INDUSTRIAL ENZYMOLOGY

The Application of Enzymes in Industry

TONY GODFREY
and
JON REICHELT

The Nature Press

Published in the United Kingdom by
MACMILLAN PUBLISHERS LTD (Journals Division), 1983
Distributed by Globe Book Services Ltd
Canada Road, Byfleet, Surrey, KT14 7JL, England

ISBN 0 333 32354 8

Published in the USA and Canada by
THE NATURE PRESS, 1983
15 East 26th Street, New York, NY 10010

Library of Congress Cataloging in Publication Data
Main entry under title:
Industrial enzymology.
 Includes index.
 1. Enzymes – Industrial applications. I. Godfrey,
Tony, 1939–   . II. Reichelt, Jon, 1949–
TP248.E51516   1983   660'.63   82–14461
ISBN 0–943818–00–1

Printed in Great Britain

CONTENTS

* Editors' note: The concluding seven sections of this chapter of industrial applications for enzymes have been separated from the preceding materials. They serve to illustrate the diversity and versatility that can be extracted from enzyme technology and translated by lateral thinking into applications representing the commercial and industrial development of biotechnology that is already taking place.

DIRECTORY OF CONTRIBUTORS

		Chapter
H. C. Barfoed	Manager for Scientific and Technical Information, Novo Industri A/S, Copenhagen, Denmark	4.7
K. Burgess	Technical Development Division, Milk Marketing Board, UK	4.6
W. D. Cowan	Technical Department, Novo Enzyme Products Ltd, UK	4.14
W. H. B. Denner	Principal Scientific Officer, Food Science Division, Ministry of Agriculture, Fisheries and Food, UK	3.1
R. I. Farrow	Marketing Director, Miles Marschall Division (Europe and Africa), France	3.3
R. Felix	Application and Development Department Swiss Ferment Company, Basle, Switzerland	4.17
P. D. Fullbrook	Business Development Manager, Imperial Biotechnology Ltd, London, UK	2
T. Godfrey (editor)	Technical Manager, Novo Enzyme Products Ltd, Windsor, UK	1, 4.5, 4.8, 4.9, 4.11, 4.13, 4.16, 4.19, 4.21, 4.22, 4.23, 5, Data Indexes 1, 2, 3, 4, 6

FOREWORD

Arthur E. Humphrey
Biotechnology Research Center
Lehigh University, Bethlehem, Pennsylvania

Considerable literature has accumulated on the laboratory production, purification, and theoretical behaviour of enzymes. Little has been written on their practical application. *Industrial Enzymology* fills this void. This text is a unique collection of practical data covering sources, performance, and specific application of various enzymes in the industrial world. As such it will be of value to a wide variety of workers in industry as well as academia. Although written almost exclusively by Europeans, the appeal of the book should be worldwide. The reader will find in the text information on industrial applications of enzymes in Western Europe, USA, Japan, and Australia, plus tabulated details of enzyme suppliers throughout the world.

The book also contains comments on current legislative thinking regarding enzymes, plus toxicological consideration and test systems for using enzymes. Of particular interest to those working with enzymes should be the material in Chapter 3 giving guidance on safety consideration when working with enzymes including the approval status of various industrial enzymes in the major countries. For example, there is a tabulated list of permitted as well as recommended industrial enzymes for several countries, including the USA GRAS listings.

In Chapter 4 the use of enzymes in 17 different and varied industries is described. This includes such industries as wine, baking, brewing, leather, paper, textiles, etc. In addition to those established industries, application in newly developing industries such as fuel, flavours, colours, and analytical are described.

However, this book is not just a source book on enzymes. Chapter 2 is devoted to the development of the basic mathematical theory of enzyme behaviour. Hence, this text should have broad based value to all those persons interested in the practical applications of enzymes.

PREFACE

The current increased demand for better utilization of regenerable resources, and the pressure on industry to operate within environmentally compatible limits, has been a stimulus to the development of new concepts in biotechnology. A parallel increase in developments using industrial enzymes is maintaining the growth of the enzyme markets that have already enjoyed more than thirty years of steadily widening applications.

This book contains a collected account of current industrial practice in the use of enzymes, several summary indexes, and comparative descriptions of industrially used enzyme preparations that is intended to provide data for use by the natural product processing industries, research teams and university teaching departments for both current understanding and their future development. It is not claimed that all industrial applications have been described, in fact some cannot yet be mentioned until industrial confidentiality restrictions can be lifted. The editors invite readers to draw their attention to such omissions to enable them to ensure that future editions may continue to be valuable up-to-date statements of the practice of industrial enzymology.

The book contains many detailed and specific examples of actual enzyme use, using named trade products, in order that meaningful treatments can be described. It is not intended that these should be considered exclusive, and by the use of simple analytical procedures it is possible to make internally valid comparisons of samples of different sources of similar enzymes. However, since most enzyme producers use their own definitions of activity, only by named products can the examples be given worthwhile meaning.

Care has been taken to ensure that the factual information in the book is accurate, but the opinions expressed are always those of the contributor and not of their employer or company. The reader is reminded that nothing is stated in this book that forms the basis of warranty or guarantee for use without prior testing, and care should be taken that patents are not infringed or local regulations violated when making use of industrial enzymes.

In the preparation of this book the editors have endeavoured to assemble a diverse account of the many applications of enzymes so that the opportunities for technology transfer become more readily apparent. Individual industry contributions may be considered as 'state of the art' for that industry, but also as the source of information and concepts for application in other industrial sectors where similar solutions to the problems of natural product processing are required. The early chapters are intended to give guidance, in the value of the mathematical concepts of kinetics as they affect industrial operations, and in the current thinking on the legal and regulatory responsibilities of both producer and user of enzymes.

The application of enzymes is a well-established part of the biotechnical society, serving both the traditional and the new industries. It is the aim of this book to bring together a substantial part of that application data to form a useful reference source.

TG
JRR

ACKNOWLEDGMENTS

Our thanks and appreciation go to all the contributors to this book, together with their employers, for their help and cooperation. We should also like to thank the Officers of the Association of Manufacturers of Food Enzyme Products for permission to reproduce their recommendations, and to Applied Science Publishers for their cooperation regarding copyright material.

TG
JRR

My grateful thanks to my wife, Eirlys, and the family, who have had to take too much of the strain; to Novo Industri A/S for allowing me to undertake this project, together with much help from my colleagues; to all at Novo Enzyme Products, for encouragement and support; to Sally James for unstinting hours of typing; to Rosemary Foster of Macmillan, who insisted the book should be written, and to her team who performed the transformation of the manuscripts.

TG

My thanks to all my colleagues at Miles Kali-Chemie, especially to Dr H. U. Geyer and Dr G. Richter, for their help and encouragement. Thanks also to Mr J. T. Brady, Dr W. Goldstein and Dr J. Marshall of Miles Laboratories Inc., Elkhart, Indiana. Special appreciation and thanks to my wife, Lesley, for her help, encouragement and typing.

JRR

Chapter 1

INTRODUCTION TO INDUSTRIAL ENZYMOLOGY

T. Godfrey and J. R. Reichelt

This book is about the industrial application of commercially available enzymes, and while no section is given to their production, some discussion of the nature of the commercial products is appropriate to introduce them.

Estimates of the 1981 world market for industrial enzymes put the sales at around 65,000 tonnes of commercial product valued at $400 million. Predictions of growth of the market forecast a rise to 75,000 tonnes, valued at $600 million by the end of 1985.

There are only a few enzyme producing companies, probably numbering about 25, of which 6 are clearly dominant in both quantity and value. Among the Western nations, almost half of all enzyme production is in Denmark, with Holland producing a further 20 per cent. American production is responsible for about 12 per cent, but a large proportion of this production is by companies who have their own 'inhouse' uses for the limited range of enzymes produced, with none of this material generally available on the free market. Japan, West Germany, France, Switzerland and the UK together with Ireland account for the remainder of enzyme production. Production in the USSR and China is significant but not readily quantified, since virtually none of the products is made available to Western industry.

1. Enzyme types and sources

More than 80 per cent of all industrial enzymes are hydrolytic in action and they are used for the depolymerization of natural substances. Almost 60 per cent of these are proteolytic, for use by the detergent, dairy and leather industries. The carbohydrases, used in baking, brewing, distilling, starch and textile industries, now represent almost 30 per cent of the total enzyme usage, leaving the lipases and highly specialized enzymes with the remainder (*see* Figure 1.1).

There are some 12 categories of enzymes used in industrial processing, with 30 different activity types in common use. When it is considered that several thousand different enzymes have been identified and characterized, the future for new processing aids is attractive. However, much new knowledge will be needed if the energy-dependent and cofactor-using enzymes are to be made

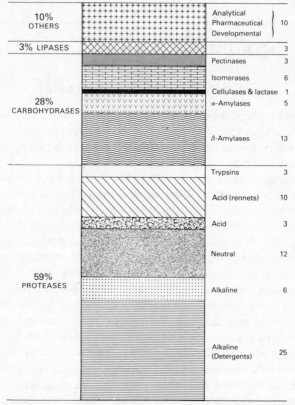

10% OTHERS	Analytical Pharmaceutical Developmental	10
3% LIPASES		3
	Pectinases	3
	Isomerases	6
28% CARBOHYDRASES	Cellulases & lactase	1
	α-Amylases	5
	β-Amylases	13
	Trypsins	3
	Acid (rennets)	10
	Acid	3
	Neutral	12
59% PROTEASES	Alkaline	6
	Alkaline (Detergents)	25

Figure 1.1   Distribution of industrial enzymes.

commercially effective. When they are available, a range of polymerizing processes will provide a completely new range of products. Some even hold out expectations that substitutes for petrochemical products will be derived by enzymic processing.

While almost any living organism can be considered a potential source of useful enzymes, in practice a limited number of plant and animal tissues are economic sources and the greatest diversity comes from microorganisms.

Plant enzymes. These include the well-known proteases papain, bromelain and ficin and the amylolytic enzymes of the cereals, soya bean lipoxygenase and specialized enzymes from the citrus fruits. Most plant enzymes are available as comparatively unpurified powder extracts, although papain is notable for being recently available as a stabilized and purified liquid. Prospects for increased supply of plant enzymes, in response to greater use in traditional applications or for new processes, depend on several factors. The influence of cultivation conditions, growth cycle and climatic requirements make new supplies comparatively long-term projects, and the influence of agricultural economics competing for culti- vated land must be coupled with national and international political forces which control agricultural activities.

Animal enzymes. These include the pancreatic trypsins and lipases and the rennets, which are produced in both ultrapure and industrial bulk qualities. Again, the prospects for large increases in supply depend on the political and agricultural policies that control the production of livestock for slaughter. Currently, these enzymes cannot meet demand on a world basis adequately, with the result that the more price sensitive user industries are increasing their interest in microbial enzymes.

Microbial enzymes. As will be seen by reference to Data Index 4 of this book, there are very few species of microorganism in use for industrial enzyme production. This is largely a limitation imposed by the desire of producing companies to have the widest possible market range for their products, which includes food processing, and the consequent high cost of obtaining approvals from legislative authorities. This cost is substantially increased if a product is produced by an organism without a recognized history of satisfac- tory approvals, and if it must be evaluated for toxicity and safety. Most industrial microbial enzymes are produced from no more than 11 fungi, 8 bacteria and 4 yeasts; producers generally seek new target enzymes from among these same species.

In response to demand fluctuations, it is relatively simple to increase or decrease the fermentation capacity devoted to the production of a particular enzyme. Within three to four months, an inventory policy change can be implemented and product availabi- lity brought into line. A full range of product qualities can be produced by the implementation of different sections of down- stream processing of fermentation biomass or liquors. Applications vary widely in the purity standards required of processing enzymes and producers generally offer two grades of product, and sometimes three or four. Food grades tend to be the dominant quality, since they can be produced in great bulk to a common standard and then function satisfactorily in most applications where lower quality would suffice. The overall economy of bulk production by well- proven fermentations, including continuous culture, provides cost effective products for most applications. It is certain that the six or so of the most frequently used enzymes now trade as commodities in the marketplace and have shown unit cost rises far below the general inflationary trends of recent years. In several cases it is now cheaper to use an enzyme than it was five years ago. This has greatly assisted the expansion of natural product processing by enzymes, especially in the alcohol producing, brewing, dairy, detergent, starch and wine industries.

Composition and stability. With such a huge variation in potency and physical presentation, it is not sensible to present specific statements about enzyme product composition. Those produced by extraction from plant and animal tissues will contain different substances, in addition to the active enzyme, than those from microbial fermentations. The table below shows the typical composition of many industrial enzyme preparations based on percentage of dry solids:

	Range of content (% dry solids)
Proteins and amino acids	10–15
Active enzyme protein	2–5
Complex carbohydrates	5–12
Sugars	2–40
Inorganic salts	3–40
Preservatives	0–0.3

Apart from noting the small proportion that is actually active enzyme protein, it should be said that sugars and inorganic salts are used as alternatives when establishing the stability of the finished product for storage and distribution, and are selected according to acceptability in the intended application. Salts and sometimes carbohydrates such as starch are used to dilute extracted enzymes to standard activity. Preservatives are generally restricted to liquid enzyme preparations and conform to regulatory information relating to the intended use and the country of destination.

Industrial enzymes are commonly used at levels of 0.1–0.5 per cent of the substrate being processed, with rare exceptions above these levels. Therefore, when the actual amount of any constituent of the preparation is evaluated as a constituent of the final processed product, it is unlikely that it will make a significant contribution in relation to other similar materials, present or introduced, in the total process.

Naturally, consultation with individual suppliers is advised when storage life of enzyme products is considered, but generally, dry products have longer shelflife than liquids, and cool conditions extend the time still further. Thus a general expectation would be

Dry products	— ambient to 25 °C (max.)	6 months
	— cooled 0–4 °C	12 months
Liquid products	— ambient to 25 °C (max.)	3 months
	— cooled 0–4 °C	6 months

These storage times would be satisfactory for most industrial stock rotations to give full performance after such storage times, providing the product remained factory-sealed and was stored according to the producer's recommendations.

Selection of processing enzymes. When the decision to investigate the use of an enzyme for a particular process has been taken, it becomes necessary to make a selection from the available enzymes with activities that appear appropriate. Many factors are involved in the selection (*see* Table 1.1).

TABLE 1.1
Key factors influencing choice of enzyme

Specificity	Sledgehammer or surgical knife?
pH	Limits to be imposed
Temperature	Rate vs inactivation vs hygiene and time
Activators and inhibitors	Control facility
Analysis method	Control/detection
Availability	Actual
	Safety
	Approved
Technical service support	Guidance on handling, dosing performance factors – new data and superceding products
Cost	Per unit of conversion
	Per cent of process

Specificity. The generally claimed specific action of enzymes is not as sharply defined as is often expected. Proteases are broad in the range of amino acid bonds they hydrolyse and exhibit only a degree of specificity. Careful investigation of the range of bonds attacked, as exemplified in Chapter 5, Table 5.1, and testing for comparable action on the actual protein target, will enable an enzyme to be chosen that has suitable performance. Some proteases are extremely narrow in their action, for example the various cheese rennets.

Among the carbohydrases, there are both highly specific enzymes, such as the components of the pectinolytic group or the lactases, and also broad acting amylases and β-glucosidases. Within any group there are small differences of action depending upon the source and type of enzyme, and these can be used to advantage in obtaining precise reaction products. Thus, the first question must be to determine what degree of specificity is required in the reaction.

pH considerations. Both the optimum operating value determined under analytical conditions, and the actual ability of the proposed industrial system to adjust away from the possibly unsuitable pH (imposed by both upstream and downstream operations) in relation to the enzyme stage, must be considered. Some limits will be dictated by the practicalities of the overall process, and these may influence the choice of a particular enzyme among alternatives. Again, the operation of an enzyme outside the anticipated best performance range may modify its specificity or sensitivity to heat in both beneficial or adverse ways.

Temperature considerations. When considering enzyme processes, the general rule is that the temperature quotient is between 1.8–2.0. (The reaction rate will increase or decrease by this order for each shift of 10°C.) By using high temperatures, the reaction may be of short duration and hygienic conditions may be maintained more easily. Conversely, the use of much greater heat than for thermolabile enzymes will be necessary to inactivate the enzyme at the end of the process. Alternatively, a significant shift of pH may be necessary to inactivate without a rise in temperature. Not all processes require the inactivation of the enzyme at the conclusion of the desired reaction, but it is important to consider this point when designing food product applications, or to prevent subsequent further modification by residual enzyme activity.

Activators and inhibitors. These are usually well-defined for specific enzymes and should be taken into account if they create expensive additional costs, or costly treatments to ensure their elimination from the reaction. Under some circumstances, the deliberate omission of a known activator will alter the pH or temperature sensitivity of an enzyme to an extent that limits the reaction, or simplifies the end process inactivation conditions.

Analysis method. This is an essential tool in process control and the detection of low levels of activity when monitoring for residual action after an inactivation treatment. Many industrial users of enzymes also maintain routine activity checks on stock enzymes to ensure good stock rotation. Therefore, when selecting an enzyme, it is helpful to have one with an established analytical method readily available (*see* Data Index 5).

Availability. Not only from the point of view of supply but also with the long term use in mind, the operator of enzyme processes will need to choose enzymes that can be supplied in consistent quality and activity. This is especially important for continuously metered enzyme dosing systems and immobilized reactor enzymes whose investment costs demand reliable and steady performance. In many processes the choice of enzyme will also relate to the safety record of industrial use, and the regulatory approval for its use, both in the process and the products made, in the countries where the finished product is to be used (*see* Chapter 3).

Technical support and service. Major improvements in the range of industrial enzymes are regular occurrences, and it is important to establish that the selected enzyme is the most up to date of its type, and that the performance data and guidance on its application are regularly updated.

Cost. Cost is a serious factor in the choice of enzyme for many processes. Clearly, no more expensive product than that necessary for the chosen duty should be selected. However, care is needed to establish that purity and activity are consistent with the cost and that any side activities have been identified, at least those that might influence the processing and product under consideration. An example to illustrate this point has been the wide availability of glucoamylases for saccharification of starches with varying levels of transglucosidase activity. The cost relationships have often been a reflection of this contaminant and the negative influence of the enzyme on the final yield of glucose is considerable. Most modern glucoamylases are quoted with information on the transglucosidase content.

Recommended literature introducing the field of enzymes

General

Dixon, M. & Webb, E. C. *Enzymes* (Longman, London, 1979).

Johnson, J. C. *Industrial Enzymes: Recent Advances* (Noyes Data Corporation, Park Ridge, New Jersey, 1977).

Lenninger, A. L. *Biochemistry* (Worth, New York, 1975).

Lilly, M. D. in *Applied Biochemistry and Bioengineering*, Vol. 2 (ed. Lemuel, B., Wyngard, J., Katchalski Katzik, E. & Goldstein, L.) (Academic, New York, 1979).

Wiseman, A. *Handbook of Enzyme Biotechnology* (Ellis and Horwood, London, 1975).

Wiseman, A. *Topics in Enzyme and Fermentation Biotechnology* (Ellis and Horwood, London, 1977–9).

Food processing

Birch, G. G., Blakebrough, N. & Parker, K. J. *Enzymes and Food Processing* (Applied Science, London, 1981).

Linko, P. & Larinkari, J. in *Food Process Engineering*, Vol. 2 *Enzyme Engineering in Food Processing* (Applied Science, London, 1980).

Pintauro, N. D. *Food Technol. Rev.*, **52** (1979).

Reed, G. C. *Enzymes in Food Processing* (Academic, New York, 1975).

Immobilized enzymes

Messing, R. A. *Immobilized Enzymes for Industrial Reactions* (Academic, New York, 1975).

Weetall, H. H. & Suzuki, S. *Immobilized Enzyme Technology* (Plenum, New York, 1975).

Bioenergy

Slesser, M. & Lewis, C. *Biological Energy Resources* (E. and F. N. Spon, London, 1979).

KINETICS

PRACTICAL APPLIED KINETICS
P. D. Fullbrook

1. Introduction

A criticism of the current developments in biotechnology is that they are often based too much on technology and not enough on basic scientific principles. The same was true at the beginning of the rapid expansion of enzyme technology, some 15 years ago. This led to confusion and frustration in the consideration and practice of industrial enzymology. It is an objective of this chapter to explain, rationalize and, hopefully, to dispel these problems.

Like most new sciences, enzymology has evolved from a descriptive phase through a quantitative phase to an applied phase. A significant part of the quantitative phase has been the development of enzyme kinetics – the study of reaction rates catalysed by enzymes, and the factors affecting them. Enzyme kinetics continues to be directed towards one of the most fundamental aspects of enzymology – an explanation of the mechanisms of enzyme action; how to account for the very high reaction rates and specific catalytic action of these remarkable proteins in terms of their chemical structure. However, to take such a restrictive view of the subject is completely to undervalue enzyme kinetics in the context of modern biotechnology. It is the object of this section of the book to show how enzyme kinetics can be used in applied enzymology in general, and in industrial biotechnology in particular.

When considering the industrial uses of enzymes – or even the purchase of a few milligrammes of a purified enzyme for academic research – the same type of questions arise: How much enzyme is needed? How long is the reaction time? What are the concentrations of substrates required? What physical conditions of temperature, pH, ionic strength should be used for optimal reaction? How much is it going to cost? Enzyme kinetics can provide some useful answers to all of these questions.

Essentially, when considering the use of enzymes on an industrial scale, one is attempting to justify their utility or viability against a background of hard economic reality. The cost-effectiveness of an

8

enzyme-catalysed process is derived directly from the relationship between the additional cost of the enzyme system involved and either the final added value of the product obtained, the cost saving achieved by overall cheaper processing, or the higher product yield. In reality this reduces to the relationship between the enzyme concentration required, the physical conditions operating and the degree of conversion obtained.

But application of enzyme kinetics can go beyond this. It can be used to design enzyme reactors for the production of kilotonne amounts of organic, biochemical, pharmaceutical or nutritive substances, and assist in the research screening for, and development of, new biotechnical products. The basic concepts and approaches of enzyme kinetics have also been developed to describe the growth patterns of populations in fermenting microorganisms and the pharmacodynamic action of drugs and other biologically active substances. They have even been used in the design of urban underground railway systems! Clearly, then, enzyme kinetics is a versatile and powerful tool, and has undergone much development since the classical work of Brown, Henri and Michaelis at the beginning of this century. My approach to a discussion of practical enzyme kinetics when applied to industrial systems will be to develop the classical enzyme kinetics models, then to criticize them by pointing out their limitations in this context, and having done this, to modify them to fit realistic conditions, and finally to introduce an alternative empirical approach which I have found useful in my experience as an industrial enzymologist. Wherever possible, actual examples with which I have been involved will be used to illustrate the points raised.

What follows, though, is intended only as a useful guide to those contemplating the use of enzymes on an industrial scale, and is not intended to replace any of the several excellent texts on enzymes and enzyme kinetics. Rather, it should be regarded as a supplement to them, based on the author's limitations and experience in applied and industrial enzymology over the past 15 years. In short, to explain what is known and accepted, and then to expand it, whilst taking into account known limitations. The points raised are obviously only my own interpretations, which are themselves open to criticism, and are not intended to be absolute or authoritarian.

2. Enzyme activity

Industrially used enzymes are characterized by the nature of the reaction they catalyse and their catalytic activity. The qualitative description of the chemical reactions they catalyse forms the basis for their classification, whilst their catalytic activity gives an indication of the magnitude of their effect, and, on a unit weight basis, an indication of their value. Since nearly all industrial enzymes are of relatively low chemical purity, they are primarily characterized as specific catalysts rather than defined active proteins. Although a consideration of their catalytic action can form a basis for their exact classification, enzyme technologists are more interested in their overall and quantitative activity. So let us start here and look at the meaning of enzyme activity in practical terms.

Enzymes are sold primarily on an activity basis – that is, a quoted cost for a specified activity. Secondary features such as degree of purity, extent of modification (stabilization, activation or physical form) and microbiological specification can modify this cost. Enzymes for analytical and medical purposes are often in a state of medium to high purity, and are sold in terms of numbers of enzyme units per lot, whilst those for industrial processing are quoted on a unit weight basis for a standardized product of guaranteed activity per unit weight. This applies to most conventionally produced solid and liquid products.

Quantitative activity of enzymes. The quantitative activity of enzymes gives an indication of how much enzyme should be used to achieve a required effect (product yield) and forms the basis for comparison of several similar enzyme products. Theoretically, as enzymes are catalysts, and therefore re-formed at the end of the reaction, a minute amount of enzyme can transform any amount of substrate. However, as is the case with any chemical catalyst, there is a practical and finite relationship between the amount of the catalyst, its initial activity and the concentration of substrate acted upon. This governs the speed of product formation, and is described by enzyme kinetics. Activity also serves as a basis for estimation of added production costs, and as a guide in assessing investment and maintenance costs, since the activity of an enzyme will have an important influence on plant throughput and hence plant size for a defined product target production rate.

Activity units. The catalytic effect of an enzyme is quantitatively expressed in terms of units of activity. These have, naturally enough, been defined under (near) optimal, or idealized, conditions so as to give a set of favourable standards for comparison. This forms the starting point for the development of classical enzyme kinetics which has evolved over the past 50 or so years, and will also be the starting point for the explanation and development of enzyme kinetics in applied enzymology.

Since enzymes were studied long before the actual chemistry of their actions was understood, enzyme unit activity has been defined in terms of (arbitrary) convenience. Hence, activity has traditionally been expressed in terms of the rate of an observed change. Often, however, a poorly defined substrate was used under arbitrary conditions, acted upon by a given amount of a crude, enzymatically active extract. For example:

'One unit of activity is the amount of enzyme that, under standard conditions (37 °C and pH 5.7), breaks down 5.26 milligrammes of starch per hour.'

'One unit of activity is the amount of enzyme which, under standard conditions (40 °C, pH 4.7 and 30 minutes), in 1 minute forms a hyrolysate equivalent in absorbancy at 275 nanometres to a solution of 1.1 microgrammes per millilitre tyrosine in 0.006 N hydrochloric acid. This absorbancy value is 0.0084.'

As a result, a large number of different methods of defining enzymatic activity have been evolved by various authors, and this has been responsible for the generation of a great deal of confusion, especially to potential enzyme users; not surprisingly, considering the above examples. This has arisen basically in three ways.

(a) Authors have not specified sufficiently the physical conditions under which they carried out and defined their enzyme units; this has as a consequence hindered repetition or development of their work.

(b) They have chosen reaction conditions which do not easily relate to their true kinetic values (or application conditions).

(c) Different units have been described for identical or similar enzyme systems, which thus makes direct comparison unnecessarily difficult. Whilst much of the earlier work in enzyme technology on activity expression was due to lack of knowledge, one cannot escape the conclusion that some authors (and enzyme producing companies) have used these shortcomings as a definite ploy to discourage enzyme users from making a too detailed comparison of activity per unit weight, and accurately assessing the cost per treatment.

(i) Definitions. In an attempt to clear up the confusion which had arisen by the 1960s, the Commission on Enzymes for the International Union of Biochemistry suggested that a standard unit definition of enzyme activity should be introduced, and proposed that:

One unit (U) of any enzyme is defined as that amount which will catalyse the transformation of one micromole of substrate per minute under defined conditions.

Wherever possible, the reaction temperature should be 25 °C. The other conditions, including pH and substrate concentration, should, where practicable, be optimal (and defined). Where more than one bond of each substrate molecule is attacked, one microequivalent of the group concerned is considered. From this basic definition, two derived expressions were suggested:

Specific activity – units of enzyme per mg protein (or $N \times 6.25$).

Molecular activity – units of activity per micromole of enzyme.

As industrially used enzymes are not pure protein, these latter terms are generally not particularly useful in enzyme biotechnology. An alternative SI unit was suggested by the Commission on Biochemical Nomenclature in 1973: the katal (kat).

One SI unit (kat) is defined as the amount of enzyme which will cause the transformation of one millimole substrate per second under specified conditions.

(*ii*) Problems. Although these recommendations are admirable in principle, they overlook several basic points when applied to enzyme technology. As is seen from other parts of this book, enzymes used industrially do not react with pure, chemically definable substrates. Usually they catalyse the conversion of complex and impure naturally occurring mixtures of compounds, the molecular composition of which is seldom known, let alone the chemical structure of the components. (What is a mole of corn starch or barley protein?) Often the observed change caused by the enzyme is measured in terms of a physical effect (e.g. a change in viscosity or filtration rate) or an approximate chemical value (e.g. production of 'reducing sugar' or α-amino nitrogen per unit time; time to reach a given optical density, or colour standard). Clearly, until the exact chemistry of the processes is known, activity, defined in terms of micromoles of substrate, is not possible. For the same reasons activities defined in terms of microequivalents of groups per unit polymer are equally impossible because very often the specificities of the enzymes used have not yet been accurately defined.

Another major problem is that of optimum conditions, and defined temperature. The solubility characteristics of many macromolecules which are industrial substrates for enzymes, prevent them from being used at optimal substrate concentration as far as enzyme unit definitions are concerned. (Starch and cellulose are insoluble, pectins and proteins often sparingly soluble at operating pH values.)

Since many industrially used enzymes are thermostable, their temperature optima are two to four times that of the recommended standard temperature (25 °C) and it makes more sense to use these optima in order to give more realistically relatable and interpretable data. Additional problems are that pH optima are sometimes affected by operation at elevated temperature and that in transforming soluble enzymes into insoluble forms, most of their 'optima' are altered anyway.

(*iii*) Solution or dilemma? Unfortunately, then, in the majority of cases of industrial enzymology, these constraints make it unrealistic to define activity in terms of the recommended universal unit basis. There are thus three options open: (*a*) To use synthetic substrates as model systems for the industrially catalysed process and define activity on the basis of these, following the Commission's recommendations. (*b*) To use standard substrates and conditions which realistically approximate to the industrial reaction being catalysed. (*c*) Totally to ignore the IUB Commission's recommendations and to carry on using one's own convenient 'special' units.

Whilst option (*a*) seems initally attractive, experience with amino acid esters as models for natural (and therefore large, high molecular weight) proteins can prove misleading for detailed development work and should only be adopted after a suitable analogue has been carefully screened for and tested in parallel with the 'natural' substrate. Use of a chemically modified form of the natural substrate is often a better alternative, although it will not have exactly the same characteristics. A useful device is to acylate the free amino groups on a protein substrate with an acidic anhydride to render it more soluble, and hence enable an optimal substrate concentration to be attained. An example of this is the assay of alkaline proteinase using a modified casein substrate, succinyl or dimethyl casein being a more soluble protein.

Option (*b*) represents the other extreme approach – to employ conditions of assay as near as possible to 'use conditions' at an industrial scale, so that data are immediately more relevant and applicable. Here again this approach often falls down on practical grounds. Different users of a standard enzyme product use different conditions for processing as well as different types of substrate. Indeed, if the substrate to be processed is a natural product – of vegetable (cereal, oil seed) or animal origin – varietal, geographic and seasonal variations will all influence the chemical nature of the substrate. Hence, it would be impossible to define a standard substrate, and pretentious to define rigid process conditions.

So one is almost forced to take up the last option (*c*). However, enzyme producers are not as obtuse as they once were, and tend to devise less arbitrary systems of enzyme unit activity which, on the one hand, at least approach assay conditions and more closely reflect the reaction conditions under which they operate, and on the other, attempt to fulfill the recommendations and criteria of the IUB Commission on enzymes. Most enzyme manufacturers will now supply customers not only with the enzyme unit definition and details of the method of assay of the enzyme under consideration, but also provide accurately standardized 'reference' enzyme samples and substrates on which it is assayed. The more resourceful will also supply conversion data to allow comparison and interconversion of their units with those of a competing product.

Essentially, then, there is no simple answer to the problem of the definition of units of activity of enzymes used industrially, and their relevance to direct industrial application requires patient interpretation, as we shall see later. The solution is a close and understanding collaboration between enzyme producer and enzyme user. As analytical techniques, digital sensors and instrumentation improve in sensitivity and reliability, they can be interfaced with a more detailed knowledge of the chemistry of catalysed reactions, to provide more relevant (and understandable) enzyme activity units. Indeed, this branch of (diagnostic) applied enzymology could prove a very important area of future biotechnology.

Enzyme dosage and productivity. Having understood the comparative value of an enzyme in terms of its stated activity, a potential user will next require to know how much enzyme is needed in a given process to achieve a desired effect in the relevant time frame. Usually the enzyme manufacturer will recommend an enzyme dosage for his product; for example, x kilogrammes of enzyme per tonne of substrate, when using enzyme of y units of defined activity per gramme. (How this is arrived at, or how this can be checked, is described below.) In other words, the customer is effectively advised of a given potential productivity per enzyme dose. Although this is only implied in the use of conventional liquid industrial enzymes, the concept is a very useful one, and one which is likely to be of greater preference in the future. This is especially so in the case of immobilized (solid state) enzymes, which are reusable. Most immobilized enzymes are specified in terms of both activity and productivity. Activities are stated in conventional terms of units of defined activity per unit weight, and this forms an expression of their commercial specification and biochemical potential. The productivity value is a fairly recent concept in applied enzymology, although it is a more valuable parameter than enzyme activity. The productivity of an enzyme is defined simply as: the weight of product per unit weight of enzyme ... formed under defined, operating conditions. In the case of an immobilized enzyme, the operating conditions definition is extended to include a 'usable life' expression, typically down to a stated fraction of its original activity or half life.

In commercial transactions, then, enzymes are sold on the following basis:

Use: Medical/analytical Form:	Industrial	
	Soluble	Immobilized
Currency units per multi-activity unit	Currency units per unit weight (volume)	
	Defined activity units per unit weight	
Currency units per lot (stated activity units per lot)		Specified productivity to defined residual activity

3. The time course of an enzyme reaction

Although industrial application of enzymes is primarily concerned with maximizing the yield of product, clearly, reaction times must be realistic. Enzymes used on an industrial scale currently operate at reaction times varying from a few minutes (e.g. liquefaction of starch) to several days (e.g. saccharification of liquefied starch, up to 80 hours). Obviously, enzyme technology looks towards speedier overall reactions, but this has to be balanced by restrictions due to processing and scale-up conditions, and must also be economically justifiable.

Generally we can recognize three phases for the overall time course of an enzyme reaction. These are:

Phase I: The transient initiation phase – a few seconds' duration

Phase II: A steady-state phase – the initial reaction rate

Phase III: A non-linear phase – the main reaction, up to completion

Industrial applications of enzymes are thus primarily concerned with phase III, although phase II is important, especially when measuring enzyme activity, or designing new enzyme processes. Phase I is also not without interest, as it provides the model on which the enzyme reaction can be considered to proceed. Let us therefore consider these three phases in turn, whilst remembering that as far as the enzyme is concerned they are one continuous process.

Phase I – reaction initiation. Application of continuous flow and stopped-flow rapid-mixing methods to enzyme reactions have provided useful information about how enzyme reactions are initiated and proceed. Such work has verified the original theories on the mechanism of enzyme catalysis first postulated by Brown (1892) and later by Henri (1902). They suggested that the reaction is initiated by the formation of an enzyme–substrate complex which subsequently breaks down to product and regenerates the original reactive enzyme. In the simplest model this may be represented:

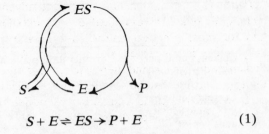

or

$$S + E \rightleftharpoons ES \rightarrow P + E \qquad (1)$$

where S, E and P represent substrate, enzyme and product, respectively, and ES represents the reversible enzyme–substrate complex. The rate limiting step is assumed to be the breakdown of $ES \rightarrow E + P$. The existence of an enzyme–substrate complex was first demonstrated experimentally in 1936 by spectrophotometric studies on peroxidase and catalase and complexes have subsequently been isolated from reaction mixtures and studied (*see*, for example, Figures 2.1 and 2.2). Detailed analysis of data obtained by these techniques allows us to produce computer generated time courses for enzyme reactions. Figure 2.3 shows such data based on experimentally obtained measurements for the above simple model. During the initiation of the reaction, the free enzyme concentration (usually only a fraction of the substrate concentration) drops to almost zero as a steady state is established between the enzyme and the substrate. The shaded portion of Figure 2.3 is enlarged in Figure 2.4 and clearly shows the induction period prior to the establishment of the steady-state phase in which $[ES]$ is constant. Similar diagrams can be generated for more complex models, such as that proposed for the isomerization of glucose to high fructose syrup by the enzyme xylose isomerase (EC 5.3.1.5). In this model the rate determining step is the breakdown of the enzyme product complex:

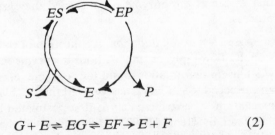

or

$$G + E \rightleftharpoons EG \rightleftharpoons EF \rightarrow E + F \qquad (2)$$

where G = glucose (substrate) and F = fructose (product).

Phase I reactions can be regarded as the step in which phase II reaction is set up.

Phase II – initial reaction. Until the advent of these modern techniques, phase II was regarded as the initial part of the reaction, and for most practical considerations, and as far as industrial enzymology is concerned, this is still so. All the reactants are in dynamic equilibrium and therefore operate at maximum efficiency, under a near-ideal steady-state system. This provides the simplest case on which to build a quantitative model for enzyme kinetics. The model can be subsequently modified to account for phase III

Figure 2.1 Formation and breakdown of an enzyme–substrate complex. The enzyme is alkaline proteinase from *Bacillus licheniformis* (EC 3.4.21.14) which is used in formulations of biological detergents. The curves show the reaction between various concentrations of enzyme, expressed in micro-Anson Units per millilitre, with 0.25 per cent casein in 0.05 M borate buffer, pH 9.0, at 50°C and is followed spectrophotometrically at 300 nanometres wavelength (author's unpublished data).

reactions, and thus used to describe the whole time course of the reaction, allowing us to calculate the product yield at any given time during the reaction.

Reaction progress curves. If we follow the transformation of substrate with time for enzyme, E, under constant physical conditions, a curve similar to that shown in Figure 2.5 is usually obtained. (Figure 2.6 shows some typical results for several industrially used enzymes, operating under their usual conditions.) The reaction velocity, v, at any point in time during the reaction, is given by the slope of the tangent to these curves at any instantaneous value of t. Hence, if S and P represent simultaneous concentrations of substrate and product respectively,

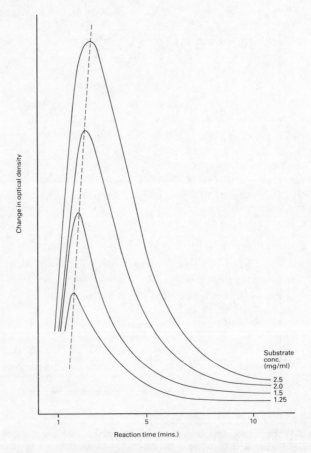

Figure 2.2   Formation and breakdown of an enzyme–substrate complex. The same system was used as that in Figure 2.1. In this experiment substrate concentration was varied as shown using 240 micro-Anson Units per millilitre of enzyme. The substrate concentrations are expressed in milligrammes per millilitre.

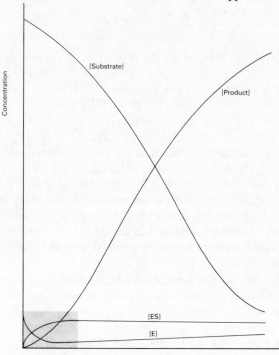

Figure 2.3   Time course of the formation of an enzyme–substrate complex and the initiation of the steady state, as derived from a computer solution of data obtained in a typical enzyme reaction (data from Lenninger, 1975).

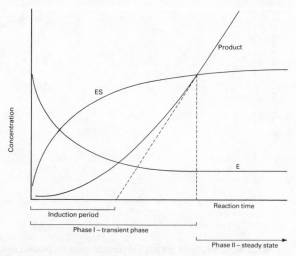

Figure 2.4   Time course of the formation of an enzyme–substrate complex. This is an enlarged portion of the shaded area of Figure 2.3.

$$v = -\frac{dS}{dt} = \frac{dP}{dt} \qquad (3)$$

Figure 2.5 shows that, where the reaction is proceeding fastest, the reaction curve is composed of an initial linear portion which gives way to a curve at longer reaction times. This linear portion is the initial reaction rate, and thus the phase II part of the reaction. Prior to the advent of computer techniques, it was often difficult to draw precise tangents to reaction curves for rapidly occurring reactions. An obvious way around this is to use low enzyme concentrations, but this requires sensitive assay systems to follow them, in order to measure the initial reaction rate accurately (Figure 2.7 shows a typical example). Nevertheless, these methods provided a relatively easy way of measuring initial reaction rates, and assisted the development of enzyme kinetics. Let us now look in more detail at the effect of substrate and enzyme concentration on the rates of enzyme reactions.

Effect of substrate concentration. For constant enzyme concentration, as the reaction proceeds available substrate becomes a contributory limiting factor in the reaction rate. It is thus important to look at the effect of substrate concentration on the initial rate of the reaction. Figure 2.8 shows the effect of different initial substrate concentrations on the rate of product formation for a typical enzyme reaction. Measurement of the initial velocity over the period $t_0 - t$ will allow a curve to be drawn of initial velocity as a

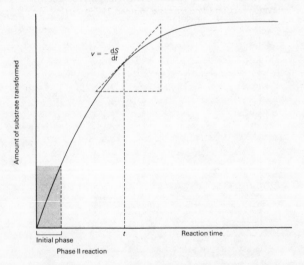

Figure 2.5 Progress curve for a typical (idealized) enzyme reaction. A tangent drawn at time t to the reaction curve gives the instantaneous reaction velocity. The phase II part of the reaction is shown as the shaded area (after Dixon & Webb, 1979).

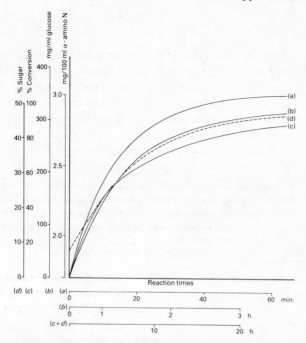

Figure 2.6   Progress curves for some important hydrolytic reactions operating on an industrial scale. Data are taken from various sources. (*a*) Production of α-amino N in Brewer's wort during conversion of barley extract using *Bacillus subtilis* neutral proteinase (EC 3.4.24.4) at pH 5.7 and 50°C. Enzyme concentration is 0.15 Anson Units per 100 millilitres. (*b*) Saccharification of liquefied starch (35 per cent dry substance) by immobilized amyloglucosidase from *Aspergillus niger* (EC 3.2.1.3) at pH 4.3 and 60°C. $E_0 = 23.7$ milligrammes per millilitre. (*c*) Hydrolysis of whey permeate containing 4.5 per cent lactose with *Kluyveromyces fragilis* β-galactosidase (EC 3.2.1.23) at pH 6.5 and 5°C. $E_0 = 9.9$ Lactase Units per millilitre. (*d*) Saccharification of liquefied starch (40 per cent dry substance, 20 dextrose equivalent) by fungal α-amylase from *Aspergillus oryzae* (EC 3.2.1.2) at pH 5.0 and 50°C. $E_0 = 160$ fungal amylase units per kilogramme.

function of initial substrate concentration. One such curve is shown in Figure 2.9. Reaction velocity units need not be expressed in terms of absolute concentration units per unit time (e.g. micromoles per second or milligrammes per millilitre per minute) but any convenient physical unit such as change in optical density or viscosity per unit time.

The shape of this curve is very significant and can give us a great deal of useful information. To begin with, it is in the form of a rectangular hyperbola and hence can be described by the mathematical expression:

$$(a - v)(S + b) = \text{Constant} \qquad (4)$$

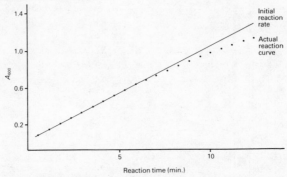

Figure 2.7 Accurate measurement of initial reaction rate using a low enzyme concentration and a very sensitive spectrophotometric assay method. The figure shows a copy of the chart recording of the hydrolysis of the medically used antibiotic ampicillin (a semisynthetic penicillin) at a concentration of 50 microgrammes per millilitre with a 0.01 per cent extract of *Klebsiella aerogenes* β-lactamase (EC 3.5.2.6) at pH 6.9 and 37°C. The assay is an iodometric assay (author's unpublished data).

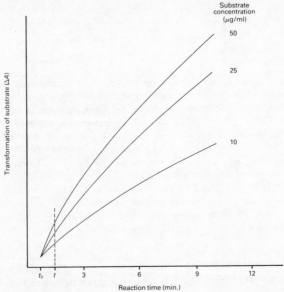

Figure 2.8 The effect of different initial substrate concentrations on reaction progress. Data are obtained using the same reaction system as for Figure 2.7.

The values of the constants a and b for a general rectangular hyperbola are shown in Figure 2.10.

(*i*) Reaction at high substrate concentration. Comparison of Figures 2.9 and 2.10 shows that a is the maximum reaction velocity, which is attained only at high substrate concentration. At this point we can write $v = V_{max}$, and note that reaction rate is now indepen-

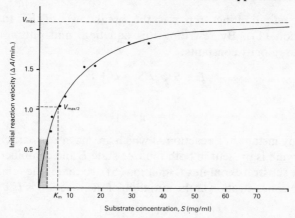

Figure 2.9 Graph of initial velocity as a function of substrate concentration. Data are obtained using the same reaction system as Figures 2.7 and 2.8. Experimentally determined points are shown exactly as they were measured. No measurements above a concentration of 40 microgrammes per millilitre were made, so this portion of the graph was calculated.

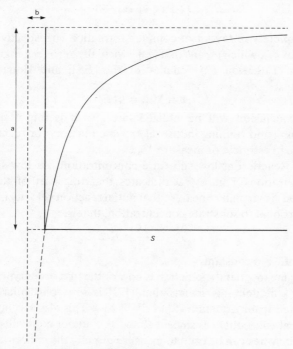

Figure 2.10 The rectangular hyperbola curve.

dent of substrate concentration. When this occurs, chemical kinetics tell us that the reaction occurring is of zero order.

In describing the phase I reaction model, it was assumed that the overall rate limiting step was the breakdown of ES to $E + P$ (Equation (1)). By rewriting this equation, and putting in the separate velocity constants:

$$E + S \underset{k_2}{\overset{k_1}{\rightleftharpoons}} ES \overset{k_3}{\rightarrow} E + P \qquad (5)$$

we see that

$$v = k_3[ES] \qquad (6)$$

At any instant in a reaction in which an enzyme is participating, the enzyme is present in both the free state E and combined as the enzyme substrate complex (Equation (5)), so that if the total molar enzyme concentration at the initiation of the reaction is E_0,

$$E_0 = \Sigma[E] + [ES] \qquad (7)$$

Usually in an enzyme reaction, and especially at high substrate concentration, $[S] \gg E_0$, so that the equilibrium governing the formation of the enzyme substrate complex:

$$[E] + [S] \underset{k_2}{\overset{k_1}{\rightleftharpoons}} [ES] \qquad (8)$$

is completely in favour of complex formation (irrespective of the value of k_2), which means that $[E]$ is virtually zero (*see* Figure 2.4), so that Equation (7) reduces to $E = [ES]$, and Equation (6) becomes:

$$v = V_{max} = k_3 E_0 \qquad (9)$$

This conclusion will be utilized later, as it is important when designing and running industrial enzyme reactors (or reactions) to be able to estimate or measure V_{max}.

(*ii*) Reaction at low substrate concentration. At low substrate concentrations Figure 2.9 indicates that this part of the curve (shaded) is virtually linear, so that initial reaction velocity is directly proportional to substrate concentration, that is,

$$v = -\frac{dS}{dt} = K'S \qquad (10)$$

where K' is a constant.

This means that the reaction is now of the first order type. On the phase I model, this means that $[ES]$ is low, particularly if (as indicated from Equation (8)) k_1/k_2 is low. This also has important practical consequences since it shows that under controlled conditions enzymes can be used to measure directly the concentration of their substrates.

The kinetics of enzyme reactions at low substrate concentration is also important when studying the control of cellular metabolism, and evaluating the efficiencies of enzymes. Such studies have led to the concept of the perfectly evolved enzyme (*see* page 33) but are not of direct relevance to industrial enzymology at present. However, these ideas may become of interest when multi-enzyme systems are being developed. Under these circumstances it will be important to have the efficiency of the second and subsequent enzymes in the reaction sequence as high as practicable, in order to steer the overall reaction in the required direction by ensuring that low concentrations of the product of the first enzyme reaction are efficiently converted by the second and subsequent enzymes to the required final product. It should be remembered that the term 'low' substrate concentration is only a relative term, and really means low only in comparison with K_m, that is, $S_0 \, K_m > 1$.

(*iii*) Reaction at intermediate substrate concentration. What we now require to know is how the reaction velocity is affected by substrate concentrations intermediate between these two extremes, in which the first order reaction gradually reduces to a zero order reaction. Since we are considering only the initial reaction rates of substrate reacting with relatively low molar enzyme concentrations (typical of most industrial enzyme reactions), the following assumptions must be valid: (1) $S \gg E_0$; (2) $P = 0$.

Using our simple phase I model, we can make a further assumption. Either the rate limiting step is the rate at which ES breaks down:

(*3a*) $\qquad ES \xrightarrow{k_3} E + P$, in which case $k_3 < k_2$

or

(*3b*) $\qquad \dfrac{\mathrm{d}[ES]}{\mathrm{d}t} = 0$, that is, $[ES]$ is constant

Assumption (*3a*) formed the basis on which the original Michaelis–Menten theory of enzyme kinetics was established. It was challenged by Haldane as being unnecessarily restrictive. This is evident from what we have seen above, that when $S \quad E_0$, $[ES] \quad E$, so that the values of k_1 or k_2 are irrelevant. Assumption (*3b*) was therefore proposed. As we have seen (Figure 2.3), this is a more realistic model, and one which we shall follow. If, then, the concentration of the enzyme–substrate complex is held constant by a steady-state situation, we can equate the rate of formation of complex with its breakdown as a simple mass balance. Using the simple single model as shown in Equation (5):

Rate of formation $\quad = k_1[E][S]$

Rate of breakdown $= k_2[ES] + k_3[ES]$

Therefore $\qquad\qquad k_1[E][S] = (k_2 + k_3)[ES]$

Hence, concentration of free enzyme

$$[E] = \frac{[ES]}{[S]} \times \frac{(k_2 + k_3)}{k_1}$$

Since k_1, k_2 and k_3 are velocity constants, they can be replaced by a single simple constant K_m, which has a characteristic value for the system. Rewriting, then,

$$[E] = \frac{[ES]}{[S]} \times K_m \qquad (11)$$

If the enzyme activity is not destroyed during the course of this part of the reaction, we can substitute Equation (11) in Equation (7) to obtain an expression relating total enzyme concentration E_0 to the variables. Thus,

$$E_0 = \frac{[ES]}{[S]} \times K_m + [ES]$$

or

$$E_0 = [ES](1 + K_m/S)$$

Therefore

$$ES = \frac{E_0}{1 + K_m/S} \qquad (12)$$

However, we know from Equation (6) that the overall (observed) reaction rate is $v = k_3[ES]$, so that by substituting Equation (12) in Equation (6) we have an expression relating a finite observable quantity, to our variables. Thus,

$$v = \frac{k_3 E_0}{1 + K_m/[S]} \qquad \text{or} \qquad v = \frac{k_3 E_0[S]}{[S] + K_m} \qquad (13)$$

This is now an expression which relates a measurable reaction velocity, v, to the substrate concentration $[S]$ and total enzyme concentration E_0, both of which are known; it also includes two constant terms, k_3 and K_m, characteristic of the reaction. This expression is extremely important in enzyme technology as it forms the basis for the design of enzyme reactors, to which we shall return later. However, Equation (13) can be simplified further since we have seen in Equation (9) that $k_3 E_0$ is an expression of the maximum velocity of the reaction, itself a useful constant. Substituting Equation (9) in Equation (13), we arrive at a simple expression

$$v = \frac{V_{max}[S]}{[S] + K_m}$$

Finally, since we have assumed that $[S] \gg E_0$, $[S] \simeq [S_0]$, where S_0 is the initial substrate concentration, this expression gives us a very useful formula, the Michaelis–Menten equation:

$$v_0 = \frac{V_{max}[S_0]}{[S_0] + K_m} \qquad (14)$$

This expression has proved extremely useful in describing classical enzyme kinetics for phase II reactions and is identical with the expression given by Michaelis and Menten (1913). Their original equilibrium assumption is now regarded as a special case of the more general steady-state situation revealed by modern techniques.

This expression should thus accurately describe Figure 2.9, and is easily checked as follows. Equation (14) can be rewritten as

$$V_{max} S_0 = v_0 S_0 + v_0 K_m$$

Adding the constant $V_{max} K_m$ to both sides of the expression and rearranging:

$$V_{max} K_m + V_{max} S_0 - v_0 S_0 - v K_m = V_{max} K_m$$

Factorizing

$$(V_{max} - v_0)(K_m + S_0) = V_{max} K_m = \text{Constant}$$

which is in the same form as Equation (4) describing a rectangular hyperbola.

The significance of K_m. A finite meaning can also be given to the Michaelis–Menten constant K_m. Rearranging Equation (11) shows that K_m has the units of concentration:

$$K_m = \frac{[E][S]}{[ES]}$$

By putting $v = V_{max}/2$, and cross multiplying, Equation (14) reduces to

$$([S_0] + K_m) V_{max} = 2 V_{max} [S_0]$$

That is

$$K_m = [S_0]$$

This means that K_m is the value of the substrate concentration which gives an initial velocity equal to half the maximum velocity at that enzyme concentration as indicated in Figure 2.9. K_m is thus independent of enzyme concentration and is a true characteristic of the enzyme under defined conditions of temperature, pH, etc., and can so be used as a genetic marker to identify a particular enzyme protein. This can be a useful parameter in specifying and characterizing an enzyme of industrial importance (e.g. in a patent), a concept which is developed in the next chapter.

The K_m values of enzymes range widely, but for most industrially used enzymes they lie in the range $10^{-1}-10^{-5}$ M when acting on biotechnologically important substrates, under normal reaction conditions.

In deriving the Michaelis–Menten equation, K_m was defined as $(k_2 + k_3)/k_1$ in order to obtain Equation (11), but it can also be written as $K_m = [E][S]/[ES]$. In this latter form, K_m is the disassociation constant of the enzyme substrate complex when k_3 k_2, that is, when k_3 is (as we have previously assumed) the rate determining factor in the overall reaction: $S \xrightarrow{\text{Enzyme}} P$. Since this is the case for most enzyme reactions, K_m is a measure of the affinity of an enzyme for a particular substrate, a low K_m value representing a high affinity and a high K_m a low affinity. This concept of affinity of an enzyme for its substrate is a useful one in applied enzymology, particularly in the context of enzyme inhibition and competing substrates, and is developed in the next chapter. It is not, however, a new concept, for the earliest attempts to account for the tendency for some chemical reactions to occur more readily than others were expressed in terms of affinities of the reacting substances as long ago as 1250 by Albertus Magnus.

K_m values are employed in applied enzymology in determining the most useful substrate concentration ranges to choose for fast conversion. As we have seen, if the substrate concentration is equal to the K_m value for the enzyme, the reaction rate will proceed at 50 per cent V_{max}. If a substrate concentration of $10K_m$ can be used, reaction will proceed at 91 per cent V_{max}. Conversely, if the K_m for the enzyme is high, and it is only possible to use substrate at a concentration of 10 per cent K_m, the reaction will proceed at only 9 per cent of V_{max}, and it might be sensible to look for another enzyme. This leads us to look at the effect of enzyme concentration.

Effect of enzyme concentration. In most industrially operated enzyme reactions, we are denied the luxury of choosing a substrate concentration suited to an enzyme, since generally manufacturers will have already developed and optimized the whole process in which the enzyme reaction is but a unit operation step. Usually, but not always, this means high substrate concentrations. Thus, we are often forced to meet required reaction velocities by adjusting the amount of enzyme. Let us look at the effect of enzyme concentration on reaction rate for a fixed initial substrate concentration.

Figure 2.11 shows a typical industrial example of some progress curves for different concentrations of enzymes, and Figure 2.12 shows a generalized curve. If tangents are drawn (computed) at t_0, t_1 and t_n for various enzyme concentrations (or dilutions), a graph of reaction velocity against enzyme concentration would appear as in Figure 2.13. As we are interested in either relating enzyme concentration to its maximum effect, or in measuring the activity of an enzyme preparation, a simple and accurate straight line relationship between enzyme concentration and observed effect is advantageous when initial reactions velocities are measured or calculated. Only under these conditions is:

$$v = -\frac{dS}{dt} = KE \qquad (15)$$

for a given initial substrate concentration, where E is enzyme activity and K is a constant for the particular system. In designing enzyme assays, therefore, velocity measurements should as far as possible be based on initial rates of reaction in the presence of adequate (or excess) substrate concentration. A knowledge of K_m is thus useful in designing enzyme assays where it would be preferable

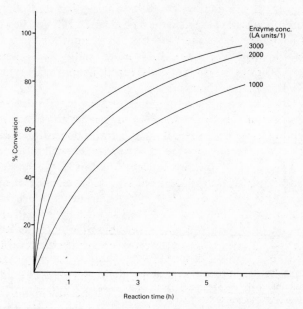

Figure 2.11 The effect of different initial enzyme concentrations on reaction progress. The figure shows the reaction of β-galactosidase on the hydrolysis of whey permeate (containing 5 per cent lactose) at pH 6.5 and 40°C. The enzyme is the same as that shown in Figure 2.6 curve (*c*), and enzyme concentration is expressed in terms of lactase units (data from Novo Industri).

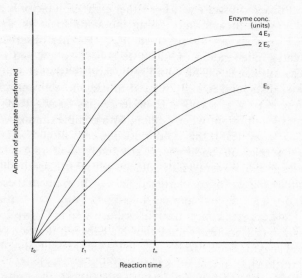

Figure 2.12 Generalized progress curves for several enzyme concentrations, to show the importance of measurement of initial reaction velocity in assay of enzyme activity (after Dixon & Webb, 1979).

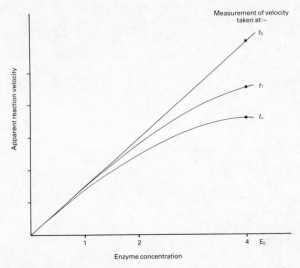

Measurement of velocity
taken at:−

t_0

t_1

t_n

Apparent reaction velocity

1   2   4 E_0

Enzyme concentration

Figure 2.13 Apparent velocity as a function of enzyme concentration. The graph shows that only in the case of initial reaction rate measurements is E_0 directly related to the observed velocity.

to have a simple straight line relationship between enzyme concentration and observed effect. As we have seen, reaction velocity is only independent of substrate concentration at high levels of substrate, that is, at V_{max}. Thus, an assay system for an enzyme should ideally utilize a substrate which is either highly soluble or has a low K_m value with the enzyme. Using the same logic as above, if $S = 100 K_m$, then $v = 99$ per cent V_{max}. Hence, observed v is particularly independent of substrate concentration and the observed assay change will be significantly large, so that enzyme units will be realistic and linearly related to enzyme concentration by Equation (9) ($V_{max} = k_3 E_0$). Experience has shown that this is particularly important when dealing with complex industrial substrates for enzymes. If this is not possible, care should be taken to ensure that reaction conditions are absolutely standard. Several test replicates should be assayed simultaneously with different dilutions of a similar enzyme of standardized activity. In many of these cases there is often a double reciprocal linearity between substrate transformed and enzyme concentration, as illustrated in Figure 2.14.

Equation (13) indicates a relationship between reaction velocity and total enzyme concentration E_0, in the presence of any substrate concentration $[S]$ and includes the constants K_m and k_3

$$v = \frac{V_{max}[S]}{[S] + K_m} = \frac{k_3 E_0}{1 + K_m/[S]}$$

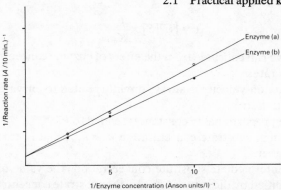

Figure 2.14 Double reciprocal relationship between reaction velocity and enzyme concentration in assay of the proteolytic activity of two commercial enzyme preparations. (*a*) *Bacillus subtilis* neutral proteinase (EC 3.4.34.4). (*b*) *Bacillus licheniformis* alkaline proteinase (EC 3.4.21.14). Assay method AF 4/5 Novo Industri (Anson method using haemoglobin substrate). Standard conditions of assay: 25°C, pH 7.5, reaction time 10 minutes.

Therefore
$$E_0 = \frac{v(1 + K_m/[S])}{k_3} \qquad (16)$$

We have ascribed a physical meaning to K_m and seen its significance in regard to industrial enzymology; can we find a similar meaning for k_3?

(*i*) Reaction at high substrate concentration. At high substrate concentration, $S \gg K_m$ and therefore Equation (13) reduces to Equation (9): that is, $v = k_3 E_0 = V_{max}$. The value of the rate constant k_3 thus governs the rate of reaction and hence under these circumstances it is often referred to as k_{cat}. k_{cat} thus reflects the number of moles of substrate converted by each active site on the enzyme per unit time, and hence is a finite constant for the enzyme system, and can be used in the same way as K_m to characterize a particular enzyme under specified conditions. Put another way, k_{cat} represents the number of re-uses of catalyst per unit time and is referred to as the turnover number of the enzyme. In practical terms this means that a high turnover number indicates the potential of an enzyme to operate at a high rate per unit of enzyme activity. Whether it actually exhibits this potential, however, is modified by the value of the K_m and the substrate concentration. Values of k_{cat} can be as high as many thousands per second.

(*ii*) Reaction at low substrate concentration. At low substrate concentration, Equation (8) indicates that [ES] is also low, so that the total enzyme concentration, E_0, is almost equivalent to the free enzyme concentration E (Equation (7)). Equation (13) thus reduces to

$$v = k_3 E[S]/K_m$$

or $$v = [E][S] \times \frac{k_3}{K_\mathrm{m}} \qquad (17)$$

Hence, we can conclude from the effect of enzyme concentration on reaction rate:

(*a*) Reaction velocity is always directly related to enzyme concentration.

(*b*) The proportionality constant is related to k_3.

(*c*) At high substrate concentration $k_3 = k_\mathrm{cat}$, that is, turnover number.

(*d*) At intermediate substrate concentrations the value of k_3 is modified by the value of K_m and the substrate concentration, and the effect of enzyme concentration is given by Equation (16).

(*e*) At low substrate concentration, enzyme is present predominantly in the free (uncombined) state and the effect of its concentration is given by Equation (17).

(*f*) Unlike k_3, V_max is not a constant for the enzyme, since it depends on E_0.

Efficiencies of enzymes. (*i*) Maximum conversion rates. We are now in a position to compare the efficiencies of enzymes in carrying out conversions of substrate. From an industrial viewpoint it is obviously desirable to be able to choose an enzyme which will have the fastest reaction rate per unit amount of enzyme (enzyme dosage), because this will mean one can attain the maximum effect for the minimum amount of added catalyst. This will generally give the best overall process, because the lower the amount of catalyst needed, the lower will be the levels of contaminants introduced into the reaction.

As we have seen above, maximum reaction rate occurs when enzyme is saturated with substrate, that is, at high substrate concentration. Under these conditions the fastest conversion per unit enzyme activity will be given by enzymes of high turnover numbers (k_cat or k_3 values). In industrial situations, however, we are often obliged to run enzyme reactions at intermediate concentrations of substrate, or have access only to enzymes which have a low affinity for the substrate to be transformed, in other words to operate systems having a low S/K_m ratio. In these cases, we have to take into account not only k_3 when relating enzyme dose to reaction velocity, but also K_m and the operating substrate concentration. Thus, given several enzymes capable of converting a substrate to a required product under defined conditions, each having a different value of K_m and k_3, how does one go about choosing the most efficient enzyme? One of the first approaches to comparing the efficiencies of several enzymes acting on a particular substrate was made by

Pollock. He was studying the resistance of bacteria to the antibiotic action of penicillin caused by the enzyme β-lactamase (EC 3.5.2.6). Despite differing K_m values for this enzyme from different species of bacteria, the physiological efficiencies of the enzymes were strikingly similar.

(ii) Physiological efficiency concepts and V_{max}. Pollock used the term 'physiological efficiency' to compare the ability of enzymes to catalyse the destruction of penicillin by extracts of bacteria, defining physiological efficiency as the ratio of V_{max}/K_m for a given amount of diluted extract (i.e. enzyme dilution obtained from a known number of bacteria). Typical results of K_m and V_{max} measurements are given in Table 2.1, which clearly shows that although the K_m of enzymes derived from differing bacterial strains have different values (and are thus different proteins and indicate different affinities for the substrates), antibiotic resistant bacteria have evolved to become strikingly similar in their effectiveness to destroy penicillin. The physiological efficiency value, since it includes the V_{max} term, incorporates both k_3 and E_0, that is, it is related to both k_{cat} and enzyme concentration, and hence provides us with a useful model for selecting the most efficient enzyme for an industrial process using a fixed initial substrate concentration. This can be illustrated as follows. Suppose nine enzymes A–I were found to be suitable candidates as catalysts for a required conversion of a substrate. If the values of K_m and k_{cat} were known, use of Equation (16) would allow us to calculate the initial likely reaction velocity v and hence choose the fastest reaction rate per unit amount of enzyme. Table 2.2 indicates such calculated values, where K_m values of the enzymes are expressed as fractions of the operating substrate concentration, S, and k_{cat} as multiples of the turnover number for enzyme A which has a value x. Clearly, the three best enzymes would be $I > F > C$, that is, the enzymes having the highest k_{cat} value. The most efficient of these is I, having the lowest K_m value, and thus the highest affinity for the enzyme. If enzymes I or F could not be used for some reason, the same reaction rate as achieved by an amount E_0 of enzyme I could be achieved using $9.52/6.67 = 1.42 \times E_0$ the amount of enzyme C.

This concept of physiological efficiency has important implications in cell metabolism and has given rise to the concept of the perfectly evolved enzyme. This is envisaged as an enzyme which has maximum reaction rate, and is characterized by having both a high k_{cat}/K_m ratio (often in the region of $10^8 - 10^9$ moles^{-1} per second^{-1}) and a K_m greater than the usual substrate concentration it encounters. Thus, the catalytic velocity is restricted only by the rate at which it comes in contact with substrate. Since most industrial

TABLE 2.1

Physiological efficiencies of β-lactamase enzymes (EC 3.5.2.6) from different species of bacteria

Substrate	Escherichia coli B-11			Klebsiella aerogenes A		
	K_m	V_{max}	PE^*	K_m	V_{max}	PE^*
Benzylpenicillin	28.9	2.92	0.10	9.0	1.18	0.13
α-amino-benzylpenicillin	36.9	3.63	0.098	22.1	1.53	0.069
p-hydroxy-α-amino-benzylpenicillin	46.0	3.75	0.082	17.9	1.09	0.061
α-carboxyl-benzylpenicillin	10.1	0.39	0.039	2.42	0.102	0.042
3-thienyl α-carboxy-benzylpenicillin	7.0	0.393	0.056	3.78	0.111	0.029

K_m values are expressed in micromoles and V_{max} in nanomoles per minute per millilitre extract. Reaction conditions were pH 7.0 at 37°C. Initial velocities were measured over a five-minute period using 1.0–100 microgrammes per millilitre substrates, and assayed spectrophotometrically using a starch-iodine microassay. (Data from Cole, Elson & Fullbrook, 1970.)
$*$ PE = physiological efficiency.

reactions are at substrate concentrations at which $S \gg K_m$, k_{cat}/K_m values are of limited use in selecting enzymes for industrial application.

TABLE 2.2

Use of enzyme kinetic data to screen and select the most suitable enzyme for a given reaction (hypothetical case)

Enzyme	K_m	k_{cat}	k_{cat}/K_m	Calculated v_0
A	0.5 S	x	$2xS$	$0.67xE_0$
B	0.5 S	$2x$	$4xS$	$1.33xE_0$
C	0.5 S	$10x$	$20xS$	$6.67xE_0$
D	0.25S	x	$4xS$	$0.8\ xE_0$
E	0.25S	$2x$	$8xS$	$1.6\ xE_0$
F	0.25S	$10x$	$40xS$	$8.0\ xE_0$
G	0.05S	x	$20xS$	$0.95xE_0$
H	0.05S	$2x$	$40xS$	$1.91xE_0$
I	0.05S	$10x$	$200xS$	$9.52xE_0$

K_m values are expressed as a fraction of the operational substrate concentration and k_{cat} values as a multiple of that for enzyme A.

Phase III – the main reactions. Figure 2.5 shows that as the reaction proceeds the initial reaction velocity decreases. Tangents drawn to the progress curve at various intervals along the reaction time abscissa will give the instantaneous reaction velocity, which will decrease as the reaction proceeds. This occurs even if complete enzyme activity is maintained, and is due to the reaction approaching equilibrium point and the reverse reaction beginning to operate. (Enzymes, like all catalysts, do not affect the position of equilibrium of a reaction – this is a function of thermodynamics – but act by lowering the activation energy for the reaction and thus speed up the process to attain equilibrium.) This point is illustrated in the hydrolysis of liquified starch by amyloglucosidase (EC 3.2.1.3) in Figure 2.15. At high substrate concentration, > 40 per cent dry substance, the yield of glucose in the reaction product is lower than at a lower initial substrate concentration. This is due to the re-synthesis of disaccharides from glucose by the reverse reaction, assisted by the low specificity of the enzyme.

(*i*) The integrated rate equation. In an industrial situation, we are particularly interested in running an enzyme reaction to near completion in order to maximize product yield, and hence require to know the enzyme concentration and the reaction time necessary to effect a desired conversion yield. Such data can be obtained by

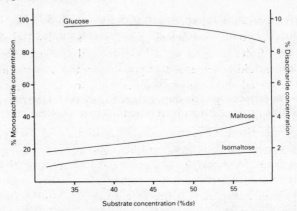

Figure 2.15 Effect of high substrate concentration on product yield. The graph shows resynthesis of disaccharide from glucose in the presence of a high initial concentration of liquefied starch substrate (18 dextrose equivalent oligosaccharide produced from maize starch). The substrate was incubated at 60°C, pH 4.5 for 48 hours using 188 Amyloglucosidase Units of *Aspergillus niger* amyloglucosidase (EC 3.2.1.3) per kilogramme dry substance (data from Novo Industri).

integrating Equation (13). (Note: S replaces the term $[S]$, for simplicity.) Since

$$v = \frac{dS}{dt} = \frac{k_3 E_0}{1 + K_m/S}$$

$$k_3 E_0 \, dt = (K_m/S + 1) \cdot dS$$

Therefore

$$k_3 E_0 \int_{t_0}^{t} dt = K_m \int_{S_0}^{S} 1/S \cdot dS + \int_{S_0}^{S} dS$$

where S_0 is the original concentration at time t_0 and S is the substrate concentration at time t. Therefore,

$$k_3 E_0 t = (S_0 - S) + K_m \cdot \ln S_0/S$$

And so

$$t = \frac{(S_0 - S) + K_m \cdot \ln S_0/S}{k_3 E_0} \qquad (18)$$

This shows that the reaction time t required to change substrate from an initial concentration of S_0 to S for a given concentration E_0 of enzyme is related by the two constants K_m and k_3, when E_0 remains at constant activity throughout the reaction. If we define conversion as:

$$X = (S_0 - S)/S_0$$

Equation (18) can be rewritten in a form relating conversion to the other parameters

$$t = \frac{S_0 X + K_m \cdot \ln\left[1/(1 - X)\right]}{E_0 k_3} \qquad (19)$$

This integrated form of the Michaelis–Menten equation thus describes the phase III portion of any enzyme reaction and is of the utmost importance in calculating required enzyme dosages to effect a given level of conversion of substrate from a stated initial concentration, in a finite time. It should be stressed, however, that this expression should be regarded as giving only the minimal level of enzyme required since, as we shall see in the next chapter, it is unlikely that the enzyme will actually maintain its full activity during the whole course of the reaction.

(*ii*) Comparison of enzymes. In medicinal chemistry and chemotherapeutic evaluations, half life concepts are often used to compare drugs, for example, to compare their resistance to inactivation by an enzyme produced by an infective organism, or detoxification by liver enzymes. $t_{\frac{1}{2}}$ is the time it takes for the organism or enzyme to reduce the drug level to half its original dose or blood level. In the case of an antibiotic the $t_{\frac{1}{2}}$ should be greater than the doubling time of the infective bacterium for it to have an effect, and the antibiotic concentration at $t_{\frac{1}{2}}$ should be much greater than the minimum inhibitory concentration.

If, in deriving Equation (18), instead of limits (S, S_0) and (t, t_0), we put $(S, S_{\frac{1}{2}})$ and $(t, t_{\frac{1}{2}})$, and put $V_{max} = k_3 E_0$, Equation (18) becomes:

$$t_{\frac{1}{2}} = \ln 2 \cdot \frac{K_m}{V_{max}} + \frac{S}{2 V_{max}} \tag{20}$$

We can use this result to compare enzymes on the basis of their physiological efficiency (*see* page 33) without necessarily knowing K_m or V_{max}. If we make S small (in relation to K_m), then in the context of Equation (20), $S/2 V_{max} \rightleftharpoons 0$. Therefore,

$$t_{\frac{1}{2}} = \frac{0.7 K_m}{E_0 \cdot k_3}$$

or
$$t_{\frac{1}{2}} \propto 1/PE$$

This allows us to use low substrate concentrations to compare the effectiveness of enzymes simply by measuring their $t_{\frac{1}{2}}$ values. The lower the value of $t_{\frac{1}{2}}$ the higher will be the (physiological) efficiency of the enzyme in the overall reaction. This thus provides us with a way of selecting the most efficient enzyme from a group of enzymes, and hence gives a valuable input in suggesting a suitable enzyme for an industrial process, simply by measuring the substrate half life in the presence of a defined enzyme concentration.

(*iii*) Immobilized enzyme reactors. One of the incidental benefits of producing immobilized enzymes is often that their stability is improved, so that Equation (19) describes their performance reasonably accurately over longer periods of time. This enables us to use Equation (19) as a basis for a design equation for flow-through, immobilized enzyme reactors. For such continuously operating reactors, reaction time t becomes residence time in the enzyme bed and can be replaced by $1/F$, where F is the flow rate of substrate through the column. If we define $E_0 = WA$, where W is the weight of enzyme of activity, A units per unit weight, Equation (19) becomes

$$W = \frac{F[S_0 X + K_m \cdot \ln 1/(1 - X)]}{Ak_3} \qquad (21)$$

Equation (21) has been used as the basis for the design of the glucose isomerase immobilized enzyme reactors, and has been shown to be valid for reactors of up to $4\,m^3$ volume, capable of producing about 30 tonnes per day of dry substance product (i.e. processing up to 3.25 tonnes substrate per hour). Such a design equation is

$$WA = \frac{F \cdot DX/100[DS(DX/100)X - K \ln 1\,(1 - X)]}{mk'\varphi(x)} \qquad (22)$$

where W = weight of enzyme (kilogrammes), A = activity of enzyme (units per gramme), DX = inlet glucose concentration (per cent dry substance), DS = dry substance content (per cent), X = required conversion per cent (= exit fructose/total sugar), K = constant (related to K_m), k' = constant (related to k_3) and m and $\varphi(x)$ = correction factors for conversion.

By substituting in Equation (21) a value for the wet bulk density, ρ, of the enzyme used, the physical size of a reactor can be calculated and designed:

$$\pi d^2 h\rho = \frac{4F[S_0 X + K_m \cdot \ln 1/(1 - X)]}{Ak_3} \qquad (23)$$

where d and h are, respectively, the diameter and height of a cylindrical column reactor. For three or more columns operating in series, it is often best to run at an h/d ratio of 1, so that the size unit, d, of 'n' identical column reactors in series can be represented by the general formula:

$$d = \sqrt[3]{\frac{4nF[S_0 X + K_m \cdot \ln 1/(1 - X)]}{\pi\rho Ak_3}} \qquad (24)$$

Determination of constants. The values of K_m and k_3 can be determined from experimental data using Equation (13). In practice, one determines the initial rates of reaction for several substrate concentrations, which can then be substituted in Equation (13) to obtain values of K_m and V_{max}, from which k_3 can be calculated, since E_0 is already known. Traditionally this has been done graphically by transforming Equation (13) into one of several possible linear forms, the most commonplace of which is the so-called Lineweaver–Burk plot. Inverting Equation (13) and putting $E_0 k_3 = V_{max}$:

$$\frac{1}{v} = \frac{([S_0] + K_m}{V_{max}[S_0]} = \frac{\cancel{[S_0]}}{V_{max}\cancel{[S_0]}} + \frac{K_m}{V_{max}[S_0]}$$

Hence

$$\frac{1}{v} = \frac{K_m}{V_{max}} \times \frac{1}{S_0} + \frac{1}{V_{max}} \qquad (25)$$

This is in the form of the equation for a straight line, $y = mx + c$, so that a graph of $1/v$ against $1/S_0$ will be a straight line of slope K_m/V_{max} and the negative intercept on the abscissa $= 1/K_m$. Similarly, the intercept on the ordinate axis will be $1/V_{max}$, as is shown in Figure 2.16.

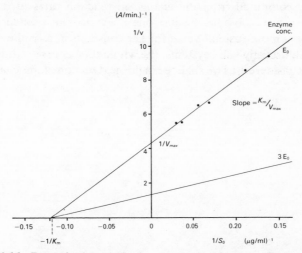

Figure 2.16 Determination of kinetic constants by Linewaver–Burk graphical method. The enzyme/substrate system used was that described in Figure 2.7. Initial velocities were measured using two concentrations of enzyme as shown (data from Cole, Elson & Fullbrook, 1979).

The accuracy of K_m and V_{max} data which can be obtained from this plot can be criticized on the ground that if an experiment is not carefully planned and an even spread of points is not obtained, the 'best fit' line is difficult to draw accurately. Hence, any inaccuracies will be magnified since the values of the constants are obtained as the reciprocals of the points read from the graph, and thus are subject to error. Among the several other plots available is S/v (ordinate) versus S ('Dixon plot'). My preference has been to use this latter method and to generate data using at least two different enzyme concentrations. This fixes K_m as the point where the lines cut the ordinate axis, allowing a direct read out (since K_m is independent of E_0). Most of the alternative plots do not give data which fit on to the lines as well as the Lineweaver–Burk plot and I suspect this is not unconnected with its popularity in the published papers in the scientific journals. My advice then is to calculate constants using a 'Dixon plot', and to use the Lineweaver–Burk plot in reports. However, the advent of the computer has changed all this, since programs are available for direct calculation of constants from raw experimental data. In fact, the availability of multi-channel, double-beam spectrophotometers which can be interfaced with computers enables values of K_m and k_3 to be estimated accurately within a few minutes, so that enzymes can actually be screened and characterized automatically for their suitability for a given process. Using these newer techniques the values K_m and k_3 can be computed from the change in reaction rates of a single progress curve obtained in just one reaction. These values could provide a more 'useful' value for the constants if, as is the case in most industrially valid systems, the whole time course of the reaction is considered, as we shall see in the next section of this chapter.

PRACTICAL LIMITS AND PROSPECTS
P. D. Fullbrook

1. Introduction

In this section, we shall concern ourselves with the factors which affect the overall enzyme action and attempt to quantify their effects, both separately and when acting together, as is the case in most industrial enzyme reactions. We shall look at ways in which to evaluate the effect of actual reaction conditions on the overall rate and progress of enzyme reactions, and conclude with practical ways of calculating the amount of enzyme required for a desired product yield under defined reaction conditions. In this way we shall be able to provide answers to the questions raised in Chapter 2.1: How much enzyme is needed? How much will it cost?

Limitations of classical kinetics. In the previous section, we looked at the relationship between initial substrate and enzyme concentrations on either initial reaction rate (Equation (13)) or conversion yield at any reaction time (Equation (19)) in circumstances in which enzyme concentration was assumed to remain constant throughout the entire reaction period. This is clearly the ideal case, and one rarely achieved in enzyme reactions of industrial importance. Figures 2.17 and 2.18 show the effect of reaction time on the residual enzyme activity of several industrially used enzymes under approximate process conditions, and indicate that kinetic equations derived for ideal conditions clearly have limited value in industrial situations. Even immobilized enzymes, which often have a higher stability than the isolated soluble enzyme, lose activity during their operation (*see* Figure 2.19), with the result that eventually they have to be discarded and substituted with a fresh charge of new enzyme.

2. The optima concept

When defining the proposed unit of activity for any enzyme, the International Union of Biochemistry stated that reaction conditions should be specified, and optimal (*see* page 11). This implies that enzyme activities are only valid within a range of physical properties, and indeed this is the case. The properties of enzymes, like those of all proteins, are modified by the prevailing physical conditions, and only within a fairly narrow range of conditions is any one property 'maximal' or optimal. A very fundamental protein proper-

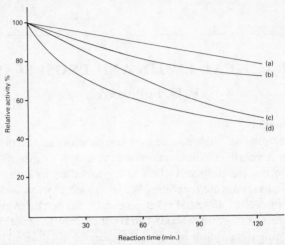

Figure 2.17 Effect of reaction time on the stability of enzymes used under industrial conditions. Data are taken from information brochures published by Novo Industri. (*a*) Residual activity of *Kluyveromyces fragilis* β-galactosidase (EC 3.2.1.23) at pH 6.6 and 40°C acting on 5 per cent lactose. (*b*) Relative activity of *Bacillus licheniformis* α-amylase (EC 3.2.1.1) at pH 6.5 and 100°C on 30 per cent dry substance starch in the presence of more than 100 parts per million calcium ion. (*c*) Relative activity of *Bacillus subtilis* α-amylase (EC 3.2.1.1) at pH 6.5 and 90°C on 30 per cent dry substance starch in the presence of more than 100 parts per million calcium ion. (*d*) Relative activity of *Aspergillus niger* γ-amylase (EC 3.2.1.3) at pH 4.5 and 70°C on 40 per cent dry substance maltodextrin.

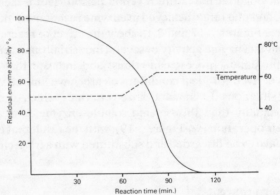

Figure 2.18 Effect of reaction time and reaction temperature on the activity of *Bacillus subtilis* n-proteinase (EC 3.4.24.4) during the hydrolysis of whole milled barley (20 percent w/w) at pH 5.7 and temperatures of 50–63°C as shown, in the presence of 0.01 M calcium chloride (author's own data).

ty, such as solubility, is dependent on such conditions as pH, temperature and the ionic strength of inorganic ions. These conditions affect the ionization of the amino acids which are the monomer building blocks of proteins, and therefore enzymes, and

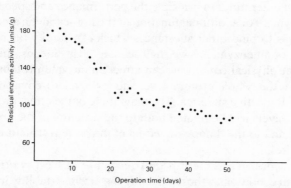

Figure 2.19   Effect of extended time on the residual activity of a continuously run immobilized enzyme reactor. The figure shows actual control laboratory assay values on a day-to-day basis during the operation of an immobilized preparation of *Bacillus circulans* glucose isomerase (EC 5.3.1.5) over a 50-day period. 93–96 per cent glucose was used as substrate at 40 per cent dry substance, pH 8.4 ± 0.1 at inlet. Reaction temperature was 62–60°C. The feed substrate contained 0.0004 M magnesium ions and the molar concentration ratio of magnesium ions to calcium ions should exceed 20 times (data from Novo Industri).

the forces of interaction between them which together are responsible for the three-dimensional structure of proteins. This, in turn, is fundamental to the properties of the protein.

Without going any deeper into the underlying chemistry involved, suffice it to say that any given protein will have a range of physical conditions within which it is able to exhibit its various properties. Outside these ranges, the particular properties will not be apparent. In the special case of enzyme proteins, the most important property as far as the industrial enzymologist is concerned, is its catalytic activity. Thus, a given enzyme will exhibit its catalytic properties only within these ranges of physical conditions. We thus say that these ranges of conditions are the optimal conditions for the enzyme and the conditions under which it expresses its maximum effect. It is therefore of great importance for enzymology in general, and for applied enzymology in particular, to be able to identify and define these optima, since outside these ranges, enzyme performance is either considerably reduced, or completely obliterated.

Activity and stability. It is generally accepted that the part of an enzyme which is actively engaged in the catalytic process is only a relatively small portion of the protein molecule. This means that a primary function of the bulk of an enzyme molecule is to maintain the catalytic area (often termed 'active site') of the protein in a limited range of configurations so that it is able to bind substrates and effect the catalytic reaction. The degree of perfection of this function determines in principle the performance and specificity of the enzyme. An additional function of the non-catalytic area of the protein is to bind other substances which can modify the catalytic activity of an enzyme, as we shall see later. This means, then, that different physical conditions can affect an enzyme protein at two levels: at the 'whole protein' level and at the micro-environmental level of the active site area. The magnitude of the effect on each of the two levels will be related both to the structure of the protein in general and to the 'knock-on' effect of the protein structure on the active site.

In practical terms we can thus recognize at least two types of optimum conditions – those affecting the enzyme stability in general, and those affecting (or effecting) the enzyme activity at particular sites. Thus, we speak of both stability and activity optima for enzymes. These are not usually identical, as we shall see. Stability optima are often broader than activity optima, as one might expect from the above discussion (*see* Figure 2.27). An important consequence of these ideas is, however, that we have the possibility of chemically altering the structure of an enzyme protein, so that we can, on the one hand modify its stability optimum range, and on the other, extend (or contract) its activity optimum – or even change its substrate specificity. Since most industrially used enzymes operate on substrates which are not their 'natural' substrates, this concept is an important one, and opens up the possibility of producing semi-synthetic enzymes, of which immobilized enzymes are but one example. Whereas activity optima are a true reflection of the structure and reactivity of the enzyme, stability optima are, from a purist point of view, somewhat artificial, since they are time dependent. However, from an applied enzymological point of view, they are just as important, since the stability of an enzyme reflects its ability to retain its activity under the various reaction conditions.

Stability and activity optima can be easily distinguished experimentally, and the separate effect of time studied. Activity optima are identified by simply measuring initial reaction velocities over the range of conditions under investigation. Stability optima are studied by exposing the enzyme to the various physical conditions for the relevant time periods and then adjusting them back to a reference value (e.g. the activity optimum) and measuring the residual activity of the enzyme in the same way as for the determination of the activity optimum.

Uses of optima data. If, then, stability and activity optima are not identical for a particular enzyme–substrate system reacting under defined conditions, we have to be able to decide which optimum to use in an industrial situation.

The criterion must be the option which gives the highest productivity. This, in turn, will be affected by at least two factors, the nature of the substrate, and the type of reaction used. Particularly when using substrates of biological origin, it is important to realize that substrates, too, have optimal ranges of stability, and this factor alone can be enough to determine the choice of conditions for the enzyme reactions. Hence, it might be pointless to use a thermostable enzyme having an activity optimum of pH 9 to produce 6-aminopenicillanic acid (an intermediate in the manufacture of semisynthetic penicillins), since penicillins are hydrolysed chemically at alkaline pH to antibiotically inactive penicilloic acids (*see* Figure 2.20).

In a batch reaction where the amount of enzyme added can easily be varied, the enzyme activity optima might be expected to be of more significance than the enzyme stability, especially if – as is usually the case in industrial reactions – no residual enzyme activity is permitted in the product. This is illustrated in Figure 2.18. For an immobilized enzyme, acting over a relatively long period in a column reactor, enzyme stability optima are more important, since less than optimal activity can be compensated for by increasing contact time (adjust flow-through rate, e.g. using Equation (21)).

Activity and stability data are also useful in screening for suitable new enzymes for a particular process, in the design of multi-enzyme processes and in the characterization of an enzyme – both in terms of its classification, and in the definition and determination of its catalytic activity.

A chemical process will usually have predetermined pH and temperature specification ranges, so that an enzyme which will act as a catalyst in the process, will have to be both stable and active

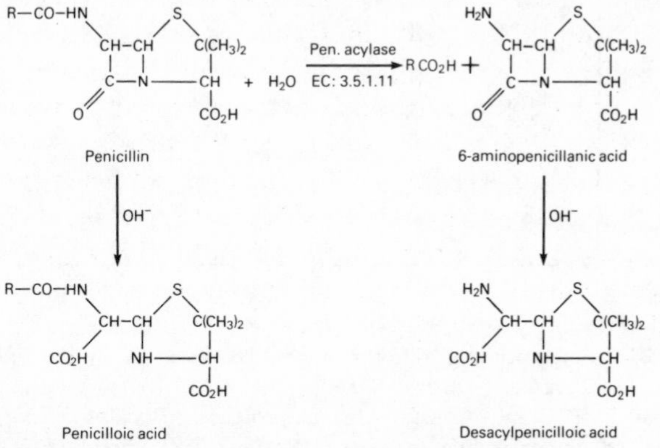

Figure 2.20 Action of penicillin acylase (EC 3.5.1.11) and alkaline pH on penicillin antibiotics. Penicilloic acids are not antibiotically active.

within these limits. In multi-step processes, such as the conversion of starch to fructose (*see* Chapter 4.15) it would be of considerable industrial advantage if all of the sequential enzyme reactions could be carried out at the same pH, since this would reduce considerably the necessity for expensive ion-exchange capacity. Failing this, if enzymes of similar optima ranges could be used, a knowledge of each separate enzyme optimum would enable a compromise pH to be adopted (the so-called 'combi-process').

In addition to specifying enzymes by their kinetic constants (K_m and k_3), optima profiles can be used to 'fingerprint' and classify enzymes. Thus, proteolytic enzymes, notoriously difficult to classify (as a glance at the International Union of Biochemistry's book *Enzyme Nomenclature* will show) are often usefully referred to as 'alkaline', 'neutral' or 'acidic' proteases, which usually are more meaningful in applied enzymology (*see* Figure 2.21). Of those enzymes of microbial origin which are currently used industrially, bacterial enzymes tend to have neutral or alkaline pH optima and are tolerant of ambient temperature or thermostable, whilst enzymes of fungal origin have neutral or acidic pH optima and are active at ambient or less than ambient temperatures, as illustrated in the examples given. However, this very general observation should only be regarded as a useful guide and not as a part of any formal enzyme classification system.

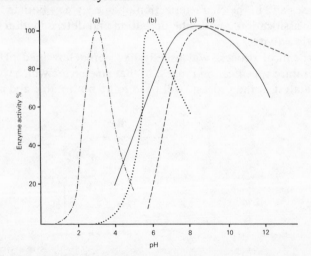

Figure 2.21 'Fingerprinting' different microbial proteases by their pH activity curves. (*a*) Acid protease – *Aspergillus niger* (EC 3.4.23.6). (*b*) Neutral protease – *Bacillus subtilis* (EC 3.4.24.4). (*c*) Alkaline protease – *Bacillus licheniformis* (EC 3.4.21.14). (*d*) Superalkaline protease – *Bacillus licheniformis* (EC 3.4.21.14). Enzymes were assayed on various substrates by standard Miles Laboratories Inc. and Novo Industri assays.

Since pH effects are more specific than temperature effects, any irregularities or tendencies to dual optima peaks in pH curves can be taken as an indication that the enzyme preparation under test probably contains more than one enzyme protein of similar catalytic effect. This is illustrated in Figure 2.22.

The advantage of defining enzyme units under their optimum conditions is both mandatory and self-evident, since it represents the maximum potential of an enzyme. Hence, it is a useful guide to its applicability and effectiveness in any given reaction (*see* Data Indexes 3 and 6).

3. Physical factors affecting optima
The main physical factors affecting activity and stability optima are pH and temperature, although other effects can have (but not usually so significantly) an effect on industrial enzyme reactions. The observed effects are due to both the general nature of pH and temperature effects on chemical reactions (ionization, dissociation, solubility, changed reaction rates, position of equilibrium etc) and their effects on enzymes as proteins and catalysts, which include effects on K_m, V_{max} and stability. In applied enzymology we are, however, not too concerned with the mechanism of the effects, only with the combined observed effects, expressed in a quantitative and practically meaningful way. Data are usually expressed graphically

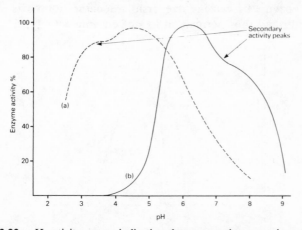

Figure 2.22   pH activity curves indication that preparations contain more than one active enzyme component. Curve (*a*) *Aspergillus oryzae* β-galactosidase (EC 3.2.1.23) acting on 5mM orthonitrophenyl β-D-galactoside synthetic substrate. Curve (*b*) an 'early' preparation of *Bacillus subtilis* having a main peak of neutral proteinase activity (EC 3.4.24.4) and a contaminating secondary peak of alkaline proteinase activity (EC 3.4.21.14) (data from Miles Laboratories Inc. and Novo Industri; both enzymes were development products).

in the form of maximum reaction rate or residual enzyme activity as a function of the variable, in order to interpret the optima values easily.

Effect of pH. As noted previously, the protein nature of enzymes means that pH will affect the ionization state of the amino acids which dictate the primary and secondary structure of the enzyme and hence control its overall activity. As change in pH will have a progressive effect on the structure of the protein, the observed enzyme activity will mirror this gradual effect on either side of the optimum range, and hence give a typical bell-shaped curve (*see* Figure 2.23), although the shapes of the curves are not always symmetrical, as indicated in Figure 2.24.

(*i*) pH activity optimum. The pH activity optimum may be due to a true reversible effect on V_{max}, to the effect of pH on the affinity of the substrate for the active site of the enzyme (in which case there will be an apparent change in K_m with pH variation), or both. Since activity data represent the maximum potential of the enzyme, it is important to know the pH activity curve in quantitative detail for the particular enzyme–substrate system under consideration, including the other prevailing physical conditions under which the reaction is to take place (such as the temperature and substrate concentration). Figures 2.21–2.24 give pH activity curves for some industrially important enzymes.

(*ii*) pH stability optimum. A change in pH may affect the stability of the enzyme by causing irreversible denaturation of its protein structure, and thus permanent loss of enzyme activity; this could

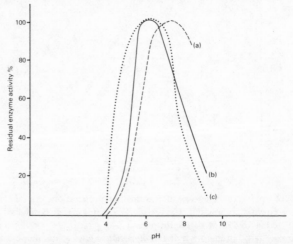

Figure 2.23 Effect of pH on the activity of enzymes. The figure shows typical bell-shaped curves for three enzymes produced by various strains of *Bacillus subtilis* assayed on suitable substrates at, or near, their optimum temperatures of reaction: (*a*) β-glucanase (EC 3.2.1.73) ; (*b*) n-proteinase and (*c*) α-amylase.

account for the shape of part of the pH activity curve, even though the reaction times during which the effects are measured are short. It is, however, likely that the effects occur in combination, so that, for example, a fall on one side of the pH optimum is due to a decrease in affinity for the substrate, and a fall on the other side of the optimum to instability of the enzyme. This is illustrated in Figure 2.25, which shows the effect of pH on monoamine oxidase (EC 1.4.3.4), an enzyme of interest in the food industry because it removes aromatic and heterocyclic amines from food materials. The effect of pH on the affinity of the substrate can often be eliminated by using a high substrate concentration (e.g. $\geqslant 10K_m$), but this may not be practical in an industrial situation. Figure 2.26 shows the stability optimum ranges for some enzymes of industrial importance. Figure 2.27 compares the stability and activity curves of bacterial neutral proteinase, and shows that, unlike monoamine oxidase (*see* Figure 2.25), this enzyme is denatured at acidic as well as alkaline conditions, giving a symmetrical pH stability curve.

(*iii*) Significance of pH optima. pH stability data, measured over a longer period of time, more relevant to the envisaged overall reaction, will reflect the effect of time on the pH activity and are thus useful in selecting enzymes for a particular reaction. It is important that (as for activity data) the effects of substrate concentration and temperature are also taken into consideration. This is because enzymes are often stabilized by the presence of their substrates, and in certain instances, temperature affects the observed pH optima, as is discussed later in this chapter (*see* page 97).

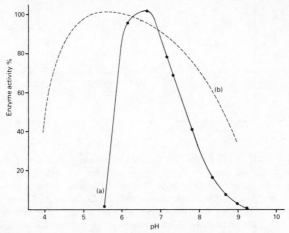

Figure 2.24 Effect of pH on the activity of enzymes; the atypical non-bell-shaped curves for (*a*) *Kluyveromyces fragilis* β-galactosidase (reaction was performed at 37°C, but over a 30-minute period); and (*b*) *Klebsiella aerogenes* pullulanase (EC 3.2.1.41) (data from Novo Industri and ABMC Food Division technical brochures).

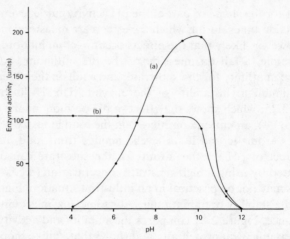

Figure 2.25   Effect of pH on monoamine oxidase (EC 1.4.3.4) showing differences between (*a*) activity and (*b*) stability. For curve (*a*) the enzyme was assayed at the pH values indicated; for curve (*b*) the enzyme preparation was exposed to the pH values given for 5 minutes and then adjusted back to pH 7.3 for assay (Hare, 1928).

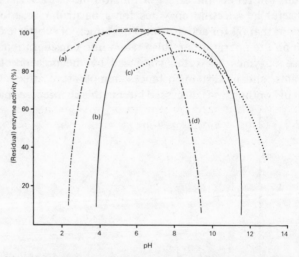

Figure 2.26   pH stability curve for four industrial enzymes. Curve (*a*) shows the influence of pH on the stability of *Aspergillus niger* α-amylase (EC 3.2.1.3) at 30°C for a holding time of 1 hour. Curve (*b*) indicates the wide stability of *Bacillus licheniformis* alkaline proteinase (EC 3.4.21.14) at 25 °C over a 24-hour period. Curve (*c*) shows the even higher alkaline stability of a different enzyme produced by a different strain of the bacterium. Note: data are residual activity not per cent maximum activity. Curve (*d*) shows the wide stability of *Aspergilus oryzae* β-galactosidase, although reaction times were measured only over a 1-hour period at 37°C (data from Miles Laboratories Inc. and Novo Industri technical brochures).

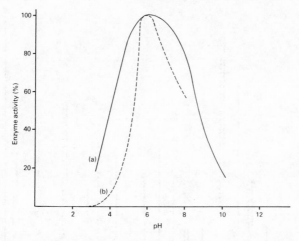

Figure 2.27 pH stability and activity curves for *Bacillus subtilis* neutral protein-ase (EC 3.4.24.4). Curve (*a*) shows stability when stored at 25°C for 21 hours. Curve (*b*) shows activity over a 10-minute reaction period at the same tempera-ture (data from Novo Industri A/S).

The various pH optima ranges of several industrially important enzymes are shown in Figure 2.28.

It is also worth remembering that many industrial enzyme reactions are not run at a fixed pH value, and slowly drift from a start pH to a terminal pH. Enzyme-assisted fermentation reactions are an example; these occur with the saccharification enzymes used to produce low-calorie beer. In other reactions, pH is assumed to be controlled by the buffering capacity of the substrate or the products of the reaction – in the mashing process in wort production, for example. Many processes do not have adequate pH control, and the pH may fluctuate around the required value. All these cases mean that pH optima data determined under the controlled conditions of the laboratory need to be applied carefully, taking such factors into consideration.

(*iv*) Reaction direction. As catalysts, enzymes cannot affect the thermodynamic equilibrium of a chemical reaction, their role being simply to speed up the rate at which equilibrium is attained. Enzymes involved in reactions which do not have a large overall free energy change ($\Delta G°$) are therefore able to catalyse both the forward and backward reactions. Thus, the enzyme glucose isomer-ase will just as easily convert fructose to a glucose–fructose mixture as convert glucose to the same product. However, pH can affect the equilibrium constant of a reaction. Thus, if one had the situation of a low free energy change reaction which could be catalysed by an enzyme having a broad pH activity/stability range, the value of the

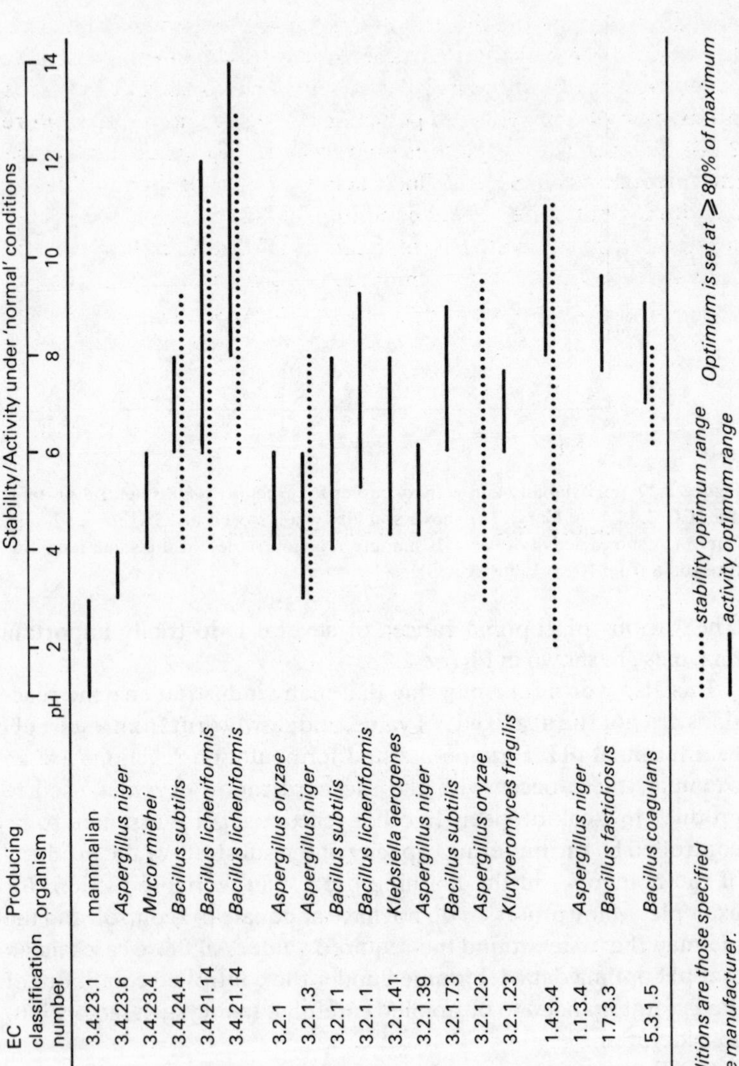

Figure 2.28   pH optima ranges for several industrially important enzymes.

pH will determine the direction in which the reaction will go. This has obvious commercial implications in the realm of using enzymes to catalyse simple synthetic reactions. One example of this is in the production of semisynthetic penicillins. We have noted (*see* Figure 2.20) that penicillin acylase is able to hydrolyse penicillins to the intermediate 6-aminopenicillanic acid at neutral or slightly alkaline conditions, but under acidic conditions it will catalyse the reverse reaction, that is, synthesis of penicillin from a carboxylic acid derivative precursor and 6-aminopenicillanic acid. The first samples of the widely used antibiotic ampicillin were made in this way:

$$C_6H_5CHCO_2H + NH_2{-}CH{-}CH \quad\quad C(CH_3)_2$$

$$\underset{NH_2}{\big|} \qquad\qquad \overset{O=C{-}N{-}CH}{\underset{CO_2H}{\big|}}$$

| Phenyl glycine | 6-aminopenicillanic acid |

$$\xrightleftharpoons[\text{pH }7.5]{\text{pH }5}$$

$$C_6H_5CHCONH{-}CH{-}CH \quad\quad C(CH_3)_2 + H_2O$$

$$\underset{NH_2}{\big|} \qquad\qquad \overset{C{-}N{-}CH}{\underset{CO_2H}{\big|}}$$

α-aminobenzylpenicillin
(ampicillin)

Effect of temperature. Bearing in mind the general remarks at the beginning of this section, the effect of temperature on the reaction rate and overall product yield of an enzyme reaction can be due to several factors, some of which are summarized in Table 2.3. The reactions will certainly be modified by the temperature stability of the enzyme, and this might be the dominant influence. Let us therefore first consider the temperature stability of an enzyme; this can best be regarded as its resilience to heat inactivation.

(*i*) Heat inactivation of enzymes. Any consideration of the inactivation of enzymes should, ideally, be related to the nature of the organism (or tissue) from which they are derived, their primary metabolic function in the living cell which produces them, and its immediate environment. The 'average' human tissue cell is replaced completely every 40–80 days, whereas the doubling time of actively growing bacteria is in the region of 20 minutes. This, then, sets the rough time frame for the required natural stability of enzymes operating under normal conditions for growth, although it should

TABLE 2.3

Some common factors affecting the effect of temperature on the reaction rate (or productivity) of an enzyme-catalysed reaction.

Effect	Due to	Determined/overcome by
Stability of enzyme	Nature of protein	Stability/exposure experiments
Stability of substrate	Chemical lability	Stability/exposure experiments
Availability of substrate (or cosubstrate)	e.g. solubility, pH	Separate experiment
Affinity of enzyme for substrate	K_m	Use high $[S]$
activators		Ensure saturation
inhibitors	K_i, (I), t_i	Avoid/purify substrate
Formation of byproducts	'Inhibition'	Increase $[E]$
Velocity of breakdown of ES complex	k_3 altered	Separate experiment
Heat of activation of reaction	Thermodynamic functions	
pH functions of reacting components	Altered pK values	Heats of ionization
Transfer of rate-limiting functions in multi-enzyme systems	Different temperature components	Fractionate and study each enzyme separately

Note: Most effects can either be studied by separate experiment, or eliminated by proper attention to detail of experimental design.

be remembered that enzymes are constantly being synthesized and destroyed during normal metabolic activity. When metabolism gets out of control, death of the cell ensues, and the catabolic degradation of cell components is due to the continued action of intracellular enzymes leading to the eventual disintegration of cellular material. Generally, then, enzymes are more than adequately stable for their natural function in a cell, and in a dehydrated or purified crystalline state are relatively very stable indeed. (A 'spring clean' of the author's laboratory refrigerator recently showed that some purified proteolytic enzymes stored at 4 °C for about 10 years, still retained over 90 per cent of their original activity.)

How, then, do enzymes stand up to unnatural conditions, particularly those of the 'industrial' conditions we impose upon them in applied enzymology?

Figure 2.29 shows a series of progress curves for a typical industrial enzyme reaction. Over the period $t_i - t_0$, true initial velocity increases as the temperature is raised, in a similar way to the use of higher enzyme concentrations (compare this with Figure 2.12 in the previous chapter). However, as the reaction proceeds over the period $t_2 - t_1$, the amount of substrate transferred per unit time (v) at any finite time falls gradually. Thus, the apparent temperature optimum is not constant since it depends on reaction time. Hence, the temperature activity determined at t_1 would be different from that at t_2, indicating that actual values of activity temperature optimum have no special significance in themselves,

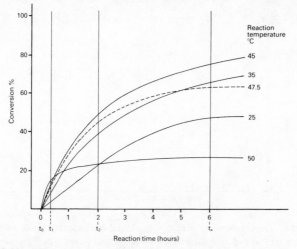

Figure 2.29   Effect of temperature on the progress of an enzyme reaction. The hydrolysis of whey permeate containing 5 per cent lactose by *Kluyveromyces fragilis* β-galactosidase (EC 3.2.1.23) is shown at various temperatures (data from Novo Industri).

and should therefore be specified and related to reaction time as well as to other reaction parameters. During the progress curve, two different effects of temperature are operating simultaneously – an increase in true catalytic activity, which causes the initial increase in initial velocity over the period $t_1 - t_0$, and an inactivation of the enzyme due to its temperature-labile protein nature. When the curve is parallel to the time axis we can assume that the enzyme has lost its activity, and the time to reach this value ($t_x - t_0$ in Figure 2.29) is a measure of the inactivation time for the enzyme under the prevailing conditions of the reaction.

(*ii*) Enzyme stability. The rate of inactivation of industrially used enzymes is thus of great importance both for conventionally used batch enzyme processes and for continuous enzyme processes. Until recently, most enzyme manufacturers have shown tempera-ture stability data for their products in the form of stability curves drawn in either of two alternative forms: (*a*) Residual per cent activity values plotted as a function of time for various temperatures (*see* Figure 2.30). Data are obtained by incubating the enzyme at various temperatures, at constant pH and in the presence/absence of the relevant substrate. Samples are withdrawn at suitable time intervals, adjusted to assay temperature and the residual activity is determined. Such data are informative but difficult to compare readily with temperature activity data. (*b*) However, using this type

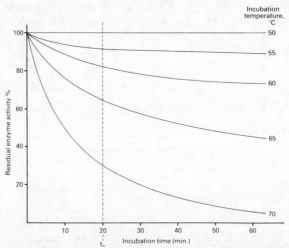

Figure 2.30   Temperature stability curves plotted in the form of residual enzyme activity as a function of time shown for various temperatures. The figure shows the (typical) results for the enzyme alkaline proteinase from *Bacillus licheniformis* (EC 3.4.21.14). Activity was estimated by the Novo Anson assay, after incubation at pH 8.5 in tris-maleate buffer. Orig nal enzyme concentration was 2.3 Anson Units per litre.

of data, and choosing a relevant time interval, a stability 'half curve' can be drawn by plotting residual per cent enzyme activity as a function of temperature at this interval (e.g. t_n in Figure 2.30). Some typical temperature stability curves obtained in this manner are shown in Figure 2.31. Moreover, if the same time interval for reaction is used, temperature stability and activity data can be readily compared, as in Figure 2.36. It should be remembered, though, that activity data are often measured in the presence of assay substrate (which may therefore have a stabilizing effect on the enzyme, as would an industrial substrate during the reaction) while stability data are usually derived in the absence of substrate, hence giving data which are not strictly comparable.

A more valuable and realistic unit of comparison of stability would be to measure the enzyme stability 'half life' ($T_{\frac{1}{2}}$) under actual reaction conditions, and it is to be hoped that this will be utilized more in the future. The half life is defined as the reaction time for the enzyme activity to drop to exactly half the initial (dosed) value under specified conditions. (It can be regarded in exactly the same way as radioactivity half life, but should not be confused with the $t_{\frac{1}{2}}$, which is the time for an enzyme to reduce substrate concentration to half its initial concentration; *see* page 37.)

$T_{\frac{1}{2}}$ is a particularly useful parameter in operating immobilized enzyme reactors and is discussed more fully on page 58. Although

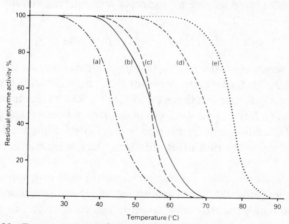

Figure 2.31 Temperature stability curves for enzymes used in the production of malt extract. Enzymes were incubated at the temperatures indicated for 45–60 minutes at pH 5.7–5.0, conditions relevant for their industrial use, prior to assay. Curve (*a*) cereal/malt β-glucanase (EC 3.2.1.73). Curve (*b*) *Bacillus subtilis* β-glucanase. Curve (*c*) *Bacillus subtilis* n-proteinase (EC 3.4.24.4). Curve (*d*) *Aspergillus niger* cellulase complex (EC 3.2.1.4). Curve (*e*) *Bacillus subtilis* α-amylase (EC 3.2.1.1) (data from Novo Industri and Miles Laboratories Inc. technical brochures and Gjertsen & Erdal, 1971).

it has been established that heat inactivation of an enzyme is a gradual process and nearly always due to denaturation of the enzyme protein (*see* Figure 2.32), there has been relatively little detailed published work on the kinetics of inactivation of enzymes, so it might be useful to illustrate how half life can be calculated, since this is likely to be an area of increased interest in the future.

Earlier in this chapter, the general effect of reaction time on residual enzyme activity was illustrated (*see* Figure 2.17) and the loss of activity of a thermostable α-amylase shown (curve *(b)* in Figure 2.17). Elsewhere in this book (Chapter 4.15) the use of this *Bacillus licheniformis* enzyme is described for the liquefaction of starch. Pilot plant studies using jet cooking of 35 per cent dry substance corn starch at 105 °C over periods of a few minutes (<7), showed that less than 15 per cent of the original enzyme activity was destroyed during this process. Laboratory tests established that under similar conditions, the mechanism of enzyme inactivation approximated to a first order decay process:

$$\frac{\mathrm{d}(DE)}{\mathrm{d}t} = aE_0 \cdot e^{-bt} f(DE) \qquad (26)$$

where DE = dextrose equivalent (titratable sugar), E_0 = initial enzyme concentration, a = enzyme activity, $f(DE)$ = correction factor, shown to be ~ 1.0 for $DE \leqslant 12$, t = reaction time, and b = decay constant = $\ln 2/T_{\frac{1}{2}}$ for constant temperature and pH. This can be integrated to give an expression of enzyme stability measured in terms of $T_{\frac{1}{2}}$:

$$\ln \left[\frac{\Delta DE}{\Delta t} \Big/ f(DE) \right] = \ln (aE_0) - t \ln 2/T_{\frac{1}{2}} \qquad (27)$$

and allows an estimation of $T_{\frac{1}{2}}$, by following the increase in DE with time. If the instantaneous slopes of the progress curves are measured at various reaction times (*see* Figure 2.33), a semi-logarithmic plot of $\ln (\Delta DE/\Delta t)$ versus t will give a straight-line graph of slope $-\ln 2/T_{\frac{1}{2}}$, enabling the half life to be calculated. Figure 2.34 shows such graphs (obtained using different start substrate concentrations).

The variation of $T_{\frac{1}{2}}$ with temperature will thus give a measure of the temperature stability of an enzyme under actual reaction conditions, and allows easy and quantitative comparison of enzyme stability. This has obvious value in assessing the operation of continuous or extended enzyme reactions, such as immobilized enzyme reactors. The effect of operating temperature on the half life stability of an immobilized enzyme is illustrated in Figure 2.35. This type of data gives us true temperature stability information having the same type of significance as pH stability data. However, to obtain such data requires much careful experimental work, which

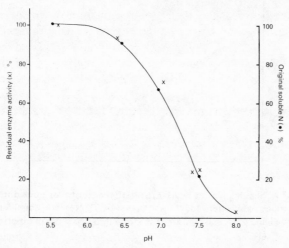

Figure 2.32 pH denaturation and inactivation of animal acid proteinase (pepsin A, EC 3.4.23.1). Decrease in soluble nitrogen indicates protein denaturation and precipitation (Northrop, 1932).

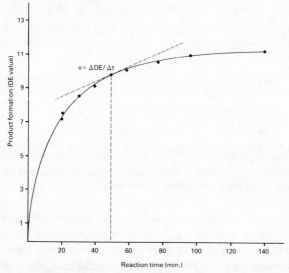

Figure 2.33 Progress curve for the hydrolysis of 26.5 per cent dry substance starch by *Bacillus licheniformis* α-amylase (EC 3.2.1.1) at 102°C and pH 5.9–6.0 in the presence of 25 parts per million calcium ion. Estimation of $\Delta DE/\Delta t$ during reaction allows $T_{\frac{1}{2}}$ to be calculated (*see* Figure 2.34) (data from Novo Industri).

is not always justifiable in a busy industrial laboratory, and this is probably the reason most temperature stability data are obtained in the form illustrated in Figures 2.30 and 2.31, for traditional enzyme systems. Some temperature stability data for commonly used industrial enzymes are shown in Figure 2.36.

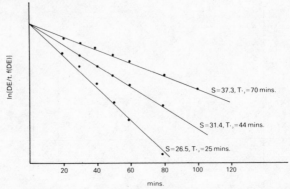

Figure 2.34 Semilog plot of $\ln(\Delta DE/\Delta t.f(DE))$ against t for three different start concentrations of substrate for the estimation of $T_{\frac{1}{2}}$. Reaction conditions are the same as in Figure 2.31, and data are taken from the same series of experiments.

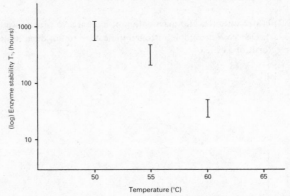

Figure 2.35 The effect of temperature on the half life of an immobilized enzyme. Approximate stability half lives of a preparation of immobilized γ-amylase (amyloglucosidase) (EC 3.2.1.3) from *Aspergillus niger* is shown for reactors run in parallel at different temperatures (data from Novo Industri).

(*iii*) Storage stability of enzymes. An important aspect of the temperature stability of commercial enzymes, is their storage stability. It has been pointed out (*see* p. 10) that enzymes are sold on an activity basis, and at the time of manufacture, their activity is measured and adjusted to a standardized value. Usually, this is some 10 per cent higher than the nominally stated value, in order to allow for inactivation during shipment and storage over the stated shelf-life conditions (usually below 20°C and for 12 months from the date of manufacture). In the development of commercial enzymes, therefore, a great deal of storage stability data are required to guarantee that the user will receive and be able to use the product at the correct enzyme level. Enzymes are most stable in crystalline or dry powder form, but due to health hazards connected with

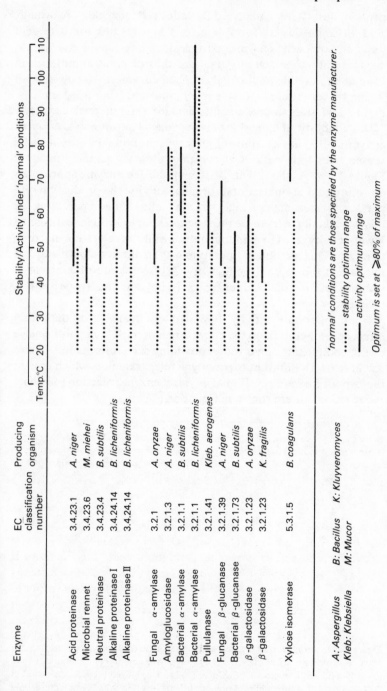

Figure 2.36 Temperature optima ranges for some industrially used enzymes.

protein dusts (*see* Chapter 3.3) industrial enzymes are usually sold in a de-dusted form (e.g. prilled enzymes for detergent use, immobilized enzyme particles) or as stabilized liquids. Temperature–time stability data are thus of great importance in evaluating potential industrially applied enzymes. A knowledge of $T_{\frac{1}{2}}$ can be very useful and save much time and laboratory analysis. Figure 2.37 gives storage stability data for a commercial enzyme.

(*iv*) Stability of immobilized enzymes. The immobilization of enzymes often has a stabilizing effect which allows the enzyme to be reused over long periods at constant, elevated temperature (*see* Figure 2.19). As the activity of an immobilized enzyme approaches one-third to one-quarter of its original activity, the production rate per reactor becomes inefficient. One method of compensating for this is gradually to increase the reaction temperature in response to lower activity so as to maintain output whilst accelerating inactivation rate. This method of operation is referred to as 'burn-off' but may affect the overall productivity. A discussion of the increased stability of immobilized enzymes is beyond the scope of this chapter.

(*v*) Temperature activity. Few data are available on the actual effect of temperature on the enzyme reaction itself, mainly because of the complexities involved in obtaining such information. There are at least 18 different thermodynamic parameters which govern the overall forward reaction of a typical enzyme reaction following the reaction pattern shown in Equation (2).

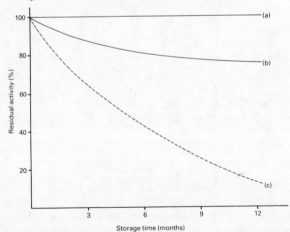

Figure 2.37 Storage stability for a typical commercial enzyme. Data are for *Bacillus fastidiosus* uricase (EC 1.7.3.3) stored over a period of 12 months. Curve (*a*) is lyophilized enzyme at 5°C. Curve (*b*) is a lyophilized preparation stored at 25°C. Curve (*c*) is a lyophilized preparation dissolved in 0.2M borate buffer, pH 9.5 and stored at 25°C (data from Novo Industri).

Since it is impossible to overcome the inactivation effects of heat on an enzyme completely, it is also impossible to attach any fundamental significance to temperature activity optima, as one can for pH activity optima curves. From the point of view of applied enzymology this does not matter too much, as an understanding and quantification of the effect of temperature and time on the inactivation of enzymes is more important than a detailed knowledge of the effect of temperature on catalysis. The important point is to be able to utilize stability data and relevant (apparent) temperature activity data, so as to ensure that an industrial enzyme reaction is run as efficiently as possible. Figure 2.38 compares temperature activity data and stability data under similar conditions.

Because, as we have seen above, immobilized enzymes are often quite temperature stable, and remembering that chemical kinetics tells us that for every 10 °C increase in temperature there will be an approximate doubling in reaction rate, small changes in temperature can have a very significant effect on the output of immobilized enzyme reactors. Figure 2.39 illustrates this point for immobilized glucose isomerase, run, in this particular case, as a batch reaction. This effect will be magnified considerably in continuous reactions,

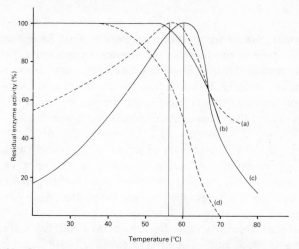

Figure 2.38 Comparison of temperature activity and stability data under similar conditions. The figure compares data for two proteinases produced by *Bacillus* spp. Activity curves show the typical bell-shaped curves and stability data, the half-curves, all measurements being made after 10 minutes reaction time using the Novo Anson assay method. Curve (*a*) shows the activity curve for *Bacillus subtilis* n-proteinase (EC 3.4.24.4) at pH 7.5. Curve (*b*) shows the stability curve for *Bacillus licheniformis* a-proteinase (EC 3.4.24.4) at pH 8.5. Curve (*c*) shows the activity curve for *Bacillus licheniformis* enzyme at pH 8.5. Curve (*d*) shows the stability curve for the *Bacillus subtilis* enzyme at pH 7.0. The temperature activity 'optima' under these conditions are indicated.

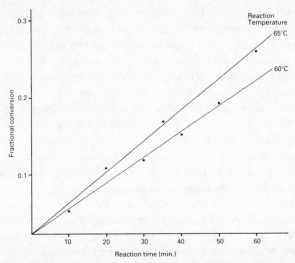

Figure 2.39 Effect of temperature on the overall reactivity of immobilized glucose isomerase in a batch conversion of glucose to glucose/fructose syrup. *Bacillus coagulans* xylose isomerase (EC 5.3.1.5) was used to react with 40 per cent dry substance commercial dextrose at the temperatures indicated. 15 grammes enzyme preparation were used per 100 millilitres syrup at pH 8.5 in the presence of 0.004 M magnesium ions (author's own unpublished data).

and a small change in operating temperature can have a dramatic effect on overall productivity of a reactor run over a period of several months, with resulting improved economic consequences. Operational data have shown that for this particular enzyme, the effect of temperature on the general design equation (Equation (22)) is to alter the value of the constant m as follows:

$$m = 1 - 0.06(65 - T), \quad T = 60 \leqslant T \leqslant 65\,°C$$

Although a 10°C increase in reaction temperature can cause a doubling in reaction rate, enzyme systems can often improve on this increase in reaction rate. Figure 2.40 shows a trebling in maximum reaction rate of the hydrolysis of penicillin by *Bacillus licheniformis* β-lactamase.

(*vi*) Heat-stable enzymes. Most biochemists still think of enzyme reactions as occurring at temperatures of 25–40 °C. In fact, the majority of industrially used enzyme reactions operate at ⩾50 °C. Thermostable α-amylase enzymes derived from strains of *Bacillus licheniformis* are used at temperatures in the region of 90–110 °C (*see* Figure 2.33). In fact, many of the species within the genus *Bacillus* are capable of producing a variety of temperature-stable enzymes, as is indicated in Table 2.4.

Most of the industrially used heat-stable enzymes were developed during the late 1960s and early 1970s on the assumption

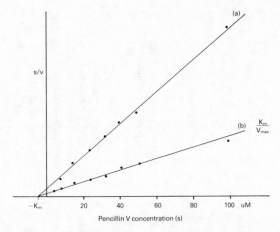

Figure 2.40   Effect of temperature on V_{max} for *Bacillus licheniformis* β-lactamase (EC 3.5.2.6) on the hydrolysis of penicillin V in 0.05M phosphate buffer, pH 6.9 (*a*) at 27°C where V_{max} is estimated at 111 µM per minute per millilitre enzyme preparation and (*b*) at 37°C where V_{max} is estimated at 333 µM per minute per millilitre (author's own unpublished data).

that since all chemical reactions proceed faster at higher temperatures, they could be speeded up still further if temperature-stable enzymes were developed. This would allow the development of continuous processing, reducing overall reaction times from days to hours or minutes, and hence reducing plant capacity size, with obvious investment savings. Heat-stable enzymes currently account for the bulk of industrially used enzymes and find application in the industries associated with the processing of starch and in domestic biological detergents, as described elsewhere in this book (*see* Chapters 4.7 ('Detergents'), 4.15 ('Starch') and 4.16 ('Textiles')).

(*vii*) Low-temperature active enzymes. Reaction at elevated temperatures requires the use of energy, and the increase in fuel and steam costs over the past years (up to 10 times in 6 years) have focused attention on enzymes which have high activity at low temperature, that need not necessarily be temperature stable. 'Low energy' proteolytic enzymes are thus in demand for the new generation of low-temperature wash, biological detergents. The use of microbial rennet in the production of cheese as a substitute for calf rennet (EC 3.4.23.6) has been hampered in taking over a larger share of this market by the fact that it was more temperature stable than the animal rennet and hence there was doubt in the industry that it might not be completely inactivated during the 'cooking' phase of the cheese-making process, and so might impair the ripening of the cheese on storage (*see*, however, page 71 and Figure 2.43; *see also* Chapter 4.6 ('Dairy')).

TABLE 2.4

Some temperature stable enzymes of current commercial and research interest produced by species of *Bacillus*

Species	Enzyme common name	EC number	Area of interest
Bacillus stearothermophilus	Amino acyl t-RNA synthetase	6.1.1.x	Protein biosynthesis
	RNA polymerase	2.7.7.6	
	Rhodenese	2.8.1.1	Research and development purposes or clinical assay
	Superoxide dismutase	1.15.1.1	
	Glyceraldehyde-3-phosphate dehydrogenase	1.2.1.12	
	Phosphofructokinase	2.7.1.11	
Bacillus amyloliquefaciens	α-amylase	3.2.1.1	Maltose production
Bacillus subtilis	n-proteinase	3.4.24.4	Brewing/protein industries
Bacillus licheniformis	α-amylase	3.2.1.1	Glucose production
	α-proteinase	3.4.21.4	Biological detergents
Bacillus cereus	β-lactamase	3.5.2.6	Penicillin resistance
Bacillus coagulans	Xylose isomerase	5.3.1.5	Fructose production
Bacillus fastidiosus	Uricase	1.7.3.3	Clinical assay of uric acid
Bacillus pumilis	β-glucanase	3.2.1.73	Brewing industry

One application of low-temperature active enzymes in a positive sense is during fermentation processes (0–15°C), when saccharification enzymes are used to produce super-attenuated worts for the production of low-calorie lagers or highly fermented distiller's wash. Figure 2.41 shows the characteristics of two amylases produced by species of *Aspergillus*, and indicates that the *Aspergillus oryzae* enzyme still has approximately one-fifth of its maximum activity even at 15°C, so that it will exert a catalytic effect at these low temperatures over the long periods required to produce the lager (3–12 weeks). The performance of immobilized enzymes for continuous processing of complex biological substances at intermediate temperature ranges (20–50°C) is easily impaired by microbiological infection and contamination, so that enzymes of high activity at low temperature are of interest in such applications. Curve (*c*) in Figure 2.6 shows how the hydrolysis of whey lactose at 4°C overcomes these problems, although the enzyme used in this particular instance can hardly be regarded as having a particularly high activity.

It has been found that enzymes active at low temperature generally have low K_m values, and we can expect these types of enzyme to become more important in the future.

Effect of pressure. In addition to temperature and pH, the other common factor affecting non-enzymatic catalysts is pressure. The effect of pressure is to reduce volume, and many catalysed inorganic or petrochemical reactions are performed at increased pressure in order to push the equilibrium in favour of the desired reaction

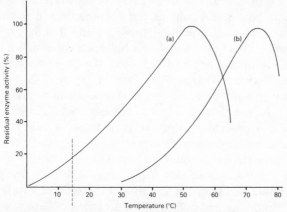

Figure 2.41   Temperature activity characteristics of two amylase enzymes produced by a different species of *Aspergillus*: (*a*) low temperature active fungal α-amylase (EC 3.2.1.2) produced by *Aspergillus oryzae* compared to the thermostable *Aspergillus niger* amyloglucosidase (EC 3.2.1.3) (data from Novo Industri).

product. The classical example is the Haber process for fixing nitrogen by catalytic reduction to ammonia:

$$N_2 + 3H_2 \rightleftharpoons 2NH_3$$

Pressures of 200–1000 atmospheres are used at 500°C in the commercial process for the synthesis of ammonia. Other points to notice are that for the reaction to occur, four molecules must come together to react on the surface of a nonspecific heterogeneous catalyst. This is in complete contrast to enzyme catalysis, which usually involves only two molecules coming together for reaction at one time ($E + S \rightleftharpoons ES$, $ES + S' \rightleftharpoons EP$), the involvement of highly selective catalysts, and overall reaction in a homogeneous phase. Hence, it is not surprising that enzyme reactions are usually indifferent to pressure, and any effect of pressure on an enzyme is usually detrimental, since it is likely to cause changes in the tertiary or quaternary structure of the enzyme, and thus impair activity.

One practical instance of this is the extraction of intracellular enzymes from bacterial cells using an X-press or similar piece of equipment, which produces high pressure and shearing forces on the cells. Enzyme yields, measured in terms of enzyme activity are often lower than when cells are disrupted by more gentle means such as enzyme digestion, although it is difficult to prove this definitively. Another specific case of the detrimental effect of pressure is in the operation of immobilized enzymes in column reactors. In order to get substrate through the enzyme layer, a certain pressure difference over the enzyme bed is necessary. This pressure difference (ΔP) depends on several factors such as the height of the bed, the flow-through rate of substrate and its viscosity (related to both solids content and temperature). If the necessary pressure difference is too large, there is a risk that the enzyme will be so compressed that it will simply block the column. Whether this happens or not depends on the physical strength of the immobilized enzyme particles. This is illustrated in Figure 2.42, where the necessary pressure difference is shown as a function of running time for three different preparations of immobilized glucose isomerase. The force to which the enzyme is exposed is indicated by the product of bed height and substrate velocity. Figure 2.42 shows that enzyme preparation A could not withstand 6 metres × metres per hour (m × m/h) without a drastic increase in ΔP, whereas preparation B could. Preparation C can withstand 24 m × m/h without pressure build-up, and industrially used enzymes have to withstand limits of between 23 and 40 m × m/h for a maximum bed height of 5 metres, which is itself a restriction on overall reactor size.

Other physical parameters. Ionic strength affects the solubility of proteins and hence of enzymes. Whilst low ionic strengths may increase enzyme solubility, high ionic strength can cause coagulation, precipitation and hence loss of enzyme activity in solution. This does not mean that the precipitated enzyme itself loses its activity, since industrially prepared enzymes often undergo a salt precipitation stage as part of their extraction/concentration processing for production. Many liquid industrial enzymes are stabilized by simple inorganic salts. The effect of particular ions is discussed on page 74.

Since enzymes are charged in solution, they are affected by ionic and electrostatic charges on other materials. This forms the basis of analytical and purification techniques such as ion-exchange chromatography and electrophoresis. It also explains why certain enzymes bind to chargeable materials (such as glass and cellulose). The deliberate use of magnetic materials as a support for enzymes can also render immobilized enzymes subject to magnetic and electromagnetic forces. These and similar effects are, however, only incidental in affecting the performance of an enzyme, although they can be very important in operating enzyme processes.

Stabilization, destabilization and inactivation. So far, we have seen that enzymes express their maximum effectiveness only within certain (narrow) limits of physical conditions, and we have already hinted that these limits can be modified either by the pressure of certain chemical substances, or by pretreatment of the enzyme with

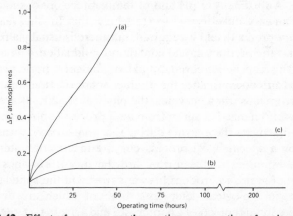

Figure 2.42 Effect of pressure on the continuous operation of an immobilized enzyme reactor. The figure shows the measured pressure difference over the enzyme bed for three development preparations of glucose isomerase. All three columns were run at 65°C with 45 per cent dry substance substrate feed. Curve (a) shows an early (unsatisfactory) enzyme preparation operating at a flow-through velocity of 3 metres per hour in a 2-metre high enzyme bed. Curve (b) is a modified product operating satisfactorily under the same constraints. Curve (c) shows a preproduction product in a 3-metre high bed operating under a flow-through rate of 8 metres per hour (data from Novo Industri).

protein-reactive agents, as in the production of immobilized enzymes. In Sections 4 and 5 we shall first look at the effect of chemical substances on the enzyme reaction, and then at the interactions of physical parameters and chemical substances on an enzyme reaction, to see how the kinetics of the overall reaction is affected. However, before moving on to this, let us look at enzyme destabilization and inactivation generally.

We have seen that although there is a finite difference between pH activity and stability, the distinction between temperature activity and stability is more tenuous. However, in practical terms, there are useful regions of overlap within which the enzyme is both stable and maximally effective (e.g. *see* Figures 2.28 and 2.36). It follows, therefore, that by deliberately moving out of these regions, enzymes will become inactivated and reaction will cease. This is precisely one of the requirements for most industrial enzyme reactions, as it provides us with an easy way of controlling an enzyme reaction, stopping it at any stage, and thus giving us enormous flexibility.

In the controlled hydrolysis of starch or protein, for example, many of the useful properties of the products obtained are derived precisely from the limited degree of hydrolysis achieved, and their production depends both on being able to measure the degree of hydrolysis effected by the enzyme reaction, and on being able to stop the reaction at a precise point in time, by inactivating the enzyme. Adjustment of pH and/or temperature out of the optima limits is an easy and effective way to achieve this. In other reactions, maximum product yield is required, but there is usually a requirement that the product should contain no residual enzyme activity. Again, this can be achieved by pH or temperature adjustment, provided this does not alter the product obtained. In several industrial conversions using enzymes, the physical conditions are either deliberately changed or suffer from poor process control, and these tend to inactivate the enzyme during reaction anyway, without the need for a specific post-process enzyme inactivation step. Two obvious examples of the former are the use of enzymes in the mashing of grains to produce brewer's wort and the application of enzymes in biological detergents, both of which are discussed elsewhere in this book (*see* Chapters 4.5 and 4.7). In fact, the skill in designing enzyme reactions is to achieve a 'dead heat' situation in which maximum product yield is achieved, just as the enzyme loses its activity. Ways in which this can be achieved are discussed in Section 5.

Failure to inactivate an enzyme completely in a product can lead to all sorts of problems, if the product of the enzyme reaction is to be used as a raw material or ingredient for another material. Some specific problems in the author's experience are bottles of runny tomato ketchup, packets of non-setting blancmanges and bottles of exploding, highly alcoholic 'low-calorie' lagers – all caused by residual amyloglucosidase used to hydrolyse the starch ingredient in their formulation. Enzyme inactivation is especially a problem if highly heat-stable enzymes are used, such as *Bacillus licheniformis* α-amylase. The problem of acceptance of microbial rennet was mentioned previously precisely on these grounds. One way this has been overcome is by producing a destabilized form of the enzyme by chemically modifying the enzyme to make it more heat labile (e.g. by selective oxidation of some of the constituent amino acids in the enzyme protein). Figure 2.43 illustrates this point. Hence, a knowledge of physical optima for an enzyme not only allows it to be characterized or used at its maximum efficiency, but also allows its effect to be precisely controlled.

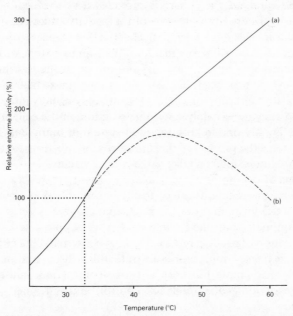

Figure 2.43 Comparison of the temperature characteristics of normal and destabilized microbial rennet produced from Mucor *miehei* (EC 3.4.23.6). Curve (*a*) shows the relative activity of the normal produce assayed by the modified Berridge method at pH 6.5 in the presence of 10 milligrammes calcium chloride per 100 millilitres. Curve (*b*) shows the relative activity of the destabilized enzyme assayed under the same conditions. At temperatures below 35°C the activity of both enzymes is similar, above this temperature – while the activity of the normal enzyme continues to increase – that of the destabilized enzyme has its maximum activity at about 45°C. At pH values above 5.6 after pasteurization at 72–74°C for 15 seconds, the normal enzyme still retains 60 per cent of its activity, while that of the destabilized enzyme is 0 (data from Novo Industri).

4. Effect of chemical substances

The effect of changes in physical conditions on an enzyme reaction is to cause a corresponding gradual change in overall enzyme reactivity, due to the nonspecific response of the protein nature of the enzyme. In contrast, the effect of chemical substances on the activity of an enzyme is often precise and specific. Absolute changes in activity result in response to even minute concentrations of substances to which the enzyme is reactive. This is because the substances react either at the catalytic site of the enzyme, or at secondary 'binding sites' which are intimately associated with the reactive catalytic site on the enzyme molecule. From an applied enzymological point of view, it is important to understand, in outline, some of these effects. On the one hand, it gives possible answers as to why an enzyme system operating on a 'natural' complex substrate, of low purity, does not always perform as expected, and on the other, it provides a powerful alternative method to 'finely tune' or control the activity of an enzyme at a molecular level, simply by adding chemical substances to or removing them from the substrate medium. In order to do this, we need to know something of the way enzymes are controlled in the living systems which produce them. An important point to bear in mind is that whilst at a technological level we are only usually interested in using an enzyme to catalyse a single reaction, in the cell, this same enzyme is only one in an integrated system of many enzymes. In such systems the product of one enzyme is the substrate for the next – or more likely, many alternative enzyme systems.

Chemical control of enzyme activity. There are several ways in which enzyme reactions are controlled or modified at a molecular (subgenetic) level, in order to ensure that enzyme systems operate in an optimal, controlled manner. A very simple one is that an accumulation of product reduces the enzyme activity, hence hindering it from transforming more substrate; this is discussed later in this section. This simple but very effective control mechanism is an example of molecular feedback control, and is termed product inhibition.

Many enzymes are synthesized as inactive proteins, and need to be activated before they can express their catalytic effect. This may require the provision of a specific (activating) chemical substance, or the removal of a portion of the protein which is occluding the active site of the enzyme. Activating substances can be simple inorganic ions (certain group II and VIII ions are common) or complex organic molecules such as the vitamins and their derivatives, or other enzymes. Hydrolytic and isomerization enzymes generally require only simple cationic activators, which therefore promotes them as candidates for cheap enzyme catalysts in industrial use, whereas oxidative enzymes require specific organic molecules to act as electron acceptors and need therefore to have a relatively complex structure. The enzymes which are produced in an inactive protein form are usually extracellular hydrolytic enzymes, whose activity is primarily towards high molecular weight substrates, such as carbohydrates or proteins. There is an obvious need for a proteolytic enzyme to be synthesized in an inactive form, since it would immediately begin to hydrolyse the surrounding tissue or cell structure. Carbohydrate-hydrolysing enzymes produced by higher organisms have a 'built-in' time-delay mechanism in addition to being produced in an inactive form. This is typified by cereal β-amylase (EC 3.2.1.2), which becomes naturally active only during germination. This feature of 'pro-active' production is a device used by living cells to control the activity of several other groups of biologically active substances such as the peptide hormones. Enzyme activation is discussed in more detail, later in the section.

A third simple way in which enzyme activity is modified is through the substrate itself. Hydrolytic enzymes acting on high molecular weight materials are often more reactive towards a modified form of the substrate molecule than unmodified (soluble) substrate. Thus, many proteinases exhibit a higher catalytic effect on certain denatured (coagulated) proteins than on a simple protein solution. Amylase enzymes are also more active towards gelatinized starch and cause more effective liquefaction after starch gelatinization. This means in industrial terms that less enzyme is required per unit time for the hydrolysis of starch to fermentable sugars, although the economic advantages of this would have to be carefully balanced against the amount of energy needed to be put into the system to effect, for example, gelatinization.

Activation and deactivation. Substances which promote the expression of maximum enzyme activity – other than the substrate itself – are known as activators. Conversely, substances which have the opposite effect and act by tending to remove the activating substance or by preventing its action, are called deactivators. Activators and deactivators are really distinct from stabilizers and inhibitors, although in practice this distinction is of lesser importance. This is evident when considering reaction at elevated temperatures – the conditions under which most industrial reactions are carried out – where the distinction between activity and stability is not precise. Activating substances exert their effect by affecting the various active sites on an enzyme. The use of activators or deactivators provides a means of coarse chemical control of enzyme reactions, although usually we are only interested in maximal enzyme effectiveness.

(*i*) Metal ion activation. More than 75 per cent of all known enzymes require the presence of metal ion activators to express their full catalytic effect. Their mechanism of action is varied, but for our purpose we need only note that the binding of the metal ion can be very loose at one extreme, so that appropriate cations need to be added with the enzyme for the reaction to occur efficiently, or at the other, very strong, as in the case of true metalloproteinases (EC 3.4.24.-). Usually, only very small concentrations of activating ions are required, and these are often present in sufficient concentration in the natural biological substrate being processed or as part of the active site of the enzyme. Thus, it is only when activating substances are deliberately abstracted by the use of deactivating substances (e.g. sequestering agents) or conditions, that their effects become apparent. Usually, restoration of the required ionic environment will restore maximal enzyme activity. Activation and deactivation are thus due to specific ion effects and deactivation is distinct from denaturation of enzyme protein (due to high non-specific ion concentration) or inhibition by other metal ions.

Metal ion activation of enzyme reactions is thus important industrially in achieving maximal catalytic efficiency. As an example, the important industrial enzyme glucose isomerase (EC 5.3.1.5) from *Bacillus coagulans* needs to be activated by both cobalt and magnesium ions to become fully active. The cobalt ion is 'tightly bound' to the enzyme protein, whilst the magnesium ion is not, and has to be added in the substrate as activator in order to ensure maximal catalytic effect. Although it is not known why cobalt has an activating effect on this enzyme, it is probably associated with the ability of this transition element to exist in several oxidation states, since the isomerization reaction itself is in effect an intermolecular oxidoreduction of an aldose sugar to a keto sugar (*see* Figure 2.44). The role of the magnesium ion is also interesting, since if Mg^{2+} is replaced by Ca^{2+}, enzyme activity is lost (*see* Figure 2.45). Activity can be restored by removing the Ca^{2+} and replacing by Mg^{2+} or by ensuring that the molar ratio of Mg^{2+} to Ca^{2+} is greater than 20. This very specific effect is interesting both academically and practically. From a theoretical point of view, of all the elements, the two which chemically resemble each other most closely are probably magnesium and calcium. From an industrial point of view, production of glucose syrup to act as substrate for the isomerization enzyme can involve an enzyme which requires Ca^{2+} as a stabilizer (*see* Figure 2.47). Thus, an ion exchange step has to be introduced in order to ensure maximal activity of the glucose isomerase enzyme which, although both inconvenient and costly, is unavoidable.

```
    C H O              C H₂O H
    |                  |
    C H O H            C O
         ⇌
  (C H O H)ₓ          (C H O H)ₓ
    |                  |
    C H₂O H            C H₂O H
```

x=3: Glucose/Fructose
x=2: Xylose/Xylulose

Figure 2.44 The intramolecular oxidoreduction of sugars by isomerizing enzymes (EC 5.3.1.5)

Another example of the consequences of deactivation is the use of enzymes in biological detergents, one of the first modern large-scale applications of enzymes. Because detergent formulations contain sequestering agents to overcome the 'hardness' of the domestic water supply, metalloproteinase enzymes could not be used (even if they had the correct pH and temperature optima characteristics). Instead, serine proteinases (EC 3.4.21.-) are used, as they are active in the presence of sequestering agents. It is also worth remembering, but nevertheless a common mistake, that buffers which contain sequestering agents, such as phosphates, should not be used when assaying enzymes which are metalloproteins or are activated by metal ions. Conversely, the activity of metalloproteinases and serine proteinases – often produced simultaneously by the same microorganism – can be distinguished and assayed by the same assay method by carrying out the reaction in the presence and absence of EDTA. In its absence, total proteolytic

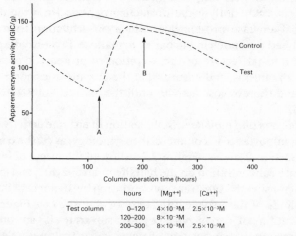

	hours	[Mg⁺⁺]	[Ca⁺⁺]
Test column	0–120	4×10⁻³M	2.5×10⁻³M
	120–200	8×10⁻³M	–
	200–300	8×10⁻³M	2.5×10⁻³M

Figure 2.45 Activation and deactivation of xylose isomerase by magnesium and calcium ions. The control column was run in the absence of the calcium deactivator while the test column had the ionic composition of the feed changed as indicated, at points A and B (data from Novo Industri).

activity is measured, and in its presence serine or 'alkaline' protein-ase activity is measured.

(*ii*) Cofactor activation. The few oxidative enzymes that are used on a large scale have in-built co-enzyme molecules, and use molecular oxygen directly as the final electron acceptor. These are the EC 1.x.3. enzymes such as glucose oxidase where $x = 1$ or monoamine oxidase where $x = 4$. Thus, the industrially used *Aspergillus niger* glucose oxidase (EC 1.1.3.4) contains two molecules of flavine adenine dinucleotide (FAD) per molecule of enzyme as activating cofactor. Removal of this (e.g. by dialysis) causes deactivation of the enzyme; this can be restored by addition of the FAD.

(*iii*) Activation of pro–enzymes. Hydrolytic enzymes which are produced naturally in their inactive (pro-enzyme) form can be activated by simple modification of their structure and represent a sophisticated method of molecular control. Cereal β-amylase, mentioned previously, can be activated by chemical agents such as cysteine or certain proteolytic enzymes which hydrolyse the protecting peptide portion of the pro-enzyme from the enzyme molecule to liberate the active enzyme. Laboratory experiments designed to investigate whether alkaline or neutral proteolytic enzymes would be more suitable in the production of brewer's wort from barley (*see* Chapter 4.5) showed unexpectedly, that as the proportion of neutral proteinase component was increased, the level of fermentable sugar increased! Closer analysis showed that this was due almost entirely to an increased maltose level. Further tests showed that only the neutral proteinase had the correct specificity to hydrolyse the protecting peptide from the dormant barley β-amylase pro-enzyme, which on activation then readily hydrolysed the starch in the barley to maltose. Pro-enzyme production is a usual feature of the secretion of proteolytic enzymes in higher organisms, and ensures that the enzyme does not hydrolyse itself and thereby lose activity until the 'correct' substrate is presented.

Stabilizers and inhibitors. Stabilization of enzyme activity is of the utmost importance in commercial enzymology, as can be concluded from Section 2 above. Figure 2.34 showed the effect of increased substrate concentration on the half life of an enzyme, and indicated that increasing the substrate concentration by 40 per cent increased the half life of the enzyme by a factor of almost 3 (*see* Figure 2.46). Stabilization of an enzyme by its substrate is an important phenomenon and means that enzymes can often operate at higher temperatures and for longer periods if high substrate concentrations are used. In other words, the presence of substrate increases the potential enzyme productivity. This is another reason why

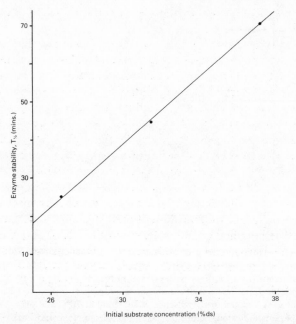

Figure 2.46   Stabilization of an enzyme by its substrate. The data from Figure 2.34 are used to show that increasing the substrate concentration 'protects' the enzyme from inactivation, as revealed by the increasing half life of the enzyme at higher substrate concentration.

temperature activity and stability data need careful interpretation: efficient enzyme reactions can often be run at temperatures in excess of their indicated maximum. Since enzymes can catalyse reactions in both directions, it also follows that enzymes are often stabilized by their products too, which means that enzymes can be more difficult to inactivate completely at the end of a reaction, and tests should be carried out to ensure that there is actually no residual enzyme activity in the product after enzyme treatment.

These properties are utilized in a positive way in order to stabilize liquid enzyme solutions during the long period between manufacture and customer use. Thus, amylases are often stabilized with starch hydrolysates, sugars or their derivatives, whilst proteolytic enzymes are stabilized by the peptides produced by their reaction on fermentation proteins. In addition to these natural stabilizing substances, enzyme solutions can be stabilized or preserved using the usual range of food preservative chemicals or simple salts.

In addition to acting as enzyme activators, certain metal ions can also act as stabilizers. An important industrial example is the effect of calcium ions on the stabilization of α-amylase. Figure 2.47 shows the effect of calcium ions on two *Bacillus* α-amylase variants, and

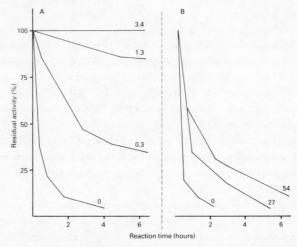

Figure 2.47   Stabilization of α-amylase enzymes by calcium ions (A) for *Bacillus licheniformis* enzyme and (B) for *Bacillus amyloliquefaciens* enzyme. The levels of calcium are shown on the figures, expressed in terms of parts per million. Both enzyme preparations had exactly the same initial concentration (60 Kilo Novo Units) and were reacted in the absence of substrate at 70°C and pH 6.5–7.0 for the time periods indicated, before the assay (data from Novo Industri).

indicates that the *Bacillus licheniformis* enzyme is much less dependent on calcium ion than the *Bacillus amyloliquefaciens* enzyme, which makes it more suitable in the production of low conductance glucose syrups or as a feedstock for the production of high-fructose syrups, since less downstream ion-exchange capacity is required. Similarly, the presence of a sequestering agent (as, for example, in the formulation of biological dishwashing detergents) would preclude use of the *amyloliquefaciens* enzyme.

Other metal ions can have an inhibitory effect on enzymes, a property they share with certain substrate analogues. This gives a clue as to the effect of inhibitors on enzymes. Inhibitors are specific substances which, at low concentration, decrease the rate of an enzyme-catalysed reaction. Inhibitors are selective agents that interact with only a limited number of sites on the enzyme protein without significantly altering the three-dimensional structure of the enzyme protein. Hence, they are distinct from general protein-reactive substances such as nonspecific chemicals (acids, alkali, salts, urea, detergents, etc) which only disrupt the general protein structure. They are also distinct from more specific protein-reactive polymers such as immuno-proteins, which are perhaps better regarded as anti-enzymes, rather than enzyme inhibitors.

In industrial enzymology, the main importance of inhibitors is that they reduce the efficiency of enzyme reactions. This dictates the way in which they are managed. A first approach should be to try to avoid them, and only if this fails, to attempt to accommodate or remove them. However, before this can be done, their effect has to be identified and quantified, and this usually requires a considerable amount of careful laboratory work. Their treatment in the context of this chapter will, therefore, be limited to these aspects. In the future it is likely that inhibitors will be used more positively in industrial enzymology, and this is discussed on page 91. In applied enzymology, the two most important considerations are inhibitor reversibility and specificity. Thus, we can regard inhibitors as falling into two groups: those whose effects are reversible, and those which have an irreversible, or permanent, inhibitory effect on the enzyme.

Reversible inhibitors. Reversible inhibitors usually combine with an enzyme in a simlar way to a substrate, but instead of being transformed to product, they stop at the enzyme complex step and form an equilibrium system with the enzyme and substrate. We can recognize a definite degree of inhibition which remains constant during at least the initial part of the enzyme reaction, and the amount of inhibition is related to the concentration of the inhibitor itself, the amount of enzyme and the concentration of the substrate (Equations (29), (32), (37)). As the name implies, the effect of inhibition can be overcome, and this can be achieved by altering slightly the environment of the reaction (*see* page 89).

Classical enzyme kinetic studies have proposed at least eight types of reversible inhibition: (*i*) competitive inhibition, (*ii*) uncompetitive inhibition, (*iii*) noncompetitive inhibition, (*iv*) mixed inhibition, (*v*) partial inhibition, (*vi*) substrate inhibition, (*vii*) product inhibition and (*viii*) allosteric inhibition. Each of these can be distinguished by slightly different initial rate reaction kinetics. However, in applied enzymology we are usually only interested in the overall reaction, and the effect an inhibitor will have on this. The mechanism by which inhibition occurs is often, but not always, irrelevant. Certain of these types of inhibition are extremely rare even for simple non-industrially important reactions (e.g. uncompetitive inhibition), whilst others (e.g. mixed inhibition) reduce under practical conditions to more simple forms of inhibition. A detailed examination of types of inhibition is superfluous to our needs in applied enzymology, and so we will limit our remarks to the most relevant points – a consideration of competitive, noncompetitive, substrate and product inhibition.

(*i*) *Competitive inhibition.* This is perhaps the most common form of reversible enzyme inhibition and is often caused by substances which closely resemble the substrate in chemical structure, or are close analogues of the substrate. In fact, it is precisely because of this close structural similarity that inhibition occurs, since both inhibitor and substrate are visualized as competing for the same binding site on the enzyme molecule. However, certain irreversible inhibitors have close structural similarity to their substrate analogues, as we shall see later. Whatever the mechanism of the inhibition, a non-productive enzyme–inhibitor complex is formed, and the inhibitor must therefore become dissociated from the enzyme molecule and be replaced by a molecule of the true substrate, before reaction can visibly take place. The kinetic model is thus visualized as:

$$E + S \rightleftharpoons ES \rightarrow E + P$$

$$-I \left\Vert \, +I \right. \tag{26}$$

$$EI$$

Using the same logic as used in deriving the Michaelis–Menten equation on the steady-state assumption (see pages 25, 26)

$$K_m = \frac{[E][S]}{[ES]} \qquad \text{or} \qquad [E] = \frac{[ES]K_m}{[S]} \tag{11}$$

In addition, we can define a value K_i, being the dissociation constant of the new enzyme–inhibitor complex

$$K_i = \frac{[E][I]}{[EI]}$$

so

$$[EI] = \frac{[E][I]}{K_i} \tag{27}$$

The total enzyme concentration, E_0, will now be distributed in this case between three species, so that

$$E_0 = [E] + [EI] + [ES] \tag{28}$$

If we substitute the value of $[EI]$ from Equation (27) in Equation (28)

$$E_0 = [E] + [E][I]/K_i + [ES]$$

Factorizing $E_0 = [E](1 + [I]/K_i) + [ES]$

Substituting the value of $[E]$ from Equation (11)

$$E_0 = \frac{[ES]K_m}{[S]} \, 1 + [I]/K_i + [ES]$$

Factorizing again $E_0 = [ES] \left(1 + \frac{K_m}{[S]} \, (1 + [I]/K_i) \right)$

Therefore

$$[ES] = \frac{E_0}{1 + \dfrac{K_m}{[S]} \, (1 + [I]/K_i)}$$

Multiplying by $[S]$,

$$[ES] = \frac{E_0[S]}{[S] + K_m(1 + [I]/K_i)}$$

And putting $[ES] = v/k_3$ as in Equation (6),

$$v = \frac{k_3 E_0 [S]}{[S] + K_m(1 + [I]/K_i)} \quad (29)$$

In other words, the only difference between Equation (29) and the standard Michaelis–Menton Equation (13) is that the presence of a competitive inhibitor of concentration I increases the apparent value of K_m by a factor of $(1 + [I]/K_i)$, thus reducing the observed initial reaction velocity. This is indicated in Figure 2.48. Thus, if Equation (29) is integrated to determine the reaction time, t_i, in the presence of inhibitor, a similar equation to Equation (18) will be obtained:

$$t_i = \frac{(S_0 - S) + K_m(1 + I/K_i) \ln S_0/S}{k_3 E_0} \quad (30)$$

Such an approach has been valuable in the screening of semi-synthetic penicillins for their ability to inhibit the penicillin-hydrolysing enzyme β-lactamase (EC 3.5.2.6), which is often responsible for the ineffectiveness of penicillin against resistant bacteria. The rationale was to look for semisynthetic penicillin analogues which had a low enough value of K_i, so that even at low concentration of I, they would increase the value of K_m (for the penicillin – β-lactamase complex) sufficiently to extend the reaction time required to decrease the concentration of the antibiotically active penicillin from S_0 (a typical blood level) to S (a value just higher than the minimum inhibitory dose of antibiotic effective against the bacterium), so that it was significantly longer than the doubling time of the bacterium, thus allowing the penicillin to inactivate the living bacterium during cell division. An example of

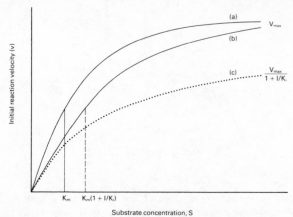

Figure 2.48 The effect of (a) substrate concentration on initial enzyme reaction velocity, in the absence and presence of (b) a competitive and (c) a noncompetitive inhibitor.

such a compound is given in Figure 2.49(a).

A similar approach was used to study the effect of sorbitol on the isomerization of glucose to fructose in an immobilized column reactor. Sorbitol is not usually a constituent of glucose syrup produced from hydrolysis of starch, but it is used as an enzyme stabilizer for the amyloglucosidase enzyme (EC 3.2.1.3) used in this process to saccharify liquefied starch. If sufficient enzyme were used (e.g. to achieve a saccharification reaction in less than 12 hours), the sorbitol concentration could significantly impair the isomerization of glucose by acting as an inhibitor.

(*ii*) Noncompetitive inhibition. In this type of inhibition, the inhibitor is regarded as forming a non-productive complex with the enzyme, irrespective of whether or not the substrate is bound. The inhibitor acts by destroying the catalytic site on the enzyme, but it

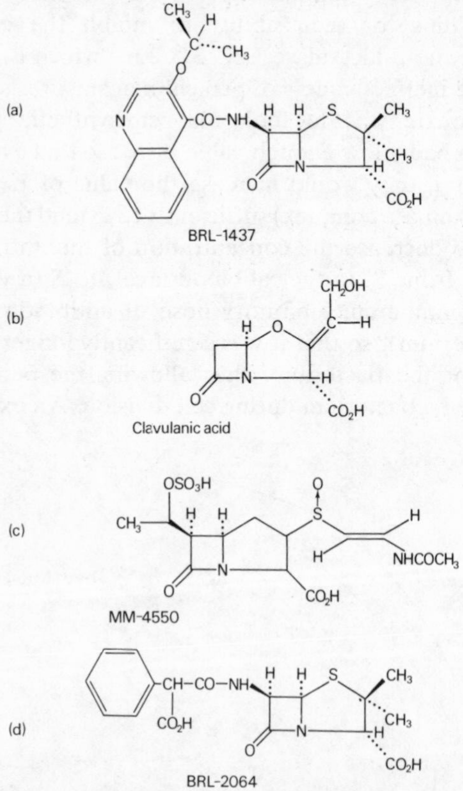

Figure 2.49 The chemical structure of several β-lactamase inhibitors. (*a*) the competitive inhibitor of gramme negative β-lactamase enzymes BRL–1437, 2-isopropoxy-1-naphthylpenicillin. (*b*) The irreversible inhibitor clavulanic acid. (*c*) The irreversible inhibitor MM–4550. (*d*) The substrate inhibitor BRL–2064, α-carboxy-penicillin (Cole, Fullbrook & Elson, 1972).

does not affect the binding of the substrate. The kinetic model for non-competitive inhibition is thus visualized as

$$E + S \rightleftharpoons ES \xrightarrow{k_3} E + P \qquad (31)$$
$$-I \Big\Vert + I \quad -I \Big\Vert + I$$
$$EI \underset{-s}{\overset{+s}{\rightleftharpoons}} EIS$$

This situation is different from those we have met previously, in that ES can be arrived at by different routes, that is, directly, or via the EIS complex, so that we cannot assume that $ds/dt = 0$ and have to invoke the assumption that the rate-limiting step is governed by k_3 and that the system is in equilibrium. In the simplest possible model (simple linear noncompetitive inhibition) the substrate does not affect the binding of the inhibitor either, so that the reactions

$$E + I \rightleftharpoons EI \text{ and } ES + I \rightleftharpoons EIS$$

have identical dissociation constants K_i. Thus,

$$K_i = \frac{[E][I]}{[EI]} = \frac{[ES][I]}{[EIS]}$$

Following the same logic as before, the total enzyme concentration E_0 will now be distributed between four species:

$$E_0 = [E] + [ES] + [EI] + [EIS]$$

Thus $\qquad E_0 = [E] + [ES] + \dfrac{[E][I]}{K_i} + \dfrac{[ES][I]}{K_i}$

Factorizing $\qquad E_0 = ([E] + [ES])(1 + I/K_i)$

Substituting for $[E]$

$$E_0 = \left(\frac{[ES]}{[S]} K_m + [ES] \right) (1 + I/K_i)$$

So that $\qquad E_0 = [ES](K_m + 1) \left(1 + \dfrac{1}{K_i} \right)$

Therefore $\qquad [ES] = \dfrac{E_0}{\left(\dfrac{K_m}{S} + 1 \right) \left(1 + \dfrac{I}{K_i} \right)}$

That is $\qquad v = \dfrac{[S]}{K_m + [S]} \dfrac{k_3 E_0}{(1 + I/K_i)} \qquad (32)$

Comparing Equation (32) with the standard (uninhibited) rate Equation (13), we see that the presence of a noncompetitive inhibitor of concentration I reduces the value of $k_3 E_0 (= V_{max})$ by the factor $(1 + I/K_i)$, while the value of K_m is unaltered. This is indicated in Figure 2.48. Since this is the same factor as affected the increase in K_m caused by a competitive inhibitor, integration of

Equation (32) to find the inhibited reaction time will give an identical expression to Equation (30). This means that it is impossible to deduce the type of inhibition occurring by simply observing the overall reaction, and one has to go back to initial reaction rate studies in order to fix the nature of the inhibition.

There are relatively few reported instances of noncompetitive inhibition for single substrate reactions, although the binding of heavy metal ions to the active sites of enzymes causing their inhibition has been claimed to follow noncompetitive inhibitor kinetics. A point to bear in mind, however, is that irreversible inhibition can often give kinetic data patterns which are similar to those of noncompetitive inhibition and hence many reported examples of noncompetitive inhibition may actually be irreversible inhibition (*see* page 88). Thus, any cases of metal ion inhibition of industrially operated enzyme processes should be examined very closely indeed if the mechanism of inhibition needs to be sought.

(*iii*) Substrate inhibition. In order to promote maximum reaction rates, it would seem logical to use as high a substrate concentration as practicable in industrial enzyme reactions. However, this can evoke its quota of problems, too, since although enzyme systems obey classical enzyme kinetics at low or intermediate substrate concentrations (as described by Equation (13)) often, at high substrate concentration the initial reaction velocity begins to fall off. This is due to substrate inhibition. Figure 2.50 shows the characteristic diagnostic curve of initial velocity versus substrate concentration for a system exhibiting this type of inhibition. It thus appears that substrate at high concentration inhibits its own conver-

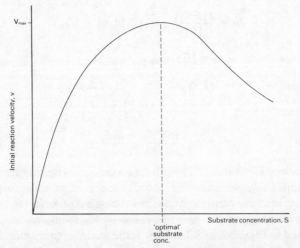

Figure 2.50   The characteristic curve for substrate inhibition.

sion to product. Substrate inhibition is envisaged to occur by the binding of one substrate to the catalytic site on the enzyme molecule, with a simultaneous binding of a second substrate molecule to a separate site on the same enzyme molecule, forming a non-productive complex. This is caused by the increased probability of double binding of substrate molecules as a direct result of their excess concentrations. The kinetic model can thus be visualized as:

$$E + S \rightleftharpoons ES \rightarrow E + P$$
$$-s \parallel +s$$
$$SES$$

(33)

The dissociation constant of this double complex is thus:

$$K_i = \frac{[ES][S]}{[SES]} \quad \text{or} \quad [SES] = \frac{[ES][S]}{K_i}$$

The total enzyme concentration E_0 will be distributed between the three species:

$$E_0 = [E] + [ES] + [SES]$$

Therefore,

$$E_0 = [E] + [ES] + \frac{[ES][S]}{K_i}$$

Factorizing

$$E_0 = [E] + [ES](1 + [S]/K_i)$$

Substituting for $[E]$

$$E_0 = \frac{[ES]}{[S]} K_m + [ES](1 + [S]/K_i)$$

So that

$$E_0 = [ES]\left(\frac{K_m}{[S]} + 1 + \frac{[S]}{K_i}\right)$$

Therefore,

$$[ES] = \frac{E_0}{\dfrac{K_m}{[S]} + 1 + \dfrac{[S]}{K_i}}$$

That is,

$$v = \frac{k_3 E_0 [S]}{K_m + [S] + [S]^2/K_i}$$

(34)

This is consistant with Figure 2.50, since at low $[S]$, $[S]^2$ will be negligible and Equation (34) reduces to the uninhibited Equation (13). Conversely, at high $[S]$, K_m is negligible so that Equation (34) reduces to:

$$v = \frac{k_3 E_0}{1 + [S]/K_i}$$

(35)

Under these conditions the value of v is reduced when S increases, as is observed in Figure 2.50. In an industrial enzyme system, which is constrained by substrate inhibition, such a graph will indicate the 'optimal substrate concentration' at which initial reaction velocity is maximal and hence the overall reaction time minimal, for a given degree of conversion. This can be calculated by integrating Equation 34 (note: S is used instead of $[S]$ for simplicity).

$$v = \frac{\mathrm{d}S}{\mathrm{d}t} = \frac{Sk_3 E_0}{S + K_m + S^2/K_i} = \frac{k_3 E_0}{1 + K_m/S + S/K_i}$$

Therefore, $k_3 E_0 \cdot \mathrm{d}t = (K_m/S + 1 + S/K_i)\ \mathrm{d}S$

$$k_3 E_0 \int_{t_0}^{t} \mathrm{d}t = K_m \int_{S_0}^{S} 1/S \cdot \mathrm{d}S + \int_{S_0}^{S} \mathrm{d}S + 1/K_i \int_{S_0}^{S} S \cdot \mathrm{d}S$$

And, $k_3 E_0 t = K_m \cdot \ln (S_0/S) + (S_0 - S)\ 1 + \dfrac{S + S_0}{2K_i}$

$$t = \frac{(S_0 - S) + (S_0^2 - S^2)/2K_i + K_m \cdot \ln S_0/S}{k_3 E_0} \tag{36}$$

K_i can be determined by plotting a graph of $1/v$ against S, and extrapolating the linear portion of the graph to meet the abscissa $(-S)$, so that when $1/v = 0$, $-S = -K_i$ (*see* Figure 2.51).

Six other mechanisms of substrate inhibition have been proposed in addition to this one, but of these only one other mechanism has been studied in depth for the sort of enzymes involved in industrial enzymology; this is illustrated on pages 35 and 94. An example of substrate inhibition that is important in applied enzymology arose during the investigation of the effect of the semi-synthetic antibiotic α-carboxybenzylpenicillin towards the β-lactamase enzyme produced by *Pseudomonas aeruginosa*, an organism which can cause severe infections following accidents involving burning of the skin. α-carboxybenzylpenicillin (*see* Figure 2.49*d*) was the first semisynthetic penicillin to be effective against *Pseudomonas* infections, and the substrate inhibition effect exhibited against the inactivating β-lactamase enzyme could explain why this particular antibiotic is effective against this organism at high (intravenous) dosage.

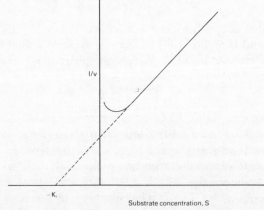

Figure 2.51   Determination of K_i for a substrate inhibitor.

(*iv*) Product inhibition. Product inhibition is not a matter of a shift in the thermodynamic equilibrium between substrate and product due to an increase in the reverse reaction, but is a specific interference with the forward reaction caused by an accumulation of product. As pointed out on page 72, this has a significant role in cell metabolism as a means of feedback control of the enzyme system, but in applied industrial enzymology it represents a serious limitation to efficient enzyme conversion. A current important example of product inhibition is in the hydrolysis of whey lactose by the enzyme β-galactosidase. The galactose produced as a product of the enzyme reaction not only acts as an inhibitor but also serves as a source of byproduct contaminant. This is illustrated in Figure 2.52. This particular problem is a current topic of research and could provide a useful model for a better understanding of this important problem in applied enzymology.

Irreversible inhibitors. An irreversible inhibitor binds to the active site of an enzyme to form an undissociable complex, often due to the formation of a chemical covalent bond: $E + I \rightarrow EI$. The breakdown of the EI complex can only be achieved by chemical reaction, and hence it is not possible to restore enzyme activity by altering the reaction conditions. Irreversible inhibitors therefore represent the most strongly inhibitory substances. This important group of inhibitors contains some of the most potent drugs, for example, the clavulanic acid antibiotics, the nerve gases and the organophos-

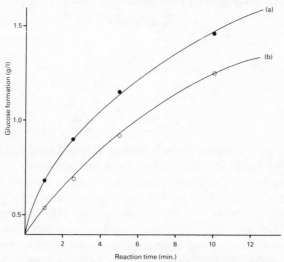

Figure 2.52   Product inhibition of the hydrolysis of lactose by galactose using the *Kluyveromyces fragilis* β-galactosidase (EC 3.2.1.23). Curve (*a*) shows the hydrolysis of 7.5 grammes per litre lactose at 37°C and pH 6.5 with time, and curve (*b*) the same reaction in the presence of 3.3 grammes per litre galactose.

phorous insecticides. It has been shown that alkylating agents such as iodoacetate combine with the —SH groups of cysteine residues located at the active sites of enzymes to form covalent linkages. Similarly, organophosphates can combine with —OH groups of serine residues in alkaline proteinase enzymes (EC 3.4.21.-):

$$E—SH + ICH_2COOH \rightarrow E—S—CH_2COOH + HI$$

Cysteine enzyme

$$E—OH + F—\underset{\underset{\displaystyle OCH(CH_3)_2}{|}}{\overset{\overset{\displaystyle OCH(CH_3)_2}{|}}{P}} = O \rightarrow E—O—\underset{\underset{\displaystyle OCH(CH_3)_2}{|}}{\overset{\overset{\displaystyle OCH(CH_3)_2}{|}}{P}} = O + HF$$

Serine proteinase

The mechanism of the clavulanic acid inactivation of β-lactamase is not fully understood, but presumably it is able to bind to the active centre of the β-lactamase enzyme, and thus protect penicillin antibiotics by preventing their activation by hydrolysis, thereby allowing them to exert their full antibiotic effect (*see* Figure 2.49*b*).

The action of irreversible inhibitors is effectively to reduce the enzyme concentration, and if there is a molar excess of inhibitor present, complete obliteration of the enzyme activity results. If, in a system containing an amount of irreversible inhibitor I_0 there is a molar excess of enzyme E_0, the effective concentration of enzyme will be reduced to $E = [E_0] - [I_0]$. This amount of enzyme will react normally with a substrate, and K_m will not be affected. If excess substrate is added, then in the absence of inhibitor,

$$V_{max} = k_3[E_0]$$

In the presence of inhibitor,

$$V'_{max} = k_3[E_0] - [I_0]$$

Therefore, $$\frac{V'_{max}}{V_{max}} = \frac{[E_0] - [I_0]}{[E_0]}$$

So $$V'_{max} = V_{max}[E_0] \left(1 - \frac{[I_0]}{[E_0]}\right) \qquad (37)$$

In other words, although the inhibitor does not affect K_m, it does reduce the value of V_{max}, just as for a noncompetitive inhibitor. However, the two types of inhibition are clearly distinct, since the value of V_{max} is reduced by different factors: by $[E_0](1 - [I_0]/[E_0])$ for the irreversible inhibitor but by $(1 + I/K_i)$ for the noncompetitive, reversible inhibitor. Another distinction (apart from reversibility tests) is that noncompetitive inhibition – or all reversible inhibition, for that matter – is set up quickly during the phase I reaction between the enzyme, its substrate and the inhibitor, whereas irreversible inhibition is slow and progressive, and often extends throughout the phase II reaction, coming to completion only when all the inhibitor I_0 or all the enzyme E_0, has been converted to EI complex. Thus, in the initial screening experiments on compound MM–4550 (the forerunner of the clavulanic acid inhibitors, *see* Figure 2.49*c*), the rate of hydrolysis of 1.5 microgrammes per millilitre benzylpenicillin by *Escherichia coli* B-11 β-lactamase at 37°C and pH 7.0 (ionic strength = M/20 phosphate) was 7.5 minutes, while in the presence of various dilutions of culture filtrates containing MM–4550, the reaction times were as follows:

Inhibitor concentration (dilution, $1/x$)	Reaction time (minutes)
2,500	14
1,875	17.5
1,000	30

Thus, provided I_0 is greater than E_0 and given enough time, complete inhibition could be achieved. If complete inhibition is required quickly, a higher inhibitor concentration would be required.

Overcoming inhibition effects. If enzyme inhibition is suspected to be the cause of poor enzyme reaction performance, and the cause is not immediately obvious, attempts should be made to determine the type of inhibition occurring by testing a concentrate of the substrate feedstock as if it were an inhibitor. Check first whether the 'inhibitor' is of the irreversible type by incubating test samples with enzyme solution. Remove samples at intervals and assay residual enzyme activity by standard assay. If the enzyme loses activity progressively with prolonged incubation time, the indication is that it is an irreversible inhibition. This should be proved by reacting different concentrations of inhibitor with a fixed concentration of enzyme until complete reaction has occurred and then use this as enzyme source in comparison with uninhibited enzyme for hydrolysis of various substrate concentrations, for example, K_m, $2K_m$, $5K_m$ and $10K_m$. Measure the initial reaction rates and estimate V_{max}. Substitute data in Equation (37) to verify. If irreversible inhibition is indicated, steps should be taken to identify the source of inhibitor in the feedstock and to attempt to avoid or remove it (it could be as simple as a heavy metal ion, requiring only ion exchange). Even minute amounts of irreversible inhibitor are potentially disastrous to the operation of an immobilized enzyme used continuously in a flowthrough reactor over a relatively long time period.

On the other hand, reversible inhibitors present in substrate feedstocks are often more of an inconvenience than a total disaster, and there is usually a way of minimizing or overcoming their effects. The effect of inhibition can be overcome by encouraging the dissociation of the enzyme–inhibitor complex. A convenient way to do this on a large scale is to increase the substrate concentration, thereby favouring the formation of *ES* complex rather than *EI*, although this may not be so effective in the case of a non-competitive inhibitor, and may even worsen the situation if inhibition is due to substrate inhibition. In this case, dilution would be a useful alternative. Dialysis or gel filtration could also be used to encourage the dissociation of the inhibitor from the enzyme molecule. If, on the other hand inhibition is only slight and it is decided that one can 'live with' the problem by accepting a longer reaction time or the need to use a higher enzyme dosage, it would be worth attempting to identify and quantify the nature of the inhibition to justify the cost-effectiveness of the operation. This could be done by measuring the initial reaction rates of a range of substrate concentrations for $0.1K_m$ to $10K_m$ using a purified or synthetic substrate preparation in the presence and absence of test inhibitor. A plot of S/V (or S/V_i) against S will quickly indicate the type of inhibition occurring, as illustrated in Figure 2.53. The values of $K_m(1 + I/K_i)$ and $V_{max}/(1 + I/K_i)$ can thus be determined and the effect of the inhibitor on the overall reaction evaluated using Equation (30).

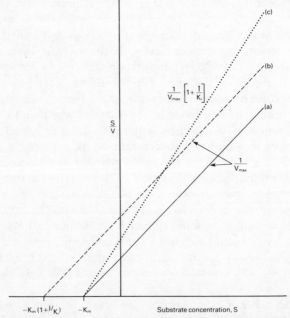

Figure 2.53 Diagnostic plot in order to determine type of reversible inhibition occuring. (*a*) Uninhibited reaction. (*b*) Inhibition due to a competitive inhibitor. (*c*) Inhibition due to a noncompetitive inhibitor.

Many cereals and oilseeds contain proteolytic enzyme inhibitors as part of their time-delay germination control system. The trypsin inhibitor in soybean is an obvious example which affects serine proteinases of the class EC 3.4.21.-. Most of these substances are thermolabile, so that they are destroyed on heating, for example. However, several industrial processes using plants as raw material require proteolytic enzyme processing prior to heat treatment. In these cases, inhibition could be prevented by using a non-serine proteinase such as a metalloproteinase (EC 3.4.24.-). An important commercial example of this is the use of neutral proteinase instead of alkaline proteinase to hydrolyse barley protein, which contains a serine proteinase inhibitor, in the production of malt extract for the food industry (*see* Figure 2.54).

Uses of inhibitors. Although enzyme inhibitors are usually re-garded as a nuisance in applied enzymology, it is likely that they will be used in a positive way in the future. Thus, one could envisage their use to control the operation of multi-enzyme reactions, or as selective binding agents for affinity chromatographic purification of enzymes. They could be used to protect active sites on enzymes during chemical modification of the enzyme protein as, for exam-ple, in the production of immobilized enzymes, or as super-

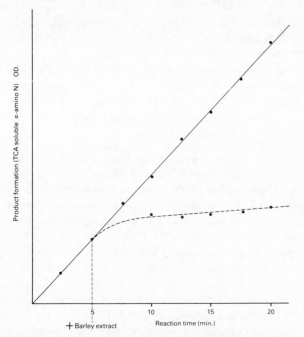

Figure 2.54 Inhibition of alkaline proteinase (EC 3.4.21.14) from *Bacillus licheniformis* by an aqueous extract of barley (Sørensen & Fullbrook, 1973).

stabilizers of these enzymes during the long period between manufacture and customer use.

Irreversible enzyme inhibitors are used as probes to investigate
the mechanism of enzyme catalysis, and these studies often have
spin-off in the area of applied enzymology. Thus, titration of the
active site of *Bacillus licheniformis* subtilidopeptidase (EC
3.4.21.14) with serine inhibitors indicates that the maximum activity of the pure enzyme is around 45 Anson Units per gramme of
enzyme protein, yet the assayed activity of the purest preparations it
is possible to make is only 30–35 Anson Units per gramme. This is
because in solution, this powerful proteolytic enzyme hydrolyses a
portion of the total protein preparation to produce stabilizing
peptides, hence resulting in apparent 'loss' of activity in solution.

It should also be remembered that a large number of therapeutic
drugs are medicinally effective, simply because they are enzyme
inhibitors. Of the currently used antibiotics, penicillins and
cephalosporins represent probably the best examples of highly
specific enzyme inhibitors used medicinally. Their antibiotic effect
is due to their competitive inhibitory effect on the enzymes which
catalyse the terminal synthesis and cross-linking of bacterial cell
walls by the enzyme D-Alanine carboxypeptidase, and they are thus
only active during cell division. Semi-synthetic penicillins are themselves synthesized from 6-aminopenicillanic acid, which is itself
produced by enzymatic hydrolysis of 'natural' penicillins, produced
by fermentation. Thus, with the advent of the clavulinic acid
antibiotics, which are irreversible inhibitors of the β-lactamase
enzymes produced naturally by bacteria to destroy penicillins, we
can truly say that modern biotechnology has produced:

> *A non-antibiotically active antibiotic*
> *That inhibits the enzyme –*
> *That hydrolyses the inhibitor –*
> *That catalyses the synthesis –*
> *Of a bacterial cell wall (!)*

This also illustrates another point in connection with the use of
inhibitors, and that is, that they can act as a convenient way of
'fingerprinting' or characterizing enzyme proteins, for use, for
example, in patent applications. One of the consequences of the
future possible (widespread?) use of genetic programming ('engineering') is that the DNA coding for the synthesis of a specific
(industrially important) enzyme protein is likely to be transposed to
several producing bacteria. Since most industrially used enzymes
have a relatively broad specificity, substrate activity is not a sufficiently precise 'handle' by which to characterize an enzyme; in
contrast, enzyme inhibitors are. Thus, it is possible to characterize
an enzyme protein by its inhibition profiles, quantified precisely as a

series of K_i or fractional inhibition values. This technique was used originally to prove that intergenic transfer of genetic material (genetic programming) was a natural phenomenon and is illustrated in Figure 2.55. Thus, the use of this technique offers a possible solution to the foreseeable legal nightmare in using genetic programming to 'get round' industrial patents for enzyme processes.

'Non-aqueous' and concentrated solution catalysis. Because most of the conventional studies in enzymology have been carried out in dilute aqueous solution, there is a tendency to overlook the possibility of 'non-aqueous' enzyme catalysis. Although absolutely anhydrous conditions are probably not feasible, some enzymes have considerable activity in organic solvents when only a very small amount of water is present. Other enzymes are able to react efficiently with aqueous insoluble substrates such as cholesterol oxidase (EC 1.1.3.6). Indeed, if we consider the environment within living cells – the environment in which the vast majority of enzyme reactions take place – a significant proportion of the enzymes are either bound to 'solid' membranes or particles, or

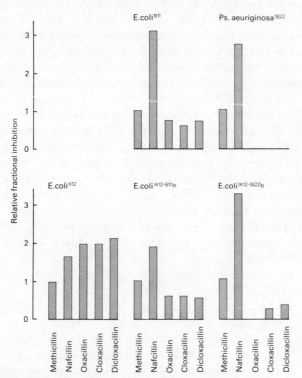

Figure 2.55 Use of inhibitors to 'fingerprint' genetic recombinants. The figure shows the original method used to follow the gene transfer of *Pseudomonas* β-lactamase into *Escherichia coli* cells (Fullbrook & Slocombe, 1970).

operate in local hydrophobic areas within the cellular constituents. Thus, we could speculate that the majority of intercellular enzyme reactions occur naturally under 'non-aqueous' conditions, even taking into consideration the low concentrations of substrates involved at this subcellular level. Without going further into this argument, perhaps a valid point can be made that, at least for non-hydrolytic enzymes, low aqueous or 'non-aqueous' enzyme catalysis is common, and could represent a useful future development in applied enzymology (currently, approximately 80 per cent of the volume of industrially used enzymes are hydrolytic applications). Of course, to the industrial organic chemist, schooled to conduct most conversions and syntheses in organic solvents, the idea of using dilute aqueous solutions as a medium for catalysis is quite foreign. Indeed, the prospect of having to remove so much water from the product, combined with low conversion yields, has perhaps been the most significant factor in discouraging serious consideration of the use of enzymes to perform organic chemical conversions.

Hydrolytic reactions can be run in concentrated solution, however, and it is perhaps significant that one of the most important applications of enzymes on an industrial scale is carried out – at least by classical enzymological standards – at 'high' substrate concentration. This is the liquefaction and saccharification of starch to disaccharides and monosaccharides, a process which operates at 30–40 per cent dry substance substrate concentration for the production of a 70–80 per cent dry substance final product. It is also possible to liquefy starch enzymatically at up to 80 per cent dry substance at high temperature (110 °C) in the presence of thermostable α-amylase, because the insoluble nature of the starch substrate means that the thermodynamic equilibrium for the reaction favours hydrolysis to soluble maltodextrins. However, the saccharification reaction needs to be carried out in a more dilute solution (*see* Figure 2.15), which is more typical of hydrolytic reactions. There are two reasons for this. The first is a fundamental reason, that water is just as much a substrate in a hydrolytic reaction as the organic substrate with which it reacts, a point often overlooked. Thus, high organic substrate concentration means low 'water substrate' concentration and hence these conditions are favourable for substrate inhibition. Figure 2.56 illustrates this point by comparing the observed variation of the rate of hydrolysis of sucrose by yeast β-fructofuranosidase (EC 3.2.1.26) with both substrate concentration and the molar water concentration. It shows that reduction in velocity is due simply to reduction in water concentration, since an almost continuous curve is evident,

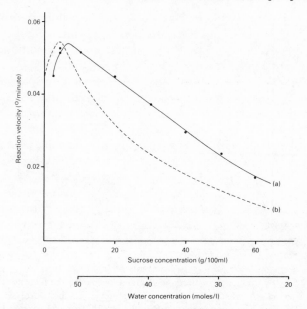

Figure 2.56 Effect of high substrate concentration (low water activity) on the initial velocity of the hydrolysis of sucrose by invertase (EC 3.2.1.26) from yeast: (*a*) water concentration lowered by sucrose; (*b*) water concentration lowered by a combination of sucrose and 10 and 20 per cent ethanol (Nelson & Schubert, 1928).

irrespective of whether sucrose itself or ethanol is responsible for abstracting water. The more practical reason saccharification/hydrolysis reactions cannot be operated at high substrate concentrations is due to equilibrium thermodynamics – that the reverse (synthetic) reaction occurs at low aqueous concentration and since the saccharification enzymes used are not absolutely specific in their catalytic effect, unwanted di- and trisaccharides by-products are formed. But at concentrations higher than 35 per cent dry substance *Aspergillus niger* amyloglucosidase (EC 3.2.1.3) will synthesize isomaltose and panose by-products from the required glucose product. A similar phenomenon occurs in the hydrolysis of a high concentration of partially demineralized whey permeate using *Kluyveromyces fragilis* β-galactosidase (EC 3.2.1.23), where trisaccharides are synthesized due to transgalactosylation of primary reaction products (*see* Figure 2.57).

This low aqueous phenomenon can, however, be put to good use in enzyme technology since it affords the attractive possibility that the direction of an enzyme-catalysed hydrolytic reaction can be controlled not only, as we have seen previously, by pH, but equally simply by the substrate/water ratio. An obvious application of this is in the hydrolysis and synthesis of oils and fats, catalysed by the

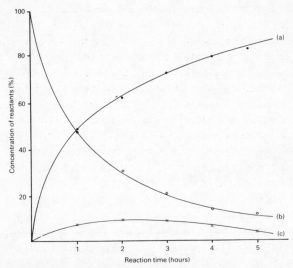

Figure 2.57 Byproduct formation and degradation during the hydrolysis of lactose by *Klebsiella fragilis* β-galactosidase. The graph shows the hydrolysis of 4.57 grammes per litre lactose at an initial pH of 6.7 at 45°C using 54 Lactose Units of enzyme per gramme lactose over a 5-hour reaction period. Curve (*a*) shows the production of product as glucose + galactose; (*b*) the disappearance of substrate; and (*c*) the formation and eventual hydrolysis of trisaccharaide (data from Novo Industri).

lipase enzymes (EC 3.1.1.3). In a highly aqueous environment hydrolysis of triglycerides occurs, whilst in a 'non-aqueous' environment synthesis of glycerides from glycerol and fatty acid is encouraged.

$$
\begin{array}{ccc}
CH_2OH & R'CO_2H & CH_2OOR' \\
| & & | \\
CHOH + R''CO_2H & \rightleftharpoons & CHOOR'' + 3H_2O \\
| & & | \\
CHOH & R'''CO_2H & CH_2OOR'''
\end{array}
$$

By choosing lipases of suitable specificity and providing the corresponding fatty acid (or derivative), a wide variety of specifically 'designed' glycerides can be enzymatically synthesized. This could have a useful potential for applied enzymology, since it would provide a synthetic route for the production of a specific glyceride from glycerol and carboxylic acids.

Cosubstrates and byproducts. Wide enzyme specificity affords the benefit of transformation of a wide variety of chemically similar substrates, but brings with it the possible penalties of increased chances of byproduct formation, as we have just seen, or inhibition by an equally wide range of inhibitors. Since most industrially used enzymes operate on substances which are probably not their natural substrates (*see*, for example, Table 2.5), they must *per se* be of wide specificity, and thus open to the above constraints. Thus, if a wide-specificity enzyme is presented with equimolar concentrations of two different substrates, the reaction that predominates will depend on the respective values of the constants K_m and k_3 for the system, and the substrates will compete with each other for reaction priority which, in turn, will determine product composition. Often in an industrial situation we are dealing not only with wide-specificity enzymes, but also with mixtures of complex biological substrates (such as protein mixtures), so that the task of controlling product specification is, to say the least, challenging.

Byproduct formation not only reduces the yield of potential product, but can also affect the stability of the enzyme during the reaction period or even its activity, by acting as an inhibitor. For efficient enzyme reaction, therefore, limitation of byproduct formation should be of high priority in the optimization of an enzyme reaction.

5. Interrelationship of effects

Having examined in some detail the various physical and chemical factors which affect both the phase II and, from our point of view more importantly, the phase III part of an enzyme reaction, it would seem appropriate to attempt to draw these factors together, to see how they affect utilization of the enzyme, and the amount of product formed during the overall course of an enzyme-catalysed process.

Interaction and compromise

It is important to realize that the variables described do not act independently, but modify, and are modified by, one or several of the other variables. This is illustrated in Figures 2.58 and 2.59. In practice, however, the picture is often more complicated. This means that each enzyme reaction has to be treated on a case-to-case basis. In turn, it explains why optimization of enzyme reactions for maximum product yield and minimum enzyme (or energy resource) use has only been attempted for a limited number of enzyme reactions, mainly those reactions which are of commercial importance. This, then, is an area of applied enzyme kinetics which has only just begun to be developed and, with the availability of better analytical methods and data processing systems, represents an area of potential interest for future development in pure and applied enzymology.

Let us, however, concentrate on what has been achieved to date. We can broadly distinguish two types of enzyme-catalysed process designed for a maximum product yield on an industrial scale:

TABLE 2.5

Exploitation of enzymes. The use of enzymes to convert 'commercial substrates' relying on their imprecise substrate specificity

Enzyme common name	EC number	'Natural' substrate	Commercial substrate → product	
Pullulanase	3.2.1.41	Pullulan	Amylopectin	α-1,4 amylose
Amyloglucosidase	3.2.1.3	Maltose (?)	Oligosaccharide	glucose
Bacterial α-amylase	3.2.1.1	Starch	Dextrins	Oligosaccharide
Fungal α-amylase	3.2.1.2 (?)	Amylose	Oligosaccharide	Maltose
Xylose isomerase	5.3.1.5	Xyl(ul)ose	Glucose	Fructose
Neutral proteinase	3.4.24.4	Protein	Pro β-amylase	β-amylase
Microbial rennet	3.4.23.6	Protein	K-casein	Milk curd
Penicillinase	3.5.2.6	?	β-lactam antibiotics (e.g. penicillin)	penicilloic acid)
Penicillin amidase	3.5.1.11	Ester(?)	Penicillin	6-aminopenicillanic acid

Figure 2.58 (*see* page 99)   Interaction of variables on the activity of an enzyme. The figure shows the combined effect of pH and temperature on the apparent activity of *Klebsiella aerogenes* pullulanase (EC 3.2.1.41). Curve (*a*) shows activity as a function of temperature at pH 5.0, and (*b*) at pH 4.0 (data from ABM Brewing and Food Group).

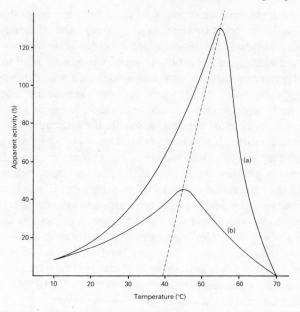

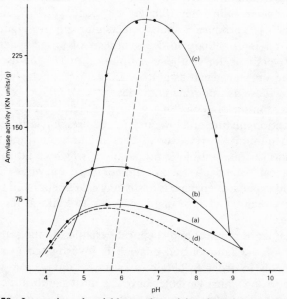

Figure 2.59   Interaction of variables on the activity of another enzyme: *Bacillus licheniformis* α-amylase (EC 3.2.1.1). Curve (*a*) shows the activity of the enzyme as a function of pH at 37°C; curve (*b*) at 60°C; and curve (*c*) at 95°C. For comparison, curve (*d*) shows the activity of *Bacillus amyloliquefaciens* α-amylase under the same reaction conditions at 95°C. Enzyme substrate was 0.46 per cent soluble starch in 4.3 M calcium ions reacted with enzymes over a 3–30 minute period (data from Novo Industri).

type A processes, in which physical conditions are accurately controlled and kept constant in order to maximize the efficiency of the enzyme; and type B processes, which specifically require no residual enzyme activity in the product. Type A processes tend to be the newer types of process, designed around the enzyme catalysis step, for high productivity, whilst type B processes are the traditional and established (bio)chemical reactions which have been recently superseded by enzyme mediation, so that the enzyme catalysis step has had to be 'fitted into' either an established process, or a process in which physical parameters are varied. This means, then, that the two types of process require radically different approaches to the study of the interactions of physical and chemical effects in order to compromise in the design of the most efficient enzyme reaction.

High productivity enzyme processes (type A processes). In many ways, these are the easiest type of processes to design for, since one is starting on a 'clean sheet' basis, and restrictions are often less, or at least can be modified. Processes are characterized by: (*i*) a requirement for an easy separation of enzyme and product, and (*ii*) minimum amount of enzyme used per process cycle.

Typical examples are (*a*) the deacylation of fermented penicillin (e.g. benzylpenicillin – penicillin G – or phenoxymethylpenicillin – penicillin V) to produce 6-aminopenicillanic acid (*see* Figure 2.20) used as an intermediate in the synthesis of broad spectrum or speciality/active semi-synthetic antibiotics, or (*b*) the use of immobilized enzymes such as glucose isomerase, amyloglucosidase, β-galactosidase or amino acid racemases.

In the example of penicillin acylase, product is solvent-extracted from reaction mixture and low enzyme requirement is preferable in order to minimize carry-over of contaminants from the enzyme preparation to the antibiotic preparation, which could cause immunological complications. For immobilized enzymes the often relatively expensive enzyme is easily physically separated from the product stream, and is re-used in order to make it more cost-effective.

The effect of physical and chemical variables on the activity and stability of the immobilized enzyme is investigated in separate experiments in order to obtain quantitative data. From this it is possible to see which variables have the most significant effect on the enzyme. The effect on the enzyme is considered in terms of the productivity of a unit amount of enzyme in a defined 'useful life' of the enzyme. The productivity is defined as the weight of product per unit weight of enzyme and is envisaged as the product of enzyme activity and enzyme stability: Productivity = Activity × Stability. If activity is defined in terms of weight of substrate transformed (or

product produced) per unit time (per unit weight of enzyme preparation), and stability in terms of half life, the productivity is simply
a number expressing weight units for a stated multiple of half lives.
Thus, if the productivity of an enzyme was given as 2000 for 3 half
lives, and the half life was stated as 1000 hours, this would mean
that 1 kilogramme of enzyme would yield 2000 kilogramme product during a working life-time of 3000 hours. This is illustrated in
Figure 2.60. The productivity is thus the accumulated weight of
product produced whilst a unit weight of enzyme was operating
during the stated operation time.

Thus, the effects of the physical and chemical variables can be
related through their influence on the activity and stability of the
enzyme, to their effect on the productivity of the enzyme. This can
be simply done in the form of a 'what-affects-what' pattern, or
'interaction logic diagram'. Figure 2.61 shows such a scheme for the
enzyme glucose isomerase. From such a diagram, and using laboratory data on the effect of each variable, a compromise reaction
process can be inferred and tested for maximum product formation.
The value of such an approach has been proved in this particular
case, and can be illustrated as follows. Laboratory data clearly
indicate that the temperature activity optimum for this particular
enzyme reaction occurs at 80 °C (Figure 2.62) yet the stability data
show that at this temperature the stability (T_1) is only approximately
10 hours. Operation at approximately 60 °C increases the stability
100-fold to around 1000 hours and this gives an improved productivity. Interaction logic diagrams also indicate how several unre-
lated effects can combine to produce a single undesirable effect such
as the formation of byproducts. These diagrams can thus form the

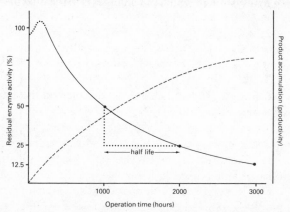

Figure 2.60   Activity decay during the operation of an immobilized enzyme reactor. These results are for glucose isomerase but are typical of most immobilized
enzymes.

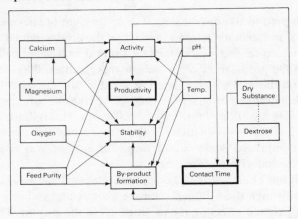

Figure 2.61 Interaction logic diagram for the operation of an immobilized glucose isomerase reactor. The diagram shows the influence of individual process parameters on the activity and stability of the enzyme in a qualitative way, and hence indicates the overall effect on the productivity of the enzyme.

basis for reaction models which can, in turn, be built into computer programs to design and optimize large-scale enzyme processes.

Varied – parameter enzyme processes (type B processes). These processes are characterized by: (i) a requirement for no residual enzyme activity in product, and (ii) maximum or controlled enzyme activity during reaction. Often physical parameters are deliberately changed during the reaction. The most common parameter change is temperature, although pH changes are sometimes involved. Typical temperature change examples are the production of malt extract or brewer's wort from an adjunct (*see* Chapter 4.5) or the use of enzymes in biological detergents. An example of an enzyme operating in a system in which pH changes is in the enzyme-assisted

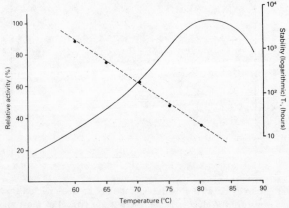

Figure 2.62 Temperature activity and stability data for glucose isomerase.

saccharification of carbohydrates during fermentation. These reactions demand that there is a 'dead heat' situation at the end of the (defined) reaction period between the activity effect of the enzyme and its stability – the last per cent conversion is being achieved just as the last surviving per cent residual enzyme activity is being lost due to instability brought about by the changed physical environment. Figure 2.18 illustrates this and shows how the activity of a neutral proteinase changes during the 'proteolytic' and 'saccharification rest' temperature programme for the production of extract from barley in the manufacture of a food product.

How, then, can we cope with the design of such a process, so that we can relate enzyme amount to required product yield for a given reaction period when activity and stability parameters are deliberately changed? A solution which has proved useful is to adopt a purely empirical approach, and to ignore the classical enzyme kinetic parameters. Provided that the magnitude and time sequences of the changes in physical parameters are faithfully repeated, their effects on enzyme and substrate will be sufficiently similar for the effect of enzyme concentration on product yield to be studied at laboratory scale. Figure 2.63 shows the results of analysis of various nitrogen fractions in an extract produced by incubating

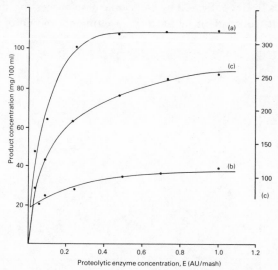

Figure 2.63 Empirical approach to relating enzyme concentration and its effect in varied-parameter enzyme processes. The figure shows the results of analysis of several nitrogenous fractions produced by incubating milled barley with *Baccillus subtilis* n-proteinase (EC 3.4.24.4) in a temperature-programmed reaction (*see* text for details) with various enzyme concentrations. Curve (*a*) α-amino N; (*b*) amino acid N fractions in the extract; and (*c*) total N assay of the partially fermented extract produced under strictly controlled conditions.

milled barley with various concentrations of neutral protease (EC 3.4.24.4) for the following reaction: barley protein → peptides + amino acids.

Temperature programme:

Time (min)	Temperature (°C)
0– 60	50
60– 73	Increase at 1 °C per minute
73–133	63
133–150	Increase at 1 °C per minute
150–180	80

Inspection of the data in Figure 2.63 indicates a hyperbolic relationship between product composition and enzyme concentration which, if tested graphically (*see* Figure 2.64), indicates a straight-line relationship between enzyme concentration E and the ratio of enzyme concentration to increase in product ($Y - Z$). Therefore,

$$\frac{E}{Y - Z} = Em + b \qquad (38)$$

where Z is endogenous product extraction, and m and b are constants; thus, $E \propto 1/\text{productivity}$.

This has been confirmed by electronic data processing statistical analysis. In subsequent laboratory experiments the nature of the

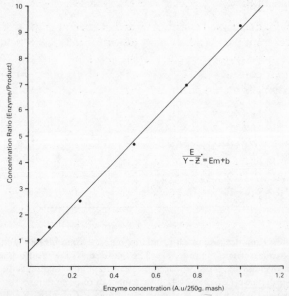

Figure 2.64   Graphical analysis of some of the data in Figure 2.63. The figure is drawn from a 'best-fit' computer printout.

complex substrate (barley) was varied. Figure 2.65 shows the effect of α-amylase concentration on the yield of reducing sugars for two types of barley, hydrolysed under identical conditions to those described above.

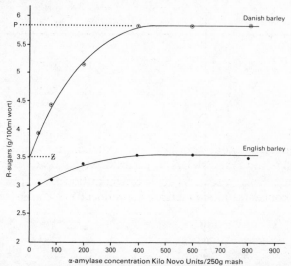

Figure 2.65   A second example of the empirical approach, this time showing the effect of *Bacillus subtilis* α-amylase concentration Kilo Novo Units per 250 g mash (EC 3.2.1.1) on the yield of reducing sugar produced from two different types of barley substrate. Reaction conditions were exactly those for Figure 2.63.

The data, if analysed statistically and plotted in graphical form as in Figure 2.66, clearly show that although the different types of substrate have different values of m and b in Equation (38), the ratio b/m is constant. Analysis reveals that the amount of product produced, Y, is related to the enzyme concentration by:

$$Y = E \left[\frac{P - Z}{1 + K} \right] + Z \qquad (39)$$

where P and Z are characteristics of the substrate, being, respectively, the maximum extract yield obtainable and the endogenous extract yield. Thus, by carrying out this type of experiment it is possible to 'fingerprint' complex substrates/raw materials by defining their aqueous extractability, Z, and maximum extract yield, P. The value $(P - Z)$ indicates its modification potential in the presence of enzyme. Use of Equation (39) also allows the calculation of the amount of enzyme required to effect any yield of product in the range $Z \rightarrow P$ as long as the reaction conditions are kept identical, and provides a simple solution to our problem. It should be borne in mind that this type of information is obtained only for laboratory conditions and needs to be tested on a plant scale before finite

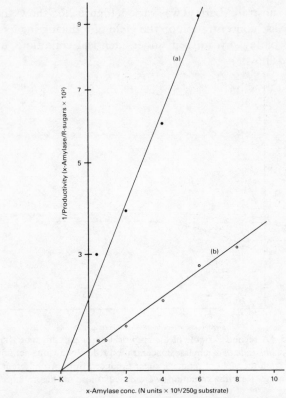

Figure 2.66 Graphical analysis of data from Figure 2.65 ((a) = English barley; (b) = Danish barley).

recommendations can be made. Experience has shown, however, that this type of modelling approach is useful in estimating enzyme dosage for varied-parameter processes, the details of which can then be adjusted in the process at plant scale. Table 2.6 illustrates this point by comparing results obtained at laboratory and plant scale using this empirical approach.

Such an approach can also usefully be extended to compare quantitatively the suitability of several different enzymes in a given process, allowing the most effective enzyme to be selected. Figure 2.67 illustrates this and compares the ability of two types of alkaline protease (EC 3.4.21.14) to assist in the cleaning ability of a detergent formulation designed as a biological washing detergent. During this enzyme process the temperature is continuously increased, whilst the enzyme stability is adversely affected by the chemical nature of the detergent formulation. In this case, then, both physical and chemical effects determine the productivity of the enzyme. From these data, Figure 2.68 shows that ΔR, the added

TABLE 2.6
Laboratory and plant scale production of malt syrups

Raw materials	Lab test Syrup 1 kg	%	Syrup 2 kg	%	Syrup 3 kg	%	Syrup 4 kg	%
				Full-scale plant trial (100 hectolitres)				
Grist								
Malt for main mash	—	—	1060	70.2	575	39.8	575	39.8
Malt for maize liquefaction	—	—	70	4.6	—	—	—	—
Barley (9.1% protein)	0.05	100	—	—	490	33.9	490	33.9
Maize grist (total solids)	—	—	380	25.2	380	26.3	380	26.3
Enzyme supplement:								
α-amylase (liquefaction)						0.4		0.4
α-amylase (main mash)		0.86				0.86		0.86
γ-amylase (amyloglucosidase)						—		1.00
Proteinase		0.08				0.08		0.08
Composition of syrup produced (expressed as per cent of total solids)								
(Total solids)	79.5		80.0		80.0		80.0	
Glucose	6.4		6.38		4.73		6.98	
Maltose	52.0		39.9		44.0		43.1	
Maltotriose	—		16.1		14.7		12.5	
Total fermentable sugar	—		62.4		63.4		62.6	
Total N	0.8		0.47		0.44		0.48	
α-amino N	—		0.12		0.09		0.11	
α-amino N total N	—		24.4		20.6		21.7	
Formol N	0.2							
Filtration time (minutes)	—		150		120		120	

Data from Novo Enzyme Information (1982).

108   Chapter 2   Kinetics

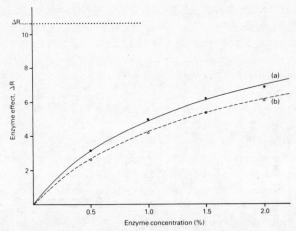

Figure 2.67 A third example of the empirical approach, this time showing the effect of *Bacillus licheniformis* a-proteinase (EC 3.4.21.14) on the removal of protein bound material from fabric for two different enzyme samples, as in a standard test for enzyme-assisted detergency in a biological detergent formulation ((*a*) = Protease-A; (*b*) = Protease-M).

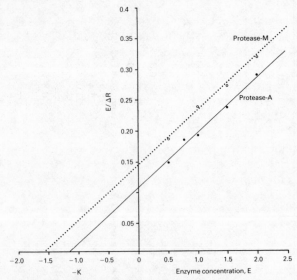

Figure 2.68 Graphical analysis of data from Figure 2.67.

cleaning effect of the enzyme (representing the increase in reflected incident light on the surface of the treated fabric) is related to the enzyme concentration E by

$$\Delta R = \frac{E \, \Delta R_{max}}{E + K} \tag{40}$$

Where ΔR_{max} is the maximum reflectance (maximum cleaning effect) and K is the enzyme constant. In other words, whereas ΔR_{max} is a characteristic of the detergent base, the value of K is a characteristic of the enzyme and will indicate which of the enzymes under test will bring about the best cleaning effect. The value $\Delta R/E$ in Equation (40) is an expression of the enzyme productivity and so in this case

$$\text{Productivity} \propto \frac{1}{K}$$

Hence, knowing the 'reciprocal productivity value', K, the amount of enzyme required to achieve a specified degree of cleaning can be estimated using Equation (40). Thus, for example, if ΔR is set at 80 per cent of ΔR_{max}, the amount of enzyme required will be $4K$.

Thus, the use of this empirical approach allows meaningful data relating enzyme concentration to its effect to be easily quantified for complex substrates and changing physical parameters. This could provide a way of studying enzymes in action, so that these experimentally derived parameter constants could be ultimately related to classical kinetic constants and the effects of reaction conditions on them.

Enzyme processing costs. Since the cost of an industrially used enzyme is quoted in currency units per unit weight (or volume) of enzyme of specified activity, if the weight of enzyme to effect a required conversion or effect is known, then the cost of its effect in a process can be calculated, and its cost-effectiveness quantified. This cost has to be significantly less than the value added to the material being processed to product by the enzyme customer, and, conversely, forms the basis for arriving at the selling price of the enzyme.

In retrospect, it is interesting to note that it was this single, simple commercial fact which led to the development of immobilized enzymes for industrial-scale operation, which is perhaps one of the most interesting developments in applied enzymology to date.

References

Birch, G. G. Parker, K. (eds) *Nutritive Sweeteners* (Applied Science, 1982).

Brown, A. J. *JCS Trans.* **61,** 369 (1892).

Brown, A. J. *JCS Trans.* **81,** 373 (1902).

Cole, M., Elson, S. & Fullbrook, P. D. *Biochem. J.* **127,** 295 (1972).

Dixon, M. & Webb, E. *Enzymes* 3rd edn (Longmans, London, 1979).

Engel, P. *Enzyme Kinetics* (Outline Series in Biology) (Chapman & Hall, London, 1980).

Fersht, A. *Enzyme Structure and Mechanism* (Freeman, San Francisco, 1977).

Fullbrook, P. & Slocombe, B. *Nature* **226,** 1054 (1970).

Gjersten, P. & Erdel, K. *EBC Proceedings* (1971).

Hare, M. L. C. *Biochem. J.* **22,** 968 (1928).

Henri, V. *C. r. hebd. Séanc. Acad. Sci., Paris* **135,** 916 (1902).

IUB Nomenclature Committee *Enzyme Nomenclature* (Academic, New York, 1978).

Nelson, J. M. & Schubert, M. P. *J. Am. Chem. Soc.* **50,** 2188 (1928).

Northrop, J. H. *Ergebn. Enzymforsch.* **1,** 302 (1932).

Michaelis, L. & Menten, M. L. *Biochem. Z.* **49,** 333 (1913).

Novo Enzymes in the Brewing Industry (Novo Enzyme Information, 1982).

Palmer, T. *Understanding Enzymes* (Ellis-Horwood, Chichester, 1981).

Pollock, M. R. *Biochem. J.* **94,** 666 (1965).

Sørensen, S. Å. & Fullbrook, P. D. *MBAA Tech. Quarterly* **9** (4), 166 (1973).

Webb, J. L. *Enzyme and Metabolic Inhibitors* (Academic, New York, 1965).

In conclusion I would like to thank my former industrial colleagues for giving me the opportunity to develop and work with these ideas, to my academic colleagues for criticizing them, and to the armies of laboratory technicians and students who have worked in the laboratory to provide the data on which these chapters depend for illustration, particularly Brenda Ridley, Janet Nalder, Merete Jensen, Ann Nielson and Laila Jensen.

LEGISLATION AND REGULATION

Chapter 3.1

THE LEGISLATIVE ASPECTS OF THE USE OF INDUSTRIAL ENZYMES IN THE MANUFACTURE OF FOOD AND FOOD INGREDIENTS

W. H. B. Denner

1. Introduction

One of the perennial problems of long-term planning in any industrial company is the accurate forecasting of legislative changes likely to affect the company's business. The food industry, perhaps more than others, finds itself particularly well circumscribed by legislation, any or all of which is subject to change. Food legislation is invariably a blend of science and politics and the relative proportions of the blend will vary from time to time and from nation to nation. The purpose of this section is not merely to provide a catalogue of the legislative status of individual enzymes in the form of a snapshot of the current scene. In addition an attempt has been made to explain the underlying factors from the legislator's viewpoint in the hope that this will promote a better understanding of the philosophy of the legislative control of enzymes in food and provide some insight into the long-term trends.

2. The perceived risk from industrial enzymes

When a traveller purchases an airline ticket, he takes a positive decision to accept a small but quantifiable risk that the aircraft in which he will travel will crash. His perception of this risk will be heightened during take-off and again on landing. On stepping out of the aircraft at his destination, however, he will experience a sense of achievement. He feels this way because his arrival signals the definite endpoint of the period of risk to which he knowingly exposed himself. But when that same traveller enters the airport restaurant and purchases food, his expectation of his exposure to risk in consuming that food approaches zero.

This desire for absolute safety of food is rooted in our instinctive

[1]The views expressed in this article are the personal views of the author and do not necessarily reflect the official UK view.

defensive misgivings about anything which is allowed to pass through the portals of our own boundary walls and which could potentially attack us from within. Further, it is now common knowledge that such an attack may not be the swift blow of the acute poison. It may come as the subtle chemical change induced by the carcinogen, the effect of which may not manifest itself until twenty or thirty years have passed. It may even affect the germ cells and not manifest itself until the next generation. In the case of substances consumed therefore, we can never experience the complete confidence of a recognizable definitive endpoint to the risk. In self-defence, the consumer tends to demand zero risk for those substances which are deliberately added to food. It is curious that the consumer appears less concerned about the relatively higher risk of acute food poisoning than about the relatively lower risk of consuming food additives. Presumably this is because food poisoning arises by chance contamination of small numbers of food items whereas food additives are deliberately added to many. The consumer's perception of the relative risk is influenced more by the 'unavoidability' of the food additive than by the higher toxicity of the contaminated food. The unease of the ordinary consumer with regard to food additives is compounded by his realization that, even if he were presented with all the relevant data on safety-in-use, he has neither the knowledge nor experience to judge the potential risk himself. Consumers tend to rank food additives high on the scale of unknown risks.[1]

Ultimately the consumer looks to the law to protect him and it is important to realize that food legislation is based not only on the desire to minimize real and quantifiable risk but also on the need to reassure the consumer from the point of view of his perception of the risk. The task of the legislator is to strike a fair and equitable balance, to maintain an equilibrium. The position of an equilibrium point depends on the relative strength and direction of the various forces prevailing at the time. As the forces change, then the pressure to alter the equilibrium point increases. However to some extent the legislator is able to act as a 'damper' to eliminate fluctuations of short duration, but when a long-term trend is detected it becomes increasingly difficult to resist legislative changes.

Certain classes of food additives (e.g. colours and artificial sweeteners) may be termed 'high-profile' additives in that their presence in food is obvious to consumers, in some cases even without labelling. Other classes are, by definition, either biologically or chemically reactive (e.g. preservatives and antioxidants). It should come as no surprise, therefore, that, in most countries, these were the first classes of food additives to be controlled by specific food legislation. In time, legislation on other classes followed until eventually many countries felt that most classes of food additives were satisfactorily controlled by statutory permitted lists. At this time, authorities were of course well aware of the use of enzymes in the manufacture of food. However, many enzymes were used as processing aids at an early stage of food manufacture or in the preparation of ingredients, as for example in the hydrolysis of starch, and the small amounts carrying over into the final food seemed unlikely to constitute a hazard to the consumer of that food. But three events were to change this rather relaxed attitude.

The first event was the marketing of enzyme-containing washing powder. Powerful advertising campaigns brought enzymes to the attention of many members of the general public for the first time. Unfortunately, the enzymes were produced at that time as fine dusty powders which produced allergic reactions among a proportion of workers who inhaled them. In addition there were many reports in the popular press of housewives experiencing skin irritation after using biological washing powders. The fact that manufacturers later altered their processing conditions to avoid the inhalation problems and that subsequent studies indicated that the enzymes themselves were not the agents primarily responsible for the skin irritation claimed by users of the washing powders, did little to assure the public. By 1969, the general public were not only aware of enzymes but were also including enzymes in the spectrum of substances which they perceived as representing a potential hazard to health.

The second event developed from the first because it arose as a result of consumer resistance to biological washing powder and the virtual collapse of the market in some countries especially the USA. A number of enzyme manufacturers were left with spare fermentation capacity and this acted as a powerful incentive for rapid development and diversification into other markets including enzymes for food manufacture. The variety of commercially available enzymes of microbial origin increased significantly and new technologies were developed (e.g. immobilized enzymes). Many of these newer enzymes were obtained from species of microorganism that were not regarded as 'traditional' from a food point of view and national authorities began to consider more seriously whether there was a need for specific legislative control of enzymes for use in food.

The final event was the development of a new scientific skill – the ability to manipulate viable DNA so that foreign genetic information could be artificially introduced into an organism and thus impart to that organism the ability to produce a substance (or substances) which it was formerly incapable of producing. When genetic engineering and the production of recombinant DNA became a practical proposition, a whole host of safety questions were raised and there were immediate demands for stringent controls over research and development of this type. In the UK, an advisory group, the Genetic Manipulation Advisory Group (GMAG), was set up in December 1976 to review the risks and to advise those undertaking activities in genetic manipulation. Initially this was a voluntary arrangement but was later given statutory effect by the Health and Safety (Genetic Manipulation) Regulations 1978 made under the Health and Safety at Work etc. Act of 1974. Industrial microorganisms and products of microbial origin were now firmly in the limelight. The industry even acquired a new all-embracing term – biotechnology. It was almost possible to forget that many of the basic foods and beverages (e.g. bread, cheese, wine and beer) had been produced with the aid of microorganisms from time immemorial.

3. Safety evaluation of enzymes – the legislative philosophy
Against this background of increasing awareness of the use of enzymes in food manufacture and of the possible hazards to health of microbial products, many national regulatory authorities questioned whether their existing legislation was adequate. A prerequisite for sound legislation is a thorough review of the safety-in-use of the substances in question. Unfortunately very few regulatory authorities publish details of how they arrive at their decisions particularly in relation to safety evaluations. However a useful barometer of toxicological opinion is the verdict of the Joint FAO/WHO Expert Committee on Food Additives (JECFA) which represents the collective view of invited experts in the field, each of whom is acting in an independent capacity.

JECFA considered the problems of the safety of enzyme preparations in food processing at its 15th[2], 18th[3] and 21st Sessions[4] in 1971, 1974 and 1977 respectively. It is interesting that in its Report of the 21st Session, JECFA acknowledged a significant shift from its previous opinion in that the possibility of mutation leading to the emergence of new, potentially toxic products was seen as being of more serious concern. At the same session, the Committee also laid down its criteria for evaluating enzymes according to the source material. Enzymes obtained from edible tissues of animals commonly used as foods, from edible portions of plants and from microorganisms that were traditionally accepted as constituents of foods or that were normally used in the preparation of foods were all regarded as foods, provided adequate chemical and microbiological specifications could be established. No toxicological studies were required in these cases. For enzymes obtained from non-pathogenic microorganisms commonly found as contaminants of food, the Committee recommended that short-term toxicity tests should be carried out. Finally, for enzymes obtained from less well-known microorganisms, more extensive toxicological studies were required including a long-term study in a rodent species. This systematic approach provided individual national authorities with very useful guidelines on the evaluation of enzymes. It has also provided industry with a warning that clearance of an enzyme obtained from an obscure microorganism might require a considerable financial investment in terms of toxicological testing.

The most recently available comprehensive evaluation of enzymes used in food is that carried out by the expert advisory committees to the UK Government. In the UK, the need for food additives is reviewed by an independent committee of experts, the Food Additives and Contaminants Committee (FACC). The FACC takes advice on safety-in-use from another independent expert committee, the Committee on Toxicity of Chemicals in Food, Consumer Products and the Environment (COT). The Committees' Reports are published together and interested parties are invited to comment on the Committees' recommendations. After these representations have been considered, proposals for regulations are issued by the Ministry of Agriculture, Fisheries and Food and, finally, new regulations are made by Parliament. The FACC's Report on the Review of Enzyme Preparations[5] was published in 1982 and the COT's Report (which forms Appendix IV of the FACC Report) makes particularly interesting reading as it departs somewhat from the earlier philosophy adopted by JECFA in 1977.

In the case of enzymes obtained from animal and plant sources that are normally consumed by man, the COT and JECFA were in accord in requiring no toxicological studies provided the specifications of purity ensured that the products met adequate microbiological standards. However the COT did not make a similar general concession in the case of enzymes obtained from microorganisms that are traditionally accepted as constituents of food or that are normally used in the preparation of food. In another respect, however, the COT was more moderate than JECFA. The COT considered that, as enzymes were essentially processing aids and very small residues of the enzyme remain in the final food and thus the daily intake for man was very low, it would be unreasonable to require long-term toxicity studies for any enzyme, unless there were special reasons in a particular case.

The COT very clearly identified three areas of potential concern in the safety evaluation of commercial enzyme preparations:

(*i*)   the catalytic activity of the enzyme itself;

(*ii*)   allergenic reactions produced by any protein present in the product including the enzyme protein itself;

(*iii*)   the presence of toxic metabolites such as mycotoxins or antibiotics.

The COT concluded that the residual catalytic activity of the enzymes added to food would be most unlikely to constitute a hazard to health. This seems a very reasonable and justifiable conclusion. After all, many foods contain a variety of active enzymes and there is no reason to suppose that these present a hazard to health. This is not to say that active enzymes are harmless under all circumstances. It is well-known that proteolytic enzymes, such as papain, in high concentrations can severely damage mucous membranes, for example when powdered papain is inhaled. However this is not relevant to consumers of food. Many enzymes added to food are required to perform their task during the preparation of the food and are deliberately denatured subsequently as the 'working-on' of active enzymes in the final food can be detrimental to the texture or flavour of the food. In such cases it is the denatured enzyme that is consumed and not the active enzyme.

Reference has already been made to the allergic response to powdered enzymes in the workplace. This is an acknowledged fact and responsible enzyme manufacturers have already taken steps to reduce the exposure of their workers. Many enzyme preparations are now marketed in solution even though this necessitates the use of preservatives. In other cases dry enzymes are produced in granular form to reduce dust. In addition, workers are screened so that atopics can be identified and excluded from certain parts of the factory such as packaging areas where the risk of exposure is highest. The allergenicity of enzymes ingested in food is quite another matter. There is no evidence to suggest that ingestion of food containing small amounts of added enzymes has given rise to any cases of allergy among consumers. Admittedly there is no evidence that anyone has looked for such cases but had this been a serious problem then it is likely that this area would have been investigated. There is no fundamental scientific reason why, from an allergy point of view, the consumption of enzymes should create more problems than the consumption of other proteins and the COT concluded that the potential allergenicity of enzymes in food was most unlikely to constitute a hazard to the health of the consumer.

The most difficult aspect of the safety evaluation of microbial enzyme preparations is convincing toxicologists that there are no potentially toxic metabolites present which might constitute a hazard to health. There are two separate but complementary aspects to this problem: firstly, are there potentially toxic metabolites present and secondly, are they present in sufficient quantities to pose a threat to human health? There is no question whatever that bacterial and fungal toxins present in food can be toxic to man. Despite the best endeavours of the food industry and of medical science, botulinum toxin continues to claim human lives every year. The number of mycotoxins of known chemical identity grows annually. This is not to say that there is cause for undue alarm but there is certainly no room for complacency either. The quantitative assessment of risk is extremely difficult as ideally it requires a reasonably accurate estimate of human exposure to the toxin as well as knowledge of the actual toxic response in man to various concentrations of the toxin.

For those toxins which may be, or are thought to be carcinogenic there are additional complications. While considerable developments have taken place in recent years in our understanding of the process of chemical carcinogenesis, the answer to the problem of the quantification of dose-response data for any such substance still eludes toxicologists. Many experts maintain there is no such thing as a safe exposure level for a chemical carcinogen. Others recognize that, for some substances it is possible to identify a level at which the induction period for the carcinogenic response exceeds, by a significant margin, the lifespan of the species upon which any studies are carried out. In such cases one presumes that in respect of carcinogenicity, the particular substance is safe below the identified level – for that species of test animal only. Even then, the toxicologist still has to decide what does this mean, for man, since assessments of carcinogenicity are based on laboratory animal experimentation.

It is true that the evidence suggests that microorganisms known to be capable of producing toxins, do not appear to do so during the phase of rapid growth. This is the phase exploited in the commercial production of extracellular enzymes but using strains of microorganisms believed not to produce toxins under any circumstances. Nevertheless, it would be foolhardy to assume we know all there is to know about the potential of microorganisms to produce toxic metabolites and not unreasonable to require positive evidence of the absence of risk. Since we are concerned about possible toxins of unknown identity as well as those which are known, chemical analysis alone would not suffice and a bioassay or biological screening test would be required.

The microorganism used to produce the enzyme is not the only source of potentially toxic metabolites. They may be present as contaminants of the raw materials used as a substrate for fermentation; they may be produced by contaminating organisms accidentally introduced into the fermenter or they may be produced by mutant strains of the source organism arising during the fermentation. Given good manufacturing practice, the first two of these potential sources should not present a problem. The third potential source is also probably not a problem in the case of batch fermentation of relative short duration in small fermenters. The chances of a mutant developing increase with the time of fermentation and the chances of detecting it decrease as the size of the fermenter increases and representative sampling of the contents becomes more difficult. Further, it is rarely made clear whether the samples of enzyme preparation used for safety evaluation tests were obtained from pilot plant production during the early stages of product development or from the fully scaled-up commercial production plant.

Most enzyme manufacturers will admit that they are constantly learning about the performance of their own strains of microorganisms. After the bringing on-stream of full scale production of a new enzyme, there will be a period of constant improvement in the yield of enzyme as the manufacturer learns by practical experience how to 'fine-tune' that particular fermentation process by subtle alterations in pH, temperature, composition of the substrate, agitation and time of fermentation. Apart from an increase in the yield of enzyme, other changes might occur that have a bearing on the safety of the commercial product relative to the product that was toxicologically tested.

Quite apart from fine-tuning of the fermentation process, manufacturers are continually trying to improve and develop the organisms themselves and the situation can turn into a taxonomist's nightmare. If, for example, a regulatory authority has authorized the use of an enzyme only from a particular named strain of organism, then the manufacturer will claim that his improved organism is a sub-strain of the original and hence still authorized. If on the other hand the manufacturer is attempting to circumvent a patent, then he will do all in his power to argue that the same 'sub-strain' is instead a new and distinct strain. Whatever the strict taxonomic ruling in each case, the fact remains that strain improvement is actively pursued by enzyme manufacturers and every strain improvement takes the product one step further away from the product that was toxicologically tested. One might assign the term 'product drift' to describe the combined effect on the composition of the product of strain improvement and fine-tuning of the fermentation conditions.

To the author's knowledge, the recent Report of the COT[6] is the only published safety evaluation of enzymes which, firstly, acknow-ledges that strain improvement is a commercial necessity and, secondly, attempts to point a way forward to reconcile product drift with safety evaluation. The Report clearly states that in the opinion of the COT the main concern is toxin production by the source organism. The recommended minimum testing requirements for a microbially-produced enzyme preparation are a well-conducted 90-day oral feeding study in a rodent species and a non-specific biological screening test for toxins carried out to a very high standard. There is no requirement for an expensive long-term feeding study. The COT further recommended that, following each strain improvement, only the non-specific biological screening test need be repeated. These recommendations reflect a genuine at-tempt to balance the degree of assurance of safety required for products of which only small residues remain in food against the cost to the manufacturers of expensive toxicological tests.

The COT also looked to the future and speculated that if suffi-ciently convenient and rapid biological screening tests for toxins could be developed then these could be included in the specification of purity and be applied to every production batch of enzyme. This would have the effect of providing additional protection against chance contamination arising in isolated production batches. The Committee was not aware that suitable tests existed as yet but recommended that research should be orientated in this direction.

4. Specifications of purity

Many chemical substances are manufactured in different purity grades and this is reflected in their cost. Broadly speaking there are four grades: analytical, pharmaceutical, food and technical. It would be a gross abuse of public money to de-ice roads in winter with highly purified analytical grade sodium chloride and, converse-ly, it would be a gross abuse of public health to add to food technical grade rock salt contaminated with significant amounts of heavy metals. The function of the statutory specification of purity is to ensure that the correct grade is used for the correct purpose. Generally the specifications of purity for food additives reflect the standard of purity of the materials that were toxicologically tested. If a food additive of a particular degree of purity has been demon-strated to be safe, then there are no public health grounds for requiring its use at higher purity. That is not to say that individual manufacturers cannot market the same substance at a higher degree of purity if they wish. This is a commercial judgement based on the quality required by their customers and what they feel the market will stand in terms of additional cost.

Drafting statutory specifications of purity for chemically synthesized substances of known chemical structure is relatively straightforward. There will be a minimum assay (say 95 per cent or even perhaps 99.5 per cent) for the substance itself and maximum limits on residues of the known starting materials, intermediates, products of predicted side-reactions and on general contaminants such as arsenic and lead. The object is not to include limits on every contaminant that does or could occur but to select purity criteria which are relevant and which should ensure that the manufacturer produces a consistent product. Many food additives are not chemically synthesized but are extracts of naturally occurring materials of both plant and animal origin (e.g. natural gums such as acacia and tragacanth, colours such as annatto and cochineal and many natural flavouring extracts). Although the identity of the chemical substance primarily responsible for the function of the food additive (i.e. the active principal) may be known, the identities of the multitude of other substances extracted along with it may not. However, provided the extract is consistently produced in the same way as the extract that was toxicologically tested, then there is no public health problem. In these circumstances the legislator resorts to a process specification which is based on a definition of the source material and the method of extraction.

From the point of view of drafting statutory specifications of purity, enzyme preparations are particularly uncooperative. Not only are the active principals present in uncommonly low proportions with respect to the extract as a whole, but also it is not possible to describe the active principals in terms of what they are, only in terms of what they do (i.e. their catalytic activity). To make matters worse, the degree of catalytic activity is highly variable depending on the conditions under which it is assayed. It may even disappear altogether after a brief period of heat-denaturation.

If one applies to a microbial enzyme preparation the normal criteria for drawing up a process specification then one requires an adequate definition of the source material and the method of extraction. But whereas the source of, for example, annatto can be simply and accurately described as the outer coating of the seeds of the tree *Bixa orellana*, the source material of a microbial enzyme preparation is the particular substrate mix after being fermented in a particular way by a particular strain of a particular species of microorganism. If the relevant system of national law requires open publication of the statutory specification of purity then this creates even more problems because the strain of microorganism used and the precise details of the fermentation conditions are often commercial secrets. In the case of enzyme preparations, there was a genuine need to depart from normal practice and to develop a unique solution to a unique problem.

The first authoritative specifications of purity for enzymes for use in food processing were drawn up at the 15th Session of JECFA in 1971 and published in 1972[7]. Similar specifications were included in the Food Chemicals Codex (FCC) in 1974[8]. Both JECFA and FCC followed the same format by laying down general specifications which in effect defined good manufacturing practice in terms of enzyme production and laid down limits for certain contaminants namely arsenic, lead, heavy metals, aflatoxin and antibiotics and for certain pathogenic microorganisms namely *Salmonella*, *Pseudomonas aeruginosa* and Coliforms. The specifications for individual enzyme preparations did little more than describe the reactions catalysed by the major enzymes present and lay down assay methods. In the case of the FCC there was a quantitative requirement that the enzyme activity was not less than 85 per cent and not more than 115 per cent of the declared activity.

The basic format has stood the test of time and has been adhered to at subsequent JECFA meetings and in the new third edition of the Food Chemicals Codex[9]. Documents published by two trade associations, The Association of Microbial Food Enzyme Producers (AMFEP)[10] and The Association of Manufacturers of Animal/Plant-Derived Food Enzymes (AMAFE)[11] to assist regulatory authorities are also in agreement with this approach. In addition AMFEP has defined the microbiological control procedures used in the cultivation of microorganisms and described tests for antibacterial activity. At the 25th Session of JECFA in 1981, the General Specifications[12] were updated to incorporate the AMFEP test for the determination of antibacterial activity and the chromatographic screening test for aflatoxin B_1, ochratoxin A, sterigmatocystin, T-2 toxin and zearelenone devised by Patterson and Roberts[13].

In recommending specifications of purity for the inclusion in UK law, the Food Additives and Contaminants Committee was fortunate to be able to draw on the past experience of others. The FACC recommendations are in line with the latest JECFA specifications with two exceptions. There has always been a reluctance to enshrine in UK law microbiological limits in terms of numbers of specified organisms. Enforcement authorities prefer to rely on the general provisions of the Food and Drugs Act so that each case can be considered on its merits. The FACC recommended specifications do not, therefore, include criteria for *Salmonella*, Coliforms and *Escherichia coli*.

The second exception is the omission of assay methods. While it is helpful for identification purposes to have assay methods for enzyme activity in advisory specifications such as those of JECFA, it is quite another to force all manufacturers by law to quote their activities in the same way. The performance of an enzyme under the conditions of food manufacture can be quite different from its apparent performance in a standardized laboratory assay. In these circumstances it was felt that statutory assay methods, far from assisting potential users, might positively mislead them.

5. Methods of legislative control

Almost all countries possess a general food law which offers protection against food which is, at worst, positively injurious to health or, at best, of poor quality. It is most unlikely, therefore, that the user of food enzymes could escape entirely from the scope of food law. But the status of enzymes with respect to subsidiary food legislation varies greatly from country to country and it is always advisable to check with the appropriate national authorities. Legislative status is important not only in terms of whether or not a substance may be used but also in terms of how the food should be labelled. Some enzymes may be regarded as foods, some as processing aids, some even as food additives. Where substances are to be subjected to individual approval undoubtedly the simplest method of legislative control is the authorization system. Here the user is required to notify the authorities of which enzyme he is using, or intends to use, and to provide details of identity, manufacture and evidence of safety-in-use. In return he usually receives a certificate of authorization. The whole system can even be operated passively by conferring automatic authorization unless the authorities object within a specified time period. The advantages to the manufacturers are speed and confidentiality and the advantages to officials are that all the necessary information can be obtained and assessed and uses can be authorized by administrative action without recourse to the cumbersome processes of amending legislation. The disadvantage is the secrecy involved and, in particular, interested parties such as consumers have no idea what is or is not permitted.

The alternative method is the usual method adopted for food additives (i.e. the published permitted list). Only substances included in the list may be used in food manufacture. The advantage is that the law is clear for all to see, for additive manufacturers, food manufacturers, importers, enforcement authorities and consumers alike. Admittedly amending the law may take some time and cause inconvenience but the process allows all interested parties to comment on proposed changes. Swift legislative action and democracy are rarely easy bed-fellows. The framework of law in some countries may not indeed allow notification or authorization systems, in which case the published permitted list is the only option.

This section concludes with a guide to the current legislative positions of enzymes in the various Member States of the European Community and in Canada and the UK. It should be interpreted with caution as the situation can change rapidly. The purpose of the guide is to indicate the general approach taken by each country and it is not intended to be a comprehensive account of all the relevant food legislation. For example, national legislation on the composition of individual food commodities may effectively prohibit the use of certain or all enzyme preparations in the production of that commodity. It is always wise to consult national authorities about the details of particular uses.

Belgium. Enzymes are regulated as processing aids and are subject to authorization. An attempt was made in 1977 to change to a published permitted list system and a draft Royal Decree was prepared in October 1977. Subsequently in June 1979, the Belgian Government decided to postpone the action and the matter still appears to be in abeyance.

Denmark. Enzymes must be notified to and registered with the Danish National Food Institute at least six months before their intended use. Within that six months the Director of the Institute may positively approve the use of the enzyme, but if after six months the director has not prohibited its use, then approval is assumed in any case. Directions on Notification and Registration of Enzyme Products were published by the Institute in May 1974[14] and these are currently being rewritten and updated.

Federal Republic of Germany. Enzymes are regarded as food additives but may be used freely and no authorization is required. However enzymes are mentioned in certain commodity regulations for cheese, fruit juices and wine-making.

France. Enzymes are regarded as food additives and the uses of named enzymes in particular commodities are authorized by Decrees. In addition, circular letters are issued from time to time giving notice that the uses of certain other enzymes are 'tolerated'. Applications for the use of enzymes must be filed with the Service de la Repression des Fraudes.

Greece. There are no specific controls on the use of enzymes.

Ireland. There are no specific controls on the use of enzymes.

Italy. Enzymes are regarded as processing aids and their use in certain commodities such as beer and wine is controlled by Decree.

The Netherlands. There are no specific controls on the use of enzymes.

UK. There are no specific controls on the use of enzymes at present and, in general, enzymes are regarded as processing aids for labelling purposes. The FACC Report on the Review of Enzymes in Food[15] recommended that in future enzymes should be controlled by permitted list and the enzymes recommended for inclusion in that list are set out in Table 3.1.1. While the recommendations are subject to comment, it is customary for regulations to follow in due course along the lines proposed by the FACC.

Canada. Enzymes are treated in a similar manner to food additives and require approval. This entails filing a Food Additive Petition in accordance with Section B.16.002 of the Food and Drug Regulations. The enzymes currently approved for use in food are listed in Table 3.1.2.

USA. The status of enzymes in the USA is complicated by the assignment of certain substances to the category of 'generally recognized as safe' (GRAS). Such substances may generally be used in food subject to good manufacturing practice. The assignment of GRAS status is not the sole prerogative of the Food and Drug Administration but may be carried out independently by any experts qualified by scientific training and experience to evaluate the safety of food ingredients. Enzymes which are not GRAS are subject to approval as food additives and these are listed in the Code of Federal Regulations. Those enzymes which are currently specifically approved in the Regulations or which are acknowledged by the FDA to be GRAS (or the subject of GRAS Petitions) are listed in Table 3.1.3. However, other GRAS enzymes may also be in use as a result of independent GRAS assessments.

References

1. Slovic P. *et al. Proc. R. Soc. Lond.* A **376,** 17–34 (1981).
2. *15th Report of the Joint FAO/WHO Expert Committee on Food Additives* (FAO Nutr. Meet. Rep. Ser. No.50; WHO Tech. Rep. Ser. No.488, 1972).
3. *18th Report of the Joint FAO/WHO Expert Committee on Food Additives* (FAO Nutr. Meet. Rep. Ser. No.54; WHO Tech. Rep. Ser. No. 557, 1974).
4. *21st Report of the Joint FAO/WHO Expert Committee on Food Additives* (WHO Tech. Rep. Ser. No.617, 1978).
5. Food Additives and Contaminants Committee *Report on the Review of Enzyme Preparations* FAC/REP/35 (includes the Report of the Committee on Toxicity of Chemicals in Food, Consumer Products and the Environment) (HMSO, London, 1982).
6. *Ibid.*
7. *Specifications for the identity and purity of some enzymes and certain other substances* (FAO Nutr. Meet. Rep. Ser. No.50B; WHO Food Additive Ser. No.2, 1972).
8. *Food Chemicals Codex* 2nd edn, 1st Suppl. (National Academy of Sciences, Washington, DC, 1974).

9. *Food Chemicals Codex* 3rd edn, (National Academy of Sciences, Washington, DC, 1981).
10 *General Standards for Enzyme Regulations* (Association of Microbial Food Enzyme Producers, Brussels 1980).
11. *Proposal for Regulation of Natural Animal/Plant-Derived Enzyme Preparations for Food Use* (Association of Manufacturers of Animal/Plant-Derived Food Enzymes, Copenhagen 1981).
12. *FAO Food and Nutrition Paper* **19** (1981).
13. Patterson, D. S. P. & Roberts, B. A. *J. Ass. Off. analyt. Chem.* **62,** (6) 1265–7 (1979).
14. *Directions on Notification and Registration of Enzyme Products* (The National Food Institute in Denmark, Söborg 1974).
15. Food Additives and Contaminants Committee *Report on the Review of Enzyme Preparations* FAC/REP/35 (includes the Report of the Committee on Toxicity of Chemicals on Food, Consumer Products and the Environment) (HMSO, London, 1982).

Table 3.1.1

UNITED KINGDOM

ENZYMES RECOMMENDED FOR INCLUSION IN
A FUTURE STATUTORY PERMITTED LIST

Enzyme	*Source*
acid proteinase (including pepsin and chymosin)	porcine gastric mucosa abomasum of calf, kid or lamb adult bovine abomasum *Mucor miehei* *Mucor pusillus*
α-amylase	porcine or bovine pancreatic tissues *Aspergillus niger* *Aspergillus oryzae* *Bacillus licheniformis* *Bacillus subtilis*
bromelain	*Ananas bracteatus* *Ananas comosus*
catalase	bovine liver *Aspergillus niger*
cellulase	*Aspergillus niger* *Trichoderma viride*
dextranase	*Penicillium funiculosum* *Penicillium lilacinum*

Enzyme	*Source*
Endothia carboxyl proteinase	*Endothia parasitica*
β-D-fructofuranosidase (invertase)	*Saccharomyces cerevisiae*
β-D-galactosidase (lactase)	*Aspergillus niger*
endo-1,3(4)-β-D-glucanase (laminarinase)	*Aspergillus niger* *Bacillus subtilis* *Penicillium emersonii*
glucose isomerase	*Bacillus coagulans*
glucose isomerase (immobilized)	*Bacillus coagulans* *Streptomyces olivaceous*
glucose oxidase	*Aspergillus niger*
exo-1,4-α-D-glucosidase (glucoamylase)	*Aspergillus niger*
exo-1,4-α-D-glucosidase (immobilized)	*Aspergillus niger*
neutral proteinase	*Aspergillus oryzae* *Bacillus subtilis*
papain/chymopapain	*Carica papaya*
pectin esterase	*Aspergillus niger*
pectin lyase	*Aspergillus niger*
polygalacturonase	*Aspergillus niger*
pullulanase	*Klebsiella aerogenes*
serine proteinase (including trypsin)	porcine or bovine pancreatic tissues *Bacillus licheniformis* *Streptomyces fradiae*
triacylglycerol lipase	edible oral forestomach tissues of the calf, kid or lamb porcine or bovine pancreatic tissues

Table 3.1.2

CANADA

ENZYMES PERMITTED FOR USE IN FOOD

Enzyme	Source	Permitted in or upon	Maximum level of use
Bovine rennet	Aqueous extracts from the fourth stomach of adult bovine animals, sheep and goats	Cheese, cottage cheese, cream cheese, cream cheese with (named added ingredients), cream cheese spread, cream cheese spread with (named added ingredients)	Good Manufacturing Practice
Bromelain	The pineapples *Ananas comosus* and *Ananas bracteatus*	(1) Ale; beer; light beer; malt liquor, porter, stout	(1) Good Manufacturing Practice
		(2) Bread; flour; whole wheat flour	(2) Good Manufacturing Practice
		(3) Edible collagen sausage casings	(3) Good Manufacturing Practice
		(4) Hydrolysed animal, milk and vegetable protein	(4) Good Manufacturing Practice
		(5) Meat cuts	(5) Good Manufacturing Practice
		(6) Meat tenderizing preparations	(6) Good Manufacturing Practice
		(7) Pumping pickle for the curing of beef cuts	(7) Good Manufacturing Practice in accordance with paragraph B.14.009(g)

Enzyme	Source	Permitted in or upon	Maximum level of use
		(8) Sugar wafers, waffles, pancakes	(8) Good Manufacturing Practice
Catalase	*Aspergillus niger* var.; *Micrococcus lysodeikticus*; Bovine (*Bos taurus*) liver	(1) Soft drinks	(1) Good Manufacturing Practice
		(2) Egg albumen	(2) Good Manufacturing Practice
Cellulase	*Aspergillus niger* var.	(1) Distillers' mash	(1) Good Manufacturing Practice
		(2) Liquid coffee concentrate	(2) Good Manufacturing Practice
		(3) Spice extracts; natural flavour and colour extractives	(3) Good Manufacturing Practice
Ficin	Latex of fig tree (*Ficus* sp.)	(1) Ale; beer; light beer; malt liquor; porter; stout	(1) Good Manufacturing Practice
		(2) Edible collagen sausage casings	(2) Good Manufacturing Practice
		(3) Hydrolysed animal, milk and vegetable protein	(3) Good Manufacturing Practice
		(4) Meat cuts	(4) Good Manufacturing Practice
		(5) Meat tenderizing preparations	(5) Good Manufacturing Practice

Enzyme	Source		Permitted in or upon	Maximum level of use
Glucoamylase (Amylo-glucosidase; Maltase)	*Aspergillus niger* var.; *Aspergillus oryzae* var.; *Rhizopus oryzae* var.	(1)	Ale; beer; light beer; malt liquor; porter; stout	(1) Good Manufacturing Practice
		(2)	Bread; flour; whole wheat flour	(2) Good Manufacturing Practice
		(3)	Chocolate syrups	(3) Good Manufacturing Practice
		(4)	Distillers' mash	(4) Good Manufacturing Practice
		(5)	Precooked (instant) cereals	(5) Good Manufacturing Practice
		(6)	Starch used in the production of dextrins, maltose, dextrose, glucose (glucose syrup), or glucose solids (dried glucose syrup)	(6) Good Manufacturing Practice
		(7)	Unstandardized bakery products	(7) Good Manufacturing Practice
	Rhizopus niveus var.	(1)	Distillers' mash	(1) Good Manufacturing Practice
	Rhizopus delemar var.;	(2)	Mash destined for vinegar manufacture	(2) Good Manufacturing Practice
	Multiplici sporus	(1)	Brewers' mash	(1) Good Manufacturing Practice

(6) Pumping pickle for the curing of beef cuts

(6) Good Manufacturing Practice in accordance with paragraph B.14.009(g)

Enzyme	Source	Permitted in or upon		Maximum level of use	
		(2)	Distillers' mash	(2)	Good Manufacturing Practice
		(3)	Mash destined for vinegar manufacture	(3)	Good Manufacturing Practice
		(4)	Starch used in the production of dextrins, maltose, dextrose, glucose (glucose syrup), or glucose solids (dried glucose syrup)	(4)	Good Manufacturing Practice
Glucanase	Aspergillus niger var.; Bacillus subtilis var.	(1)	Ales; beer; light beer; malt liquor; porter; stout	(1)	Good Manufacturing Practice
		(2)	Corn for degermination	(2)	Good Manufacturing Practice
		(3)	Distillers' mash	(3)	Good Manufacturing Practice
		(4)	Mash destined for vinegar manufacture	(4)	Good Manufacturing Practice
		(5)	Unstandardized bakery products	(5)	Good Manufacturing Practice
Glucose oxidase	Aspergillus niger var.	(1)	Soft drinks	(1)	Good Manufacturing Practice
		(2)	Liquid whole egg; egg white (albumen); and liquid egg yolk destined for drying	(2)	Good Manufacturing Practice in accordance with paragraphs B.22.034(b), B.22.035(b) and B.22.036(b)

Enzyme	Source	Permitted use	Limit
Glucose isomerase	*Bacillus coagulans* var.; *Streptomyces olivochromogenes* var.; *Actinoplanes missouriensis* var.; *Streptomyces olivaceus* var.	(1) Glucose (glucose syrup) to be partially or completely isomerized to fructose	(1) Good Manufacturing Practice
Hemicellulase	*Bacillus subtilis* var.	(1) Distillers' mash	(1) Good Manufacturing Practice
		(2) Liquid coffee concentrate	(2) Good Manufacturing Practice
		(3) Mash destined for vinegar manufacture	(3) Good Manufacturing Practice
Invertase	*Saccharomyces* sp.	(1) Soft-centred and liquid-centred confections	(1) Good Manufacturing Practice
		(2) Unstandardized bakery foods	(2) Good Manufacturing Practice
Lactase	*Aspergillus niger* var.; *Aspergillus oryzae* var.; *Saccharomyces* sp.	(1) Lactose-reducing enzyme preparations	(1) Good Manufacturing Practice
		(2) Milk destined for use in ice cream mix	(2) Good Manufacturing Practice
Lipase	*Aspergillus niger* var.; *Aspergillus oryzae* var.; edible forestomach tissue of calves, kids or lambs; animal pancreatic tissue	(1) Dairy based flavouring preparations	(1) Good Manufacturing Practice
		(2) Liquid and dried egg white (liquid and dried albumen)	(2) Good Manufacturing Practice
		(3) Romano cheese	(3) Good Manufacturing Practice

Enzyme	Source	Permitted in or upon		Maximum level of use
Lipoxidase	Soyabean whey or meal	(1)	Bread; flour	(1) Good Manufacturing Practice
Milk coagulating enzyme	*Mucor miehei* (Cooney and Emerson) or *Mucor pusillus lindt* by pure culture fermentation process; *Endothia parasitica* by pure culture fermentation processes	(1) (1)	Cheese; cottage cheese; sour cream Emmentaler (Emmental, Swiss) cheese	(1) Good Manufacturing Practice (1) Good Manufacturing Practice
Pancreatin	Pancreas of the hog (*Sus scofa*) or ox (*Bos taurus*)	(1) (2) (3)	Liquid and dried egg white (liquid and dried albumen) Precooked (instant) cereals Starch used in the production of dextrins, maltose, dextrose, glucose (glucose syrup), or glucose solids (dried glucose syrup)	(1) Good Manufacturing Practice (2) Good Manufacturing Practice (3) Good Manufacturing Practice
Papain	Fruit of the papaya *Carica papaya* (L) (Fam. Caricaceae)	(1) (2) (3)	Ale; beer; light beer; malt liquor; porter; stout Beef before slaughter Edible collagen sausage casings; water-soluble edible collagen films	(1) Good Manufacturing Practice (2) Good Manufacturing Practice (3) Good Manufacturing Practice

	(4) Hydrolysed animal, milk and vegetable protein	(4)	Good Manufacturing Practice
	(5) Meat cuts	(5)	Good Manufacturing Practice
	(6) Meat tenderizing preparations	(6)	Good Manufacturing Practice
	(7) Precooked (instant) cereals	(7)	Good Manufacturing Practice
	(8) Pumping pickle for the curing of beef cuts	(8)	Good Manufacturing Practice
Pectinase	*Aspergillus niger* var.; *Rhizopus oryzae* var.		
	(1) Cider; wine	(1)	Good Manufacturing Practice
	(2) Distillers' mash	(2)	Good Manufacturing Practice
	(3) (Naming the fruit) juice	(3)	Good Manufacturing Practice
	(4) Natural flavour and colour extractives	(4)	Good Manufacturing Practice
	(5) Skins of citrus fruits destined for jam, marmalade and candied fruit production	(5)	Good Manufacturing Practice
	(6) Vegetable stock for use in soups	(6)	Good Manufacturing Practice

Enzyme	Source	Permitted in or upon	Maximum level of use
Pentosanase	Aspergillus niger var.; Bacillus subtilis var.	(1) Ale; beer; light beer; malt liquor; porter; stout	(1) Good Manufacturing Practice
		(2) Corn for degermination	(2) Good Manufacturing Practice
		(3) Distillers' mash	(3) Good Manufacturing Practice
		(4) Mash destined for vinegar manufacture	(4) Good Manufacturing Practice
		(5) Unstandardized bakery products	(5) Good Manufacturing Practice
Pepsin	Glandular layer of porcine stomach	(1) Ale; beer; light beer; malt liquor; porter; stout	(1) Good Manufacturing Practice
		(2) Cheese; cottage cheese; cream cheese; cream cheese with (added named ingredients); cream cheese spread with (added named ingredients)	(2) Good Manufacturing Practice
		(3) Defatted soya flour	(3) Good Manufacturing Practice
		(4) Precooked (instant) cereals	(4) Good Manufacturing Practice
Protease	Aspergillus oryzae var.; Aspergillus niger var.; Bacillus subtilis var.	(1) Ale; beer; light beer; malt liquor; porter; stout	(1) Good Manufacturing Practice
		(2) Bread; flour; whole wheat flour	(2) Good Manufacturing Practice
		(3) Dairy based flavouring	(3) Good Manufacturing

	(4) Distillers' mash	(4) Good Manufacturing Practice
	(5) Edible collagen sausage casings	(5) Good Manufacturing Practice
	(6) Hydrolysed animal, milk and vegetable protein	(6) Good Manufacturing Practice
	(7) Industrial spraydried cheese powder	(7) Good Manufacturing Practice
	(8) Meat cuts	(8) Good Manufacturing Practice
	(9) Meat tenderizing preparations	(9) Good Manufacturing Practice
	(10) Precooked (instant) cereals	(10) Good Manufacturing Practice
	(11) Unstandardized bakery foods	
Rennet Aqueous extracts from fourth stomach of calves, kids, or lambs	(1) Cheese; cottage cheese; cream cheese with (added named ingredients); cream cheese spread; cream cheese with (added named ingredients); sour cream	(1) Good Manufacturing Practice
	(2) Unstandardized milk based dessert preparations	(2) Good Manufacturing Practice

Table 3.1.3

USA

ENZYMES SPECIFICALLY PERMITTED FOR USE
IN FOOD BY THE US CODE OF FEDERAL
REGULATIONS 1981 (CFR) OR ACKNOWLEDGED
BY THE FOOD AND DRUG ADMINISTRATION TO
BE GENERALLY RECOGNIZED AS SAFE (GRAS)
OR THE SUBJECT OF GRAS PETITIONS (GRASP)

Enzyme	Source	Regulatory status
α-amylase	Aspergillus niger	GRAS
	Aspergillus oryzae	GRAS
	Rhizopus oryzae	GRAS
	Bacillus subtilis	GRAS
	Barley malt	GRAS
β-amylase	Barley malt	GRAS
Cellulase	Aspergillus niger	GRAS
α-galactosidase	Morteirella vinaceae	CFR. 173.145
Glucoamylase (Amyloglucosidase)	Aspergillus niger	GRAS
	Aspergillus oryzae	GRAS
	Rhizopus oryzae	CFR. 173.145
	Rhizopus niveus	CFR. 173.110
Invertase	Saccharomyces cerevisiae	GRAS
Lactase	Aspergillus niger	GRAS
	Aspergillus oryzae	GRAS
	Saccharomyces fragilis	GRAS
Pectinase	Aspergillus niger	GRAS
	Rhizopus oryzae	CFR. 173.130
Glucose isomerase	Streptomyces rubiginosus	GRASP-4G0042
	Actinoplanes missouriensis	GRASP-6G0060
	Streptomyces olivaceus	GRASP-7G0080
	Streptomyces olivochromogenes	GRASP-7G0084
	Bacillus coagulans	GRASP-7G0086
	Arthrobacter globiformis	GRASP-7G0087
Lactase	Kluyveromyces lactis	GRASP-6G0077
	Kluyveromyces lactis	GRASP-7G0088
α-amylase	Bacillus licheniformis	GRASP-3G0026
Catalase	Aspergillus niger	GRAS

Enzyme	Source	Regulatory status
	Bovine liver	GRAS
	Micrococcus lysodeikticus	CFR.173.135
Glucose oxidase	*Aspergillus niger*	GRAS
Bromelin	Pineapples; *Ananas Comosus; Ananas Bracteatus*(L)	GRAS
Ficin	Figs: *Ficus* spp.	GRAS
Papain	Papaya: *Carica papaya*(L)	GRAS
Rennet	4th stomach of ruminants	GRAS
Pepsin	Porcine and bovine stomachs	GRAS
Trypsin	Pancreas	GRAS
Rennet	*Endothia parasitica*	CFR.173.150
	Bacillus cereus	CFR.173.150
	Mucor miehei	CFR.173.150
	Mucor pusillus	CFR.173.150
Protease (general)	*Aspergillus niger*	GRAS
	Bacillus subtilis	GRAS
Lipase	Edible forestomach tissues of calves, kids, lambs	GRAS
	Pancreatic tissue	GRAS
	Aspergillus niger	GRAS
	Aspergillus oryzae	GRAS
Lipase-esterase	*Mucor miehei*	GRASP-6G0067

Toxicology
J. R. Reichelt

1. Introduction

In the UK the Ministry of Agriculture, Fisheries and Food (MAFF), through the actions of the Food Additives and Contaminants Committee (FACC) and the Committee of Toxicology (COT), have undertaken a review of enzyme preparations and drawn up recommendations governing their use in food processing.[1] In addition, a second association has been formed to bring together the manufacturers of animal and plant enzymes. This is the Association of Manufacturers of Animal and Plant Food Enzymes (AMAPFE). At the time of writing this section, no formal documented comments similar to those of AMFEP have been published.

An outline of the proposed UK regulations is given in Chapter 3.1 (pp. 125–6) and concerns toxicological aspects and areas of manufacture where Good Manufacturing Practice (GMP) are essential. The enzyme manufacturers were asked by MAFF to make submissions which would include confidential information on manufacturing procedures, source organisms, toxicity testing, quality criteria and methods of analysis. Manufacturers had made earlier submissions to the US Food and Drug Authority (FDA)[2] for GRAS recognition ('generally recognized as safe') and the Food and Agriculture Organization of the United Nations World Health Organization (FAO/WHO) through the Joint Expert Committee on Food Additives (JECFA).[3–5]

In the USA an enzyme manufacturers' association, the Ad-Hoc Enzyme Committee, was formed to prepare data for these earlier submissions. Within the European Community the enzyme manufacturers set up their own association, the Association of Microbial Food Enzyme Producers (AMFEP),[6] to draw up general guidelines to assist regulatory authorities.

Discussions held with legislative bodies indicate that this approach is both helpful and informative. Further evidence for this comes from the acceptance by JECFA in 1981[7] of the AMFEP test for the determination of antibacterial activity and the chromatographic screening tests for aflatoxin B_1, ochratoxin A, sterigmatocystin, T-2 toxin and zearalenone by Patterson and Roberts.[8]

Guidelines for the safety evaluation of enzymes are given below and reflect both the AMFEP approach and that of the FDA for

GRAS approval. The following text, together with shorter extracts covering the classification of enzyme producing microorganisms, Good Manufacturing Practice, safety testing and the basic data of the tables in this section are reproduced with the kind permission of AMFEP.

2. AMFEP general standards for enzyme regulations

The purpose of this preamble is to present for microbial food enzymes such facts as are relevant in a discussion of whether or to what extent their application should be regulated by authorities.

Throughout living nature, from its simplest to its most complicated beings, including Man, the dependence on enzymes for performance is one of the important common features.

The mere fact that enzymes are natural substances, of course, does not guarantee that they are always harmless. Many natural substances of small molecular size are known to be active as physiological agents. Some of these are used in medical treatment, but very often this activity is only beneficial at a low dosage and detrimental at high dosage level. The reason for this is that these compounds exert their activity by reacting chemically with appropriate reaction partners. The activity of such natural substances is explained to be due to their primary molecular structure.

The activity of an enzyme, on the other hand, is due to its catalytic nature. The enzyme exerts its activity without being consumed in the reaction, but the reaction takes place at a much higher rate when the enzyme is present.

Each enzyme is extremely specific in its catalytic properties. It only works on a certain type of compound, called its substrate. Even a very minute amount of enzyme will make a large (in theory unlimited) amount of substrate react.

The catalytic activity of an enzyme is not only due to its primary molecular structure, but also to the intricately folded configuration of the enzyme molecule as a whole. This configuration conveys to the enzyme its activity, and when the configuration is disturbed, the activity is lost and what remains is a protein composed of the amino acid building blocks found in all other proteins. In some enzymes the additional presence of an equimolar amount of a low molecular weight cofactor is necessary for the catalytic activity. The most common cofactors are magnesium, calcium or zinc ions and nucleotides structurally related to the building blocks in nucleic acids.

It follows from this description that the effect of an enzyme is the same no matter how much is used, although the higher the dosage the faster the result is obtained. It also follows, that adding an enzyme is not adding a reactive chemical compound but rather adding a catalytic tool built from innocuous parts.

Since enzymes will work in a very specific way and under moderate conditions of temperatures and pH, they make it possible to replace known processes with better-controlled processes having fewer byproducts, lower energy consumption and less environmental pollution. They even make possible new processes upgrading the nutritional value of natural substances.

Being the result of a delicate configuration, the activity of an enzyme is easily lost. This, in fact, makes the enzyme as a tool even more attractive from a technological point of view, since it can easily be dispensed with after use.

Many examples of processes made possible or improved by the use of enzymes can be given, but for the purpose of this discussion it is more important to point out, that we are just in the beginning of a new technological development and the use of enzymes is expected to expand into many more areas than can be illustrated by examples today. One important field of new applications is the use of immobilized enzymes. A list of commercially used enzymes with sources and applications is given in Table 3.2.1.

The enzymes described are the active ingredients in microbial enzyme products. Microbial enzyme products are manufactured by fermentation of selected strains of nonpathogenic microorganisms. The delicate nature of enzymes makes it necessary to use very gentle methods of extraction and purification in order to isolate the enzyme from the fermentation. Besides the enzyme itself the enzyme products will normally contain other soluble parts of the microorganism and the fermentation medium as well as residues of harmless amounts of materials used for purification and food grade materials used as fillers and diluents to obtain a product of standardized enzymic activity.

From this description of the way microbial enzyme products are made, it is not surprising that animal feeding trials have not demonstrated any harmful effects. Furthermore, because of their catalytic nature they are always used at very low dosage levels in food processes. The consumption of food processed with enzymes can therefore hardly present a hazard to the consumers and might be considered too unimportant to merit the attention of regulating authorities.

From a purely scientific and technical point of view we would support this reasoning. However, since in today's consumer society it is not to be expected that any area can be left unregulated, we have felt it more realistic to anticipate a future demand for enzyme legislation. We have therefore directed our efforts towards contributing to a regulatory framework, which complies with the best of our present knowledge about food enzyme technology and which would not inhibit future developments.

Leaving for a moment the enzyme technical field in which we find ourselves to be the experts, we would like to make a few comments on how to regulate enzymes. While for food additives a detailed positive listing system specifying permitted applications seems to be universally adopted, this is not necessarily the best way to regulate the use of processing aids (*auxiliaires technologiques*) like enzymes. In fact we would propose that certain microorganisms as such are approved as sources of food enzymes to be used in all foods and with no other limitations than Good Manufacturing Practice. This in our opinion should be the content of the positive list in a horizontal enzyme directive, leaving any other limitations (if found desirable in special cases) to vertical directives for the food in question.

Irrespective of the way in which enzymes (if at all) are eventually regulated, we believe that approval should be based upon certain kinds of information from the enzyme producers (and enzyme users) to the authorities, and in the following text we have outlined the extent of this information.

A dossier should contain the following information, some of which may have to be confidential.

Soluble enzyme preparations. (*i*) The source of the enzyme. Classification criteria are given on page 150 and in Table 3.2.2.

- (*ii*) The main activity and the EC number following IUPAC designation.
- (*iii*) Animal toxicity studies. The extent of the studies is given in Table 3.2.3.
- (*iv*) Method of production and controls in production. The definition of Good Manufacturing Practice (GMP) should comply with FCC and JECFA recommendations. Raw materials should be food or feed grade quality. The microbiological part of GMP is outlined on page 151.
- (*vi*) Food application, including dosage and residual level in food.
- (*vii*) Activity of the commercial preparation in manufacturers' units as well as in any common or reference units.
- (*viii*) Product specification based upon JECFA and FCC recommendations, but including total viable count (maximum 5×10^4), and *Escherichia coli* (not detected). To be in agreement with the recent FCC specifications, *Pseudomonas aeruginosa* is deleted. Aflatoxin is also deleted as a general test. Testing for antibiotic activity should be done by means of a panel of microorganisms as indicated on page 152.

TABLE 3.2.1

The use of enzymes in food

Biological origin	Principal enzymatic activity	IUB number	Application examples*
Bacteria			
Actinoplanes missouriensis	Glucose isomerase	5.3.1.5	H, I, J, L
Arthrobacter sp.	Glucose isomerase	5.3.1.5	H, I, J, L
Bacillus cereus	Protease	3.4.24.4	J, O
	β-amylase	3.2.1.2	H, J
	Iso-amylase	3.2.1.68	H, J
Bacillus circulans	Endo-β-glucanase	3.2.1.6	J
Bacillus coagulans var.	Glucose isomerase	5.3.1.5	H, I, J, L
Bacillus licheniformis	α-amylase	3.2.1.1	B, C, H, I, J, L, N, O
	Protease	3.4.21.14	E, F, K
Bacillus megaterium	β-amylase	3.2.1.2	H, J
Bacillus subtilis	α-amylase	3.2.1.1	B, C, H, I, J, L, N, O
(including strains known under the	β-amylase	3.2.1.2	H, J
name *Bacillus mesentericus* and	Endo-β-glucanase	3.2.1.6	J
Bacillus amyloliquefaciens)	Hemicellulase	3.2.1.78	M
	Protease	3.4.24.4	J, O

Organism	Enzyme	EC number	Codes
Klebsiella aerogenes	Dextranase	3.2.1.11	L
	Pullulanase	3.2.1.41	L, H, J
Leuconostoc oenos	Malic acid decarboxylase	1.1.1.39	J
Micrococcus lysodeikticus	Catalase	1.11.1.6	A, B, G, J, P
Streptomyces albus	Glucose isomerase	5.3.1.5	H, I, J, L
Streptomyces olivaceus	Glucose isomerase	5.3.1.5	H, I, J, L
Streptomyces olivochromogenes	Glucose isomerase	5.3.1.5	H, I, J, L
Streptomyces sp.	Xylanase	3.2.1.32	H, J, M
Fungi			
Aspergillus melleus	Protease	3.4.23.4	B
Aspergillus niger var.	α-amylase	3.2.1.1	H, I, L, N, O
	Glucoamylase or amyloglucosidase	3.2.1.3	I, J, L, N, O, R
	Catalase	1.11.1.6	A, B, G, J, P
	Cellulase	3.2.1.4	I, J, R
	Endo-β-glucanase	3.2.1.6	J
	Glucose oxidase	1.1.3.4	G, J, P
	Hemicellulase	3.2.1.78	M, Q
	Lipase	3.1.1.3	B, C
	Maltase or α-glucosidase	3.2.1.20	H
	Pectinase	3.2.1.15	C, I, J, Q

Biological origin	Principal enzymatic activity	IUB number	Application examples*
	Protease	3.4.23.4	B, E, F, H, I, J, K, O, R
	Cellobiase	3.2.1.21	I
	Tannase	3.1.1.20	J
	Xylanase	3.2.1.32	H, J, M
	Lactase	3.2.1.23	A, B, D, R
	α-galactosidase	3.2.1.22	L
	Invertase	3.2.1.26	N
Aspergillus oryzae var.	α-amylase	3.2.1.1	C, H, I, J, L, N, O
	Glucoamylase or amyloglucosidase	3.2.1.3	C, I, J, L, N, O, R
	Cellulase	3.2.1.4	I, J, R
	Endo-β-glucanase	3.2.1.6	J
	Hemicellulase	3.2.1.78	M
	Lipase	3.1.1.3	B, C
	Maltase or α-glucosidase	3.2.1.20	H
	Pectinase	3.2.1.15	C, I, F
	Protease	3.4.23.4	B, E, F, H, I, J, K, O, R
	Tannase	3.1.1.20	J
	Lactase	3.2.1.23	A, B, D, R
Candida lipolytica	Lipase	3.1.1.3	B, C
Endothia parasitica	Protease	3.4.23.6	B
Mortierella vinacea sp.	α-galactosidase	3.2.1.22	L

Mucor javanicus	Lipase	3.1.1.3	B, C
Mucor miehei	Lipase	3.1.1.3	B, C
	Esterase	3.1.1.1	B, C
	Protease	3.4.23.6	B
Mucor pusillus	Protease	3.4.23.6	B
	Lipase	3.1.1.3	B, C
Penicillium emersonii	Endo-β-glucanase	3.2.1.6	J
Penicillium funicolosum	Dextranase	3.2.1.11	L
Penicillium lilacinum	Dextranase	3.2.1.11	L
Penicillium simplicissium	Pectinase	3.2.1.15	C, I, J, Q
Rhizopus arrhizus	Lipase	3.1.1.3	B, C
	Glucoamylase or amyloglucosidase	3.2.1.3	I, J, L, N, O, R
Rhizopus delemar	α-amylase	3.2.1.1	H, I, J, L, N, O
	Glucoamylase or amyloglucosidase	3.2.1.3	I, J, L, N, O, R
	Cellulase	3.2.1.4	I, J, R
	Endo-β-glucanase	3.2.1.6	J
	Hemicellulase	3.2.1.78	M

Biological origin	Principal enzymatic activity	IUB number	Application examples*
Rhizopus niveus	Lipase	3.1.1.3	B, C
	Glucoamylase or amyloglucosidase	3.2.1.3	I, J, L, N, O, R
Rhizopus oryzae	α-amylase	3.2.1.1	H, I, J, L, N, O
	Glucoamylase or amyloglucosidase	3.2.1.3	I, J, L, N, O, R
	Cellulase	3.2.1.4	I, J, R
	Endo-β-glucanase	3.2.1.6	J
	Hemicellulase	3.2.1.78	M
	Maltase or α-glucosidase	3.2.1.20	H
	Pectinase	3.2.1.15	C, I, J
Sporotrichum dimorphosporum	Cellulase	3.2.1.4	I, J, R
	Hemicellulase	3.2.1.78	M
	Xylanase	3.2.1.32	H, J, M
Thielavia terrestris	Cellulase	3.2.1.4	I, J, R
Trichoderma reesei (*Trichoderma viride*)	Glucoamylase or amyloglucosidase	3.2.1.3	I, J, L, N, O, R
	Cellulase	3.2.1.4	I, J, R
	Hemicellulase	3.2.1.78	M
	Cellobiase or β-glucosidase	3.2.1.21	I
	Pectinase	3.2.1.15	C, I, J, Q

Yeasts

Kluyveromyces fragilis	Lactase	3.2.1.23	A, B, D, R
	Inulinase	3.2.1.7	L
	Invertase	3.2.1.26	N
Kluyveromyces lactis	Lactase	3.2.1.23	A, B, D, R
Saccharomyces carlsbergensis	Invertase	3.2.1.26	N
	α-galactosidase	3.2.1.22	L
Saccharomyces cerevisiae	Invertase	3.2.1.26	N

* A, milk; B, cheese; C, fats and oils; D, edible ice; E, meat; F, fish; G, egg; H, cereal and starch; I, fruit and vegetables; J, beverages (soft drinks, beer, wine); K, soups and broths; L, sugar and honey; M, cocoa, chocolate, coffee and tea; N, confectionery; O, bakery; P, salads; Q, spices and flavours; R, dietary food.

TABLE 3.2.2

Microorganisms used for enzyme production

Group A: Microorganisms that have traditionally been used
            in food or in food processing
            Bacillus subtilis (including strains known
                under the names *mesentericus, natto* and
                amyloliquefaciens).
            Aspergillus niger (including strains known
                under the names *awamori, foetidus,
                phoenicis, saitoi* and *usamii*).
            Aspergillus oryzae (including strains known
                under the names *sojae* and *effusus*).
            Mucor javanicus
            Rhizopus arrhizus
            Rhizopus oligosporus
            Rhizopus oryzae
            Saccharomyces cerevisiae
            Kluyveromyces fragilis
            Kluyveromyces lactis
            Leuconostoc oenos

Group B: Microorganisms that are accepted as harmless
            contaminants present in food
            Bacillus stearothermophilus
            Bacillus licheniformis
            Bacillus coagulans
            Bacillus megaterium
            Bacillus circulans
            Klebsiella aerogenes

Group C: Microorganisms that are not included in groups A
            and B
            Mucor miehei
            Mucor pusillus
            Endothia parasitica
            Actinoplanes missouriensis
            Streptomyces albus
            Bacillus cereus
            Trichoderma reesei (T. viride)
            Penicillium lilacinum
            Penicillium emersonii
            Sporotrichum dimorphosporum
            Streptomyces olivaceus
            Penicillium simplicissium
            Penicillium funiculosum

(*ix*) Carriers, diluents, preservatives, anti-oxidants, anticaking
        agents and stabilizers in the enzyme product. We agree to
        follow FCC recommendations on carriers and diluents. On
        preservatives we agree that liquid enzyme products should

TABLE 3.2.3

Safety testing of food enzymes based on the AMFEP classification

Group Tests (X = to be performed)	A, Microorganisms that have traditionally been used in food, or in food processing	B, Microorganisms that are accepted as harmless contaminants present in food	C, Microorganisms that are not included in A or B
Pathogenicity	In general no testing required		X
Acute oral toxicity, mouse and rat		X	X
Subacute oral toxicity, rat four weeks		X	X
Three month oral toxicity, rat		X	X
In vitro mutagenicity		X	X
Teratogenicity, rat			(X)*
In vivo mutagencity, mouse and hamster			(X)*
Toxicity studies on the final food			(X)*
Carcinogenicity, rat			(X)*
Fertility and reproduction			(X)*

*Only to be performed under exceptional conditions.

be permitted to contain the amount necessary of commonly accepted preservatives like benzoate, sorbate and parabens and no more than that amount.

(x)   Typical average composition of the enzyme product.

Immobilized enzyme preparations. Since it was found that immobilized enzymes presented different problems, they have been discussed separately from soluble enzymes. The following conclusions have been reached regarding regulatory aspects of immobilized enzymes:

(i)   Source: should be restricted to just the range of species suitable for production of soluble enzymes.

(ii)   Carrier: the materials chosen as carriers should either be inert or generally acceptable for use in food or food manufacture. When new materials are being considered, they should be tested to prove that no harmful residues will leak out into the food.

(iii) Immobilization agents: as such only agents approved for relevant uses should be used, but not those banned for such uses. Agents neither banned nor approved should be accepted on the basis of suitable toxicological and/or analytical testing on the food product.

(iv)   Toxicity studies: certain immobilization techniques should be approved so that when used with an enzyme approved by itself no additional toxicicity testing is needed. In any case the ratio between enzyme and food is lower than for the soluble enzyme. However, for new immobilization techniques and/or new enzymes (possibly never used in soluble form) some toxicity testing would be necessary.

(v)   Leakage: tests will have to be performed showing that leakage of immobilization agent or enzyme is kept within certain limits.

(vi)   Product specifications: as for soluble enzymes.

(vii) Assay methods: ideally, an assay method should both give a good measure of productivity and be a short term method.

Enzyme use in food. The AMFEP guidelines on enzymes used in food processing and their application within the industry are summarized in Table 3.2.1. The table shows the biological origin and principal activity of the enzymes together with their IUB classification number. A key to the examples used to illustrate the application of industrial enzymes is included.

Classification of enzyme producing microorganisms. In Table 3.2.2 microorganisms are classified according to their use in food or enzyme manufacture. The names used are those recommended in Bergey 8th edition (Raper and Fennell), *The Genus Aspergillus*

(Baltimore, 1965) and J. Lodder's *The Yeasts* (Delft, 1970).

When new developments in taxonomy are generally and officially accepted, the names should be changed accordingly, but it is recommended that reference is made to the original names.

In cases where an organism is known under several names, the most common synonyms are mentioned. *Aspergillus niger* is used in the broad meaning which is referred to in *The Genus Aspergillus*.

The list is not exhaustive; each of the three groups contains organisms not mentioned in Table 3.2.2.

Tests for antibacterial activity. The method described below is used with the following organisms: *Staphylococcus aureus* ATCC 6538; *Escherichia coli* ATCC 11229; *Bacillus cereus* ATCC 2; *Bacillus circulans* ATCC 4516; *Streptococcus pyogenes* ATCC 12344; *Serratia marcescens* ATCC 14041. The test is made in the absence of preservatives.

Determination of antibacterial activity

Culture plates. Using the organisms mentioned above make a test plate of each organism by preparing a 1 : 10 dilution of a 24-hour Trypticase Soya Broth culture in Trypticase Agar (TSA) (for *Streptococcus pyogenes* a 1 : 20 dilution). Pour 15 millilitres of plain TSA into a Petri dish and allow the medium to harden. Overlay with 10 millilitres of seeded TSA and let solidify. Place a paper disk of the tested enzyme on each of the six inoculated plates.

Disk preparation. Make a 10 per cent solution of the enzyme by adding 1 gramme of enzyme to 9 millilitres of sterile distilled water. Mix thoroughly with a Vortex mixer to obtain a homogeneous suspension. Autoclave the paper disks (S&S Analytical Filter Papers No 740-E, 12.7 millimetres in diameter), then saturate them with the enzyme by application of 0.1 millilitres (about three drops) of a 10 per cent solution of the enzyme to the disk surface. Prepare six disks (one for each of the six organisms) for each enzyme; place a disk on the surface of the six inoculated agar plates.

Incubation. Keep the six plates in the refrigerator overnight to obtain proper diffusion. Incubate the plates at 37 °C for 24 hours. Examine the plates for any inhibition zones that may have been caused by the enzyme preparation.

Interpretation. A visually clear zone around a disk (total diameter greater than 16 millimetres) indicates the presence of antibacterial components in the enzyme preparation. If an enzyme preparation shows obvious antibacterial activity against three (or more) organisms it is concluded that antimicrobial agents are present.

Good Manufacturing Practice (microbiological aspects of enzyme production). The methods used to control microbial growth during enzyme manufacture are designed to ensure that the enzyme production is performed by a pure culture of the producing microorganisms. The number of foreign microorganisms should be insignificant (i.e. they should not be able to influence the fermentation process or the final enzyme product). The tests given below are performed on all batches. If a contamination is suspected, more extensive tests will be made.

Microbiological control procedures used in the cultivation of microorganisms for production of food grade enzymes. Food grade enzymes should be prepared by a pure culture of the enzyme-forming microorganism in question. This objective is secured by performing the cultivation under conditions of controlled microbial purity. Tests for purity of the culture are made on samples drawn aseptically in the following points (*see* below) of the manufacturing process: (*i*) (*iii*) (*v*) and (*vi*). On uninoculated medium [(*ii*) and (*iv*)] the samples are observed for absence of growth.

   (*i*)   Stock culture.
   (*ii*) Uninoculated seed culture medium.
   (*iii*) Seed culture before transfer to fermenter.
   (*iv*) Uninoculated fermenter medium.
   (*v*)   Fermentation liquid at regular intervals during the fermentation.
   (*vi*) Final sample immediately before transfer to the recovery plant.

The tests are performed in the following ways: (*i*) By visual inspection, macroscopic and microscopic, and (*ii*) by streaking on plates and observation for the presence of foreign microorganisms.

In the more traditional methods of semi-solid cultivation, control procedures are the same as in the submerged fermentation processes as far as points (*i*) to (*iii*) of the list above are concerned.

In this technology it is part of good manufacturing practice to avoid significant contaminations by choosing appropriate process conditionsns such as composition of media, sterilization and large inoculation.

Due to the impossibility of taking representative samples from large volumes of semi-solid masses, observations for the absence of undesired growth are restricted to visual inspections.

Final sampling is done at the extraction step, corresponding to point (*vi*) of the list above.

If a significant contamination develops in a batch, the batch is discarded. In addition to these tests, a close check is kept on aberrations from normal of fermentation parameters like enzyme activity, pH etc, which might signal a contamination before it can be detected microbiologically.

The stock culture is stored as a pure culture in such a way that degeneration is avoided that is, by lyophilization, storage under liquid nitrogen or by other methods which have been found suitable for the organisms in question. In revival of the culture strict aseptic technique is used.

Safety testing of food enzymes. The guidelines from AMFEP on the classification of safety testing for use with enzymes produced by microorganisms in food processing are summarized in Table 3.2.3. AMFEP recommend that safety testing is not necessary except under special conditions for microorganisms that have traditionally been used in food or food processing. This reflects a different approach from that proposed by the FDA guidelines.

3. FDA requirements for safety studies for enzyme preparations
The US Federal Food, Drug and Cosmetic Act and the FDA regulations, place the burden of proof of safety of food ingredients on the manufacturer and/or user of the substance. Level of proof is dependent on many factors, including whether or not the substance was in use in foods in the USA prior to 1958; the year this portion of the FD&C Act was amended.

In general, it is necessary to conduct adequate and appropriate animal and other biological experiments to show whether or not the substance will be safe for its intended use. This is equally true for regulated food additives and GRAS (generally recognized as safe) substances. The FDA announced,[9] but never formally adopted, some general guidelines concerning principles and procedures for food additive (or GRAS substances) safety evaluations. Food additives fall into one of three categories depending on human exposure and toxicological properties.

(*i*) An additive whose molecular structure is not associated with any known toxic potential and whose presence in the daily diet is less than 0.05 parts per million would initially be assigned to the lowest level of concern. Tests likely to be associated with this level are: a 28-day continuous feeding study in a rodent species, and a genetic toxicity screen.

(*ii*) An additive with a molecular structure that is not associated with any known toxic potential, but which is present in the daily diet at levels between 0.05 and 1.0 parts per million would be assigned to an intermediate level of concern. Tests likely to be required at the level are: a 90-day (subchronic) feeding study in a rodent species; a subchronic feeding study in a non-rodent species; a reproduction study with a teratology phase, in the rodent; and a genetic toxicity screen.

(*iii*) Tests likely to be required for an additive with a presence in the daily diet of 1.0 parts per million or more are: lifetime feeding studies for carcinogenicity in two rodent species; a chronic toxicity study in both a rodent species and a non-rodent species (one year dog, for example); a reproduction study with a teratology phase in the rodent; and a genetic toxicity screen.

Latest thinking in possible safety studies for microbial enzyme preparation. Despite these announced guidelines, recent contacts with the FDA and current industry practices in submission of animal safety data indicate specific requirements for microbial enzyme preparations to be: studies or data to demonstrate that the microorganism used is nonpathogenic and nontoxic in man or other animals and does not produce antibiotics; and multigeneration feeding studies at three levels (1.24, 2.50 and 5 per cent of diet) in a rodent (rat normally, but mouse probably acceptable), reproductive study with *in utero* exposure and with teratology.

Animal numbers: At least 20 litters per generation per dose for reproduction studies, 20 litters per dose for teratological studies and 20 surviving animals per dose per sex for histology assessment. This is usually accomplished by using:

F_0: 25 to 30 males and 25 to 30 females per dose to ensure 20 litters per dose are delivered.

F_1: Select 50 males and 50 females per dose. Of these animals, 25 males and 25 females should be studied in the growth and development phase and undergo histological examination following the delivery of the F_2 a litter. Use of 25 animals will ensure that 20 animals per sex/dose are available for histology (only 20 need undergo histological examination provided they are selected randomly). Use the remaining 25 males and 25 females that did not undergo histological examination for breeding to produce the 20 F_2b generation litters delivered by caesarian section for teratological analysis. One half of these fetuses should undergo visceral and the other half skeletal examinations.

A genetic toxicity study may or may not be deemed necessary due to the obvious problems with testing substances such as enzymes.

Enzyme preparations from plant or animal sources do not fall under the above scheme. Requirements are unclear as to my knowledge, no new plant or animal enzyme preparations have been introduced into the American, food industry since 1958. It should be emphasized, however, that the American, FDA does not have any published guidelines or criteria by which to prejudge what data are needed and how they should be developed to prove the safety of enzymes. Each such substance will be judged on a case by case basis, on its own merits.

The above information, together with manufacturing materials, methods and controls, intended use levels and effectiveness in foods and other pertinent information, is then submitted to the FDA in support of a petition for a 'generally recognized as safe' affirmation or for a 'food additive regulation'.

4. Conclusions

Enzyme manufacturers have through the formation of industry associations endeavoured to furnish data for legislative bodies to apply as guidelines on the use of enzymes and assist in the framing of regulatory controls for their use in food processing. A clarification of enzyme producing microorganisms has been given together with indications of their applications in food processing at the present time. Indications on quality assurance criteria – antibacterial activity and microbiological aspects of Good Manufacturing Practice by AMFEP – reflect the industry's concern and dedication to product quality and safety.

Manufacturers have also indicated a specific rationale for soluble and immobilized enzyme preparations to clarify the differences between them. Immobilized enzyme production techniques by the use of immobilization agents and support media present different problems from those of soluble enzyme products, and should therefore be considered separately for regulatory purposes.

Proposals for safety evaluations of food enzymes produced from microorganisms have been made by AMFEP, together with the FDA guidelines for GRAS approval. These two separate views have been given in an effort to show the latest thinking on possible future safety evaluation requirements for microbial food enzymes, in the USA and in the European Community.

Unfortunately, no information has been given on the possible safety evaluations required for enzymes derived from plant or animal sources, as they are not contained within the published guidelines for enzymes from microorganisms, and none are available at this time.

References

1. Food Additives and Contaminants Committee *Report on the Review of Enzyme Preparations* FAC/REP/35 (includes the Report of the Committee on Toxicity of Chemicals in Food, Consumer Products and the Environment) (HMSO, London, 1982).
2. *Food Chemicals Codex* 2nd edn, 1st Suppl. (National Academy of Sciences, Washington, DC, 1974); *Food Chemicals Codex* 3rd edn (National Academy of Sciences, Washington, DC, 1981).
3. *15th Report of the Joint FAO/WHO Expert Committee on Food Additives* (FAO Nutr. Meet. Rep. Ser. No. 50; WHO Tech. Rep. Ser. No. 488, 1972).

4. *18th Report of the Joint FAO/WHO Expert Committee on Food Additives* (FAO Nutr. Meet. Rep. Ser. No. 54; WHO Tech. Rep. Ser. No. 557, 1974).
5. *21st Report of the Joint FAO/WHO Expert Committee on Food Additives* (WHO Tech. Rep. Ser. No. 617, 1978).
6. *General Standards for Enzyme Regulations* (Association of Microbial Food Enzyme Producers, Brussels, 1980).
7. *FAO Food and Nutrition Paper* **19** (1981).
8. Patterson, D. S. P. & Roberts, B. A. *J. Ass. Off. analyt. Chem.* **62**, (6) 1265–1267 (1979).

SAFE HANDLING
R. I. Farrow and J. R. Reichelt

1. Introduction
This chapter takes a prospective view with regard to the requirements of the Health and Safety at Work Act 1974, together with
perceived legislative changes. Observations have been made on
labelling and disclosures of product constituents in the light of
future EEC legislation for packaging and transportation of substances. Enzyme production criteria and product composition are
discussed, together with sensitivity reactions which can occur
through misuse. Current enzyme production trends are reviewed in
the light of future regulatory constraints with the development of
concentrated liquid enzymes and powder agglomeration techniques. Finally, recommendations for the safe handling and use of
industrial enzymes are given together with practical guidelines for
use in cases of accidental spillage or body contact.

The authors have used their experience of the operation and
interpretation of the current UK Health and Safety at Work Act
(1974) to provide this account in the belief that it may be considered
a sound example for the development of good industrial practice.

2. Legislative constraints
Microbial and animal or plant-derived enzymes supplied as food
processing aids are subject to two major regulatory constraints at
the time of writing: (*a*) use in the factory and (*b*) enzyme product
compositions and description.

The Health and Safety at Work Act (1974). The Health and
Safety at Work Act (HSW Act) is most relevant to health and safety
considerations in the factory environment. Part I of the Act aims to
secure the health, safety and welfare of persons at work – including
self-employed persons – and to protect other people from the
results of the activities of persons at work. The Act sets forth
general duties of employers to their employees, and covers maintenance of plant, safe storage of raw materials, provisions of training
and creation of a safe and healthy working environment. The
general duties of employers, and the self-employed, to persons
other than their employees, are defined, along with the general
duties of the persons concerned with certain premises in relation to
harmful emissions into the atmosphere. Atmosphere in this case
appears to mean the external environment rather than the 'in plant'

atmosphere which is covered by other provisions of the Act.

Special attention should be given to Part II, Sections 2–9, which with perceived future legislation will be given extra emphasis and new powers by additional regulations aimed at strengthening the statutory powers of the HSW Act. Section 2 defines the general duties of employers to their employees. Section 4 defines the general duties of persons concerned with premises to persons other than their employees to ensure there is no hazard. Section 6 defines the general duties of manufacturers etc, relating to articles and substances for use at work. This section of the Act lays upon manufacturers the obligation to ensure that: (a) the article is safe and without risks to health; (b) tests are carried out to demonstrate the safety of materials in use; (c) adequate training and information relating to the material is provided; (d) any necessary research is carried out with a view to the elimination of any risks to health or safety to which the substance may give rise.

Section 7 defines general duties of an employee at work who should: (a) take reasonable care for the health and safety of himself and of other persons who may be affected by his acts or omissions at work; and (b) as regards any duty or requirement imposed on his employer or any other person by or under any of the relevant statutory provisions, to cooperate with him so far as is necessary to enable that duty or requirement to be performed or complied with.

In future, matters relating to processes carried out in the course of business and under manufacturers' control, will ultimately be governed by interpretation viewed as a result of judicial proceedings. As a result, legal action could be taken against individual employees or company officers, and not just the company as has been the case with proven infringements in the past.

Regulatory control for enzymes in food processing. The Ministry of Agriculture, Fisheries and Food (MAFF), through the actions of the Food Standards Committee (FSC), the Food Additives and Contaminants Committee (FACC), and the Committee of Toxicology (COT), have undertaken a review of enzyme preparations with a view to the creation of regulations governing their use in food processing.

An outline of the proposed UK regulations concerning the use of enzymes in food processing has been extensively reviewed in Chapter 3.3, 'The Legislative Aspects of the Use of Industrial Enzymes in the Manufacture of Food and Food Ingredients'.

Labelling and packaging. Further prospective legislation regarding labelling and packaging of chemical substances during the next few years is inevitable; the universal adoption of the sixth amendment to European Commission Directive 67/548/EEC concerning the classification packaging and labelling of chemical substances confirms this.

In the opinion of the authors these changes would be linked into the HSW Act in the UK and Northern Ireland, and in turn there are no clearly defined exemption guidelines for food processing aids or enzyme preparations within the published guidelines. Prospective regulations would make recommendations regarding the labelling of enzyme preparations. The majority of industrial enzyme labels currently in use could be greatly improved.

It has been suggested that the following information should be made available on product labels (future transport legislation may require special labelling for all products) to provide adequate identification for both government bodies (e.g. Customs) and users:

(*i*)   Trade name and generic name (IUB Classification Number)
(*ii*)   Enzyme source
(*iii*) Activity of enzyme(s)
(*iv*)   Contents by volume or weight
(*v*)   Major ingredients by percentage to include diluents, stabilizers, preservatives, colours and salt composition
(*vi*)   Manufacturers' name and address – country of origin
(*vii*) Warning notice

CAUTION

Product contains enzymes which may cause allergic reactions in sensitive individuals. Handle the product with care and in accordance with instructions given in the manufacturers' technical literature.

Avoid contact with eyes, skin and mucous membranes. In case of accidental spillage, wash with large volume of water. If accidental contact with skin or eyes wash immediately with tap water.

Technical literature provided by enzyme suppliers is generally of a high standard, providing ample scientific advice regarding enzyme performance in the process for which the enzyme is supplied. Health hazards could be made more obvious, and it is suggested that a cautionary statement similar to that outlined for product labels could be added.

3. Enzymes

Enzyme preparations supplied to the food processing industry are offered in liquid, powdered, prilled or granulated forms. They are commonly derived from the controlled fermentation of microorganisms or by extraction from plant and animal tissues.

Sources of industrial enzyme. Bovine and porcine animals are the primary source for animal derived enzymes, for example pepsin, trypsin, chymotrypsin, pancreatic lipase and pancreatin. Papain, bromelain and ficin are examples of plant extracted enzymes from *Carica papaya* (L), *Ananas comosus*, *Ananas bracteatus* (L) and *Ficus* sp. Fungi as moulds and yeast together with bacteria form a ready source of microorganisms for enzyme manufacture, *Bacillus subtilis*, *Aspergillus niger* var. and *Aspergillus oryzae* are the most commonly known examples.

Enzymes are often referred to as 'biological catalysts' which are found in, or closely associated with, living cells. Their structure is based primarily on protein, but other components such as carbohydrates, lipids and metal ions may also be present. Some enzymes have special nonprotein groupings or moieties called 'cofactors' (of which haems and flavins are examples) which are essential for correct catalytic functioning of the enzyme.

Production criteria and product composition. Production processes have been effectively described exhaustively in many texts, but they may be summarized into three major operations: (*a*) culture selection and maintenance of purity criteria; (*b*) preparation of inoculum; (*c*) large scale fermentation and product extraction. Successful enzyme production may be achieved only by the careful maintenance of strain purity, asepsis during media preparation and fermentation together with 'fine tuning' for maximum enzyme yield coupled with fermentor efficiency. Manufacturers are constantly striving to improve production criteria and the current interest in large scale fermentation processes has resulted in the application of the latest genetic engineering techniques to assist in product development.

Enzymes produced by fermentation must be extracted from the fermentation media or the harvested cells after disruption of the cell membranes. The extraction procedures are varied and further details can be obtained in Neckhorn *et al.* (1965).

The final stage of production usually involves precipitation or concentration, together with assays and standardization for products. In the case of liquid enzymes, stabilizers and preservatives must be added to ensure adequate stability. Typical compositions of liquid and powdered enzyme products are shown in Table 3.3.1.

Liquid and powdered enzyme preparations used for food processing are usually single addition procedures. Process operators dispense enzymes with measuring equipment using gravimetric or volumetric procedures. It is at this point that process workers may be exposed to an airborne microfine powder, or an aerosol, or splashes from liquid enzyme products if simple safety precautions are ignored. In the majority of individuals, such exposure has no obvious adverse effects (providing exposure has not been at very high concentration or long duration) and it must be stressed that

TABLE 3.3.1

Analysis of powdered and liquid industrial enzymes to show their major constituents

	% w/w	
Constituent	Powder	Liquid
Inactive protein	1–5	2–5
Enzyme-active protein	1–8[1]	0.5–5[3]
Diluents	up to 90+[2]	—
Preservative	—	0.2–1.0[4]
Colour	—	0.05–0.1[5]
Electrolytes	2–5	up to 16–18[6]
Water (potable)	3–5	up to 90+

[1] Indicates the high potency of powdered enzymes over liquids.
[2] Corn starch, lactose, dextrins, sodium and potassium chloride can be used.
[3] Indicates activity of enzymic proteins.
[4] Benzoic acid or mixture of benzoic acid esters – usually expressed as benzoates or methyl, ethyl and propylparabens.
[5] Caramel or annato.
[6] Rennet contains up to 16–18% sodium chloride.

enzyme preparations are considerably less harmful than many industrial chemicals – ammonia, chlorine, acids, alkali and organic solvents.

Immune response. Unfortunately all foreign protein is antigenic and enzymes are protein. Some individuals show a high degree of sensitivity to 'foreign protein' of many kinds. Enzyme preparations form a small part of this category and thus can present a degree of health risk.

The risk inherent in this exposure and subsequent contact concerns a highly complex body defence mechanism which is termed the 'Immune response'. In the vast majority of cases, this is a complicated, subtle but trouble-free mechanism which protects the body against invasion by viruses, bacteria and alien proteins.

Classification of hypersensitivity. A malfunction of this 'Immune response' is known as 'Hypersensitivity disease', and results in injury or death to the host cells. The defence mechanisms of the human body are comprehensive and diverse, ranging from 'macro-effects' – such as inflammation, coughing, sneezing, vomiting, diarrhoea, reflex muscular movements and skin exfoliation – to intrinsic enzyme-mediated reactions in blood, on the surface of mucous membranes, and in the intracellular matrix. These may be termed 'microeffects', although the bodily symptoms can sometimes be-

TABLE 3.3.2

Generalized hypersensitivity classification table

Prognosis	Skin tests	Antibody type	Tissue response
Type I Hay fever Allergic rhinitis Allergic asthma Anaphylaxis Some urticaria migraine Some gastro-intestinal disorders	Immediate wheal and flare Maximal at 10–15 mins Duration 1½–2 hours	Circulating and cell-fixed in certain tissues (nose, eyes, lungs, skin) Cells are passively sensitized by 'Reagins', IgE-type antibody produced elsewhere Non-precipitating	Antigen–antibody complex sets off reaction at cell membrane resulting in release of pharmacologically active substances causing increased capillary permeability, arteriolar dilation, and smooth muscle contraction Eosinophilia
Type II Incompatible blood transfusion Haemolytic disease of newborn Drug-induced purpura Drug-induced haemolytic anaemia	Not applicable	Circulating antibodies IgC and IgM	Antigen is present in or on a tissue cell, or is formed by combination of a hapten with a tissue cell. Antibody reacts with antigenic component of cell membrane to cause

Drug-induced
granulocytopoenia
Transplant rejection

Type III
Arthus reaction
Bronco-pulmonary
aspergillosis
Farmer's lung, bird-fancier's
lung etc
Serum sickness
Polyarteritis nodosa
Disseminated lupus
erythematosus
Other auto-immune diseases

Circulating antibodies IgG
and IgM
'Precipitins'

Dual reaction
Immediate wheal followed
by oedema at 3 hours
Maximal at 6–7 hours
Duration 24 hours

Antigen reacts with
antibody in tissue-spaces,
forming micro-
precipitates in and
around small vessels
Intravascular thrombosis,
fibrinoid necrosis,
haemorrhage
Neutrophil debris

Type IV
Contact dermatitis
Tuberculin reaction
Some auto-immune diseases

Non-circulating antibody
present in on
lymphocytes

Induration, oedema,
erythema, bullae, vesicles,
necrosis and
pigmentation developing
after 24 hours
Maximal at 48–72 hours

Antibody-carrying
lymphocytes infiltrate site
where antigen is located.
Perivascular lymphoid
cell granulomata with
epithelioid and giant cells

come major. The latter reactions have been evolved to prevent ingress of foreign materials. One of the results of exposure to foreign proteins is widely apparent – in approximately 10–15 per cent of the population who suffer from allergy at some time in their lives – hay fever and asthma, allergic reactions to animal hairs and danders (exfoliated skin cells) and housedust mites. Other antigens exist, some of which are recognized as occupational diseases, for example Farmer's Lung (an allergic reaction to mouldy vegetable compost).

Some individuals who exhibit allergic reactions which are categorized as 'Type I' responses (*see* Table 3.3.2) are known to have a built-in genetic susceptibility or weakness. These individuals may become extremely sensitive to foreign proteins, where the large majority of the population is unaffected. The growth of industry over the last century has led to a huge increase in the use of industrial chemicals which inevitably have found their way into the environment.

These environmental changes have been caused largely by poor industrial housekeeping, the dumping of waste and poor air pollution laws. Legislators under pressure from the general public and trade unions have started to examine these problems and legislation drafted over the last decade has seen dramatic improvements.

TABLE 3.3.3

Estimated protein molecular weights for certain selected industrial enzymes

Molecular weight	Enzyme	Application
27000	Protease (*Staphylococcus aureus*)	Protein sequencing
44000	Peroxidase	Diagnostics
58000	Glucoamylase (Amyloglucosidase)	Starch; brewing; distilling; baking; animal feeds; textiles; paper; pharmaceuticals; fermentation industries
186000	Glucose oxidase	Diagnostics; food processing; soft drinks
244000	Catalase	Diagnostics; dairy; food processing; textiles

Hypersensitivity reactions are only thought to occur where the molecular weights of foreign proteins exceed 8000–10000 (data from Miles Laboratories Ltd).

The HSW Act represents part of this legislation and the new measures currently envisaged will strengthen the Act which in turn will help to reduce chemical emissions to both the working environment and the environment as a whole.

In view of these facts a rise in sensitivity reactions among individuals is therefore not surprising. Many hypersensitive people suffer from nasal discharge, sneezing, watery eyes, eye redness, asthma-type respiratory distress and skin reactions such as contact dermatitis. Identical symptomatic effects can also occur from exposure to proteins in fractionated enzyme preparations of the type shown in Table 3.3.3.

Further detailed information on the Immune response and Hypersensitivity disease is to be found in Farrow (1981).

4. Safe handling

For many years enzyme manufacturers have advised enzyme users about the correct way to use enzymes and the possible effects proteins could have on hypersensitive individuals. Individuals who have known history of asthma or related sensitivity problems should not work directly with enzymes.

Enzyme product types and characteristics. Most manufacturers now offer 'liquid' enzyme preparation in more concentrated forms than in the past. Liquids of higher viscosity exhibit reduced splashing and aerosol formation. Increases in concentration (activity per millilitre) also leads to a reduction in preservatives levels in the final products, as less preservative is required. However, more stabilizers may be required with liquid proteases preparations as they tend to 'digest' themselves over long periods of time.

The major reasons for the use of higher concentration liquid products are: (*a*) user safety – easier handling, less splashing, lower aerosol formation; (*b*) reduction in enzyme costs, transport and storage costs; (*c*) lower preservative and stabilizer levels in final food products.

Powdered products, particularly microfine protease enzymes, are the most difficult to handle and manufacturers have responded and largely overcome these problems by using the latest 'prilling, encapsulation and granulation (agglomeration)' techniques during manufacture. Immobilization of certain enzymes and their use in columns for substrate processing have significantly reduced these problems. Process workers are not exposed to free enzyme protein, because none exists – therefore health risks diminish.

Unfortunately immobilized enzymes cannot yet be employed in all processes, and indeed many are not practicable anyway. Immobilized enzymes can only be employed where a cleaned liquid substrate is available, and process conditions can be optimized to a similar or higher product/efficiency level than that achieved with free enzyme systems. The major problem of many immobilized enzyme systems is their failure to meet these minimum requirements for industrial use.

Now, with the results of legislative action, Good Manufacturing Practice (GMP) is widely used within industry. The pharmaceutical industry has traditionally been at the forefront in this respect, in view of the high potency of pharmaceuticals and inherent risks to patients from faulty products. Enzyme manufacturers therefore lay down strict plant procedures and criteria during the production of enzyme products. Further guidelines for Good Manufacturing Practice in enzyme manufacture are given on page 139 by The Association of Microbial Food Enzyme Producers (AMFEP).

It should be fully appreciated that the major enzyme producers only use food grade or standard materials as their minimum requirements for raw materials in enzyme manufacture and in many cases higher purity ingredients are used together with strict asepsis procedures. Quality control criteria of raw materials and final products are at a very high level and generally on a par with that of the pharmaceutical industry.

General safety precautions. The following is a guide to general safety precautions on the handling of enzymes, whether in liquid, powder, slurry or agglomerized forms. If these simple precautions are followed there is a minimal risk of allergic response in the vast majority of individuals handling enzymes.

Please note, however, that *no person with asthma or other respiratory complaint, or who has suffered from eczema, dermatitis or hay fever should use enzymes.* Please remember that enzymes are biologically active proteins and can cause allergic reactions in sensitive individuals. The following precautions should be observed:

(i) Avoid the inhalation of powdered enzymes and contact with eyes, skin and mucous membranes.

(ii) Wear suitable protective clothing, masks, goggles and gloves and work in a well-ventilated – but not draughty – environment.

(iii) Avoid contact between liquid enzymes, skin and mucous membranes.

(iv) Warn personnel handling enzymes – by suitable means such as notices, literature and training – to avoid contact with enzymes as far as possible by suitable procedures and use of protective clothing.

(v)   Warn personnel to watch for and report immediately skin reactions like reddening and itching and any respiratory problems like wheezing and shortness of breath.

The above simple precautions should enable enzymes to be handled safely in your laboratory or plant.

Processors should ensure that adequate ventilation, dust extraction and protective clothing is provided for users irrespective of legislative and trade union pressures.

Dust extractors and ventilation (clean air systems) should have effective filter systems and be regularly maintained, for maximum effect.

Protective clothing should be comfortable and not rub or chaff the skin. Regular changes of protective clothing should be provided together with a regular laundry service.

Safety showers and adequate changing room facilities should be provided for workers.

Regular safety courses should be initiated with follow-up courses at regular intervals.

Employees should be screened both on joining the company and periodically for allergic responses.

These procedures carried out by a qualified doctor appointed by the company with the agreement of both employees and trade unions would aid industrial relations. The company would be seen as safeguarding the interests of its employees by using regular health screening and together with the trade union protect their interests.

Protective clothing and plant should be practical, easily serviced and maintained, easy to use and comfortable. Incentives should be given where necessary to ensure these procedures are complied with where employees are using full protective clothing with breathing equipment when handling microfine powdered protease enzymes.

Action in the case of accidental spillage. In all cases of accidental spillage the material should be cleaned up using the following guidelines:

(i)   Liquids should not be allowed to dry to prevent powder formation.

(ii)   Powders or granules should be removed by vacuum collection into filter bags, with an exhaust air system external to the plant environment. The exhaust should also have a filter system, but not one capable of creating positive back pressure for obvious reasons.

(iii)   Where vacuum collection is unavailable, damp down with water.

(*iv*) All waste should be burned or incinerated.

Action in the case of accidental contact. In any accidental contact with skin or eyes the area of contact should be washed with water immediately using eye-wash bottles, safety shower hosepipe or total immersion in water. All enzymes are soluble in water and accidents should only be treated with water.

Contaminated clothing should be changed and the whole body washed with water and clothing laundered.

Any spillage should be rinsed off equipment and floors into drains with water to prevent dust generation from drying.

In the case of mouth contact rinse the mouth with plenty of water and drink plenty of water.

If symptoms develop in the respiratory passages, on the skin or in the gastro-intestinal tract see a doctor immediately. All intoxication cases with enzymes should be treated in hospital as a matter of course.

5. Conclusions

Health risks created by the use of enzymes exist, but should be kept in perspective in relation to other industrial chemicals. Product improvements by enzyme suppliers together with Good Manufacturing Practice by users and the adoption of simple safety precautions can restrict health risks to a very low level.

Biotechnology is a growth area for modern industry and enzymes, with their very high substrate specificity, low temperature and atmospheric requirements, represent the ideal natural catalysts for future industrial processes. Enzymes are environmentally clean, and do not represent a biological hazard to the ecosystem, as the vast majority of products are naturally-occurring ones.

Enzymes are the key to the future for industrial processing.

References

Chan-Yeung, M. & Grzybowski, S. *Canad. Med. Assoc. J.* **114,** (5) 433–6 (1976).

Farrow, R. I. *Enzymes: Health and Safety Considerations, Enzymes and Food Processing.* Ed. Birch, G. G., Blakebrough, N., Parker, K. J. (Applied Science, London, 1981).

HMSO Health and Safety Commission *Proposals for Classification Packaging and Labelling of Substances Regulation* (Consultative Document HMSO).

HMSO Health and Safety at Work etc Act (HMSO, London, 1974).

Kirk, R. E. & Othmer, D. F. *Encyclopaedia of Chemical Technology* (Interscience, New York, 1965).

Neckhorn, E. J., Labbee, M. D. & Underkofler, L. A. *J. Agr. Food Chem.* **13,** (1), 30–4 (1965).

Roitt, I. M. *Essential Immunology*, 3rd edn (Blackwell Scientific, Oxford, 1977).

Simson, R. E. & Simpson, G. R. *Med. J. Austr.* **1,** (23) (1971).

Weill, H., Waddell, L. C. & Ziskind, M. *J. Am. Med. Assoc.* **217,** (4) (1971).

Zetterstrom, O. *Clinical Allergy* **7,** 355–63 (1977).

Zweiman, B., Green, G., Maycock, R. L. & Hildreth, E. A. *J. Allergy* **39,** (1) 11–16 (1967).

INDUSTRIAL APPLICATIONS

Chapter 4.1

ALCOHOL – POTABLE
P. B. Poulson

1. Introduction

Potable alcohol is the term used for all distilled spirits (ethanol content higher than around 20 per cent) intended for human consumption. The term covers an enormous number of different spirits produced by many different processes, and consequently this chapter aims only at covering the most important parts of potable alcohol production.

Potable alcohol has been produced industrially as well as domestically for many hundreds of years. In fact, it is one of the oldest industries. The industry presumably grew from chance fermentation of sugar-containing juices followed by natural distillation by the sun. Examples like the fermentation of cactus juice (which has now been developed as Tequila a.o. (Mexico)) and the fermentation of palm juice (e.g. Ogogoro (Nigeria)) can be mentioned. The industry later became more sophisticated with the finding that barley could be transformed to malt and thereby be used in the processing of starch-containing crops to alcohol (e.g. whisky (Scotland)). A further improvement was made in China and Japan where special microorganisms were grown on cooked rice in order to produce starch-fermentable agents (koji). Around 1890, the enzymes present in koji were extracted and concentrated by Takamine and sold as takadiastase. This was the beginning of the modern enzyme industry. Thus it can be seen that the history of the development of the potable alcohol industry is in fact also the history of the growing importance of industrial enzymes.

When sugar-containing crops are to be used as raw materials for the production of spirits, there is only a need for ethanol-producing agents. However, when starch materials are to be used, it is essential that the starch is hydrolysed to fermentable sugars. In some areas, this was done by means of enzymes of vegetable origin (malt), and in others, by means of enzymes of microbial origin (koji). Both methods have survived, although there has been a tendency – especially during the past 20 years – towards the replacement of malt (and originally koji) by industrially produced enzymes (from microbes). The reason for this replacement has been partly economical and partly the need for a product of consistent quality.

2. Raw materials

Many varieties and mixtures of sugar-containing and starch-containing raw materials are used in the production of potable alcohol. Tables 4.1.1 and 4.1.2 give some examples, while Table 4.1.3 shows the relative proportions of the raw materials used.

TABLE 4.1.1

Examples of potable alcohol produced from sugar-containing raw materials

Raw materials	Products
Molasses (sugar cane)	Caribbean rum
	Brazilian cachaça
Wine (grapes)	Cognac
	Pisco (Peru)
Agave azul tequilana	Tequila (Mexico)
Cherry	Kirsch (Switzerland)
Pear	Pear brandy (Switzerland)
Plums	Slivovice (Balkans)
Palm juice	Ogogoro (Nigeria)

TABLE 4.1.2

Examples of potable alcohol produced from starch-containing raw materials

Raw materials	Products
Barley	Whisky
Maize and rye	Bourbon whiskey
Potatoes and barley	Aquavit
Potatoes, rye, wheat (etc)	Vodka
Rice	Chinese brandies

3. Applied cultures and enzymes

Starch-hydrolysing agents include the following.

Malt. Malt is germinated barley. During the germination, enzymes are formed or activated. Enzymes of special interest for brewing processes are:

Starch-hydrolysing enzymes (α- and β-amylases)

Protein-hydrolysing enzymes (proteases, peptidases)

Hemicellulose-hydrolysing enzymes (cytases)

Phytin-hydrolysing enzymes (phytases)

When barley is to be transformed to malt, it must be made to germinate. This requires a water content of 42–46 per cent in the barley. The malt process is initiated by soaking (steeping) the barley in water for two or three days, and then allowing it to germinate, a process which takes six or seven days. The temperature (10–22 °C) during this step is very important in determining which type of malt is produced. At this stage, the malt is called green malt, and the final step is to dry to dry the green malt. The drying step causes:

Reduction in the water content from around 45 per cent to 1.5–4 per cent.

Cessation of germination and digestion.

The formation of colouring and aromatic compounds.

On average, 100 kilogrammes barley will give around 80 kilogrammes malt.

Koji. The manufacture of koji can be performed as follows. Dehusked brown rice (unpolished) is pounded briefly with a wooden pestle in order to scratch the surface of the outer epidermis. The rice is then washed thoroughly, soaked in water overnight and cooked in live steam for about an hour. It is then taken out of the steamer, cooled, and mixed manually and evenly with burnt wood ash (two per cent by weight). When the temperature of the mixture is low enough so as not to be harmful to the microbes, the powdery koji seed, which is carefully preserved by successive transplantations, is sprinkled over the rice and rubbed with the fingers so as to distribute the spores and to bring them into contact with the surface of the rice. The whole mass is incubated overnight in a warm koji chamber until the temperature of the mass has risen to 35 °C. The mass is then divided among small shallow wooden trays which are piled up in the chamber. The temperature and humidity are controlled by changing the type of piling and by opening the ceiling window.

The propagation of mould as well as the abundant formation of spores can be completed within five or six days, after which the trays are taken out into the open air and each tray is covered with thin paper and exposed to direct sunlight for one day. The mass is then carefully dried again at 40°C in an indirectly heated drying chamber, and the final product is wrapped in a paper bag and stored.

Microbial enzymes. The market share of the abovementioned enzyme products has been declining, especially over the past ten years, as an increasing number of distilleries have decided to use industrially produced microbial enzyme products. The reasons for this have been both economical and technical, the technical reasons being that the microbial enzyme products have a known standardized activity and are much more concentrated, and thus require less handling.

The microbial enzymes are produced by special, carefully optimized mutants in very large vessels – typically 100 – 200 cubic metres. The enzyme products on the market are either single enzyme products (α-amylase, β-amylase, glucoamylase, protease etc) or enzyme mixture products.

Ethanol-producing organisms. Whereas starch hydrolysis has almost become an art, the selection of ethanol-producing organisms has been rather casual. For example, in the fermentation of grapes, many wineries are quite satisfied with yeasts of natural origin. Generally speaking, yeasts, such as *Saccharomyces cerevisiae*, *Saccharomyces carlsbergensis* and *Schizzosaccharomyces* types, are used in temperate climates for the production of alcohol, while in tropical countries, bacteria (e.g. *Xymomonas mobilis*), are used.

4. Processes involved
Main processes in distilling
The manufacture of alcohol from starch-containing raw materials is based on the following main processes (*see* Figure 4.1.1).

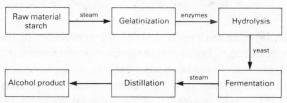

Figure 4.1.1   Main processes

Gelatinization. The dissolution of the raw materials into a mash by steam cooking to make the starch available for enzymatic attack. Normally, a concentration of 15 – 20 per cent starch in the mash is aimed at.

Hydrolysis. The breakdown of the dissolved starch to fermentable sugars by means of enzymes.

Fermentation. Conversion of the sugars to alcohol by the action of yeast.

Distillation. Separation and purification of the alcohol.
The enzymatic hydrolysis consists of two stages:

Liquefaction. The gelatinized starch is broken down into short molecule fragments (dextrins) by means of α-amylase, resulting in a rapid reduction in mash viscosity.

Saccharification. The dextrins formed during liquefaction are further hydrolysed to fermentable sugar (glucose) by means of glucoamylase.

American batch process. In the American batch process (*see* Figure 4.1.2) milled corn is slurried with water and fed into the cooker at a concentration of about 25 per cent, where it is kept under agitation. A thermostable, liquefying bacterial α-amylase (e.g. Novo *Termamyl*® 60L at 0.15–0.3 kilogrammes per tonne) is added, and the temperature of the 'mash' is gradually increased by the injection of live steam to about 150°C. During cooking, the starch becomes gelatinized and the mash viscosity increases. The liquefying α-amylase partially hydrolyses the gelatinized starch and reduces the mash viscosity sufficiently to allow agitation to continue. This is referred to as the pre-liquefaction stage.

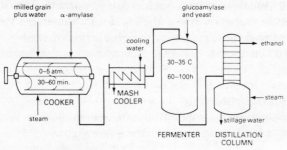

Figure 4.1.2   American batch process

As heating continues up to 150°C, the remainder of the starch becomes gelatinized, but the enzyme has by now been inactivated. However, the starch granules have been completely disrupted and the starch is fully susceptible to enzyme attack. The mash is cooled and further bacterial α-amylase added (e.g. Novo *Termamyl*® 60L at 0.35–0.7 kilogrammes per tonne). The enzyme partially hydrolyses the gelatinized starch and reduces the viscosity of the cooling mash and prevents starch retrogradation. This is referred to as post-liquefaction. After the post-liquefaction stage, the mash is cooled to about 60°C and the saccharifying enzyme (glucoamylase) added (e.g. Novo *SAN* 150L at 1.5–2.0 litres per tonne). Yeast is added when the temperature has been lowered to about 30°C. The glucoamylase converts the partially hydrolysed starch to dextrose which is fermented to ethanol. Simultaneous saccharification and fermentation continues for 60–100 hours, after which the ethanol is distilled off.

German batch process. In the German batch process (*see* Figure 4.1.3), the raw material is gelatinized without previous milling by cooking with live steam in a Henze cooker. No addition of enzyme or mechanical agitation are necessary in the cooking stage. The cooked mash is blown through a strainer valve into the mash tub where the liquefaction takes place according to one of two procedures:

(i) High temperature liquefaction: The blow-down is carried out within the shortest possible time, after which the mash is cooled in the mash tub to 80°C. At 80°C, α-amylase (typically 0.15–0.6 kilogrammes Novo *Termamy*® 60L per tonne of starch) is added and the temperature is maintained for 20 minutes before further cooling.

(ii) Low temperature liquefaction: Before blow-down of the cooker, the mash tub is filled with cold water, sufficient to cover the lowest part of the cooling coil, and α-amylase is added (e.g. Novo *Fungamyl*®800L at 0.1–0.2 kilogrammes per tonne). The mash is blown down into the mash tub under agitation and cooling at such a rate that a temperature of between 55 and 60°C is maintained in the mash tub. When the blow-down is complete, the mash may be further cooled to the fermentation temperature.

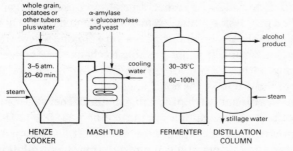

Figure 4.1.3   German batch process

Saccharification. Glucoamylase (e.g. 1.1–2 litres Novo *SAN* 150L per tonne of starch) is added to the liquefied mash at 60°C or lower. This will bring about complete saccharification within the normal period of fermentation. Yeast is added to the mash after cooling to 30°C or lower.

Continuous cooking. Cooking and liquefaction may be carried out continuously, thus giving better process control and more efficient use of equipment. Such a process is shown in Figure 4.1.4. A milled corn slurry to which a thermostable bacterial α-amylase has been added is heated by direct steam injection to about 150°C in a jet cooker. The mash is then flash-cooled to 80–90°C and the second addition of α–amylase made. The cooked mash is held at this temperature for 30–60 minutes to complete liquefaction before it is transferred to the saccharification/fermentation tanks. (Continuous cooking followed by continuous fermentation has also been described by Rosén (1978).) Enzyme dosages are the same as for the American batch process.

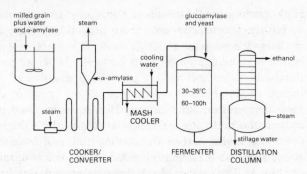

Figure 4.1.4   Continuous cooking process

Newer process layouts and future process developments. What will happen to the technical processes used in the potable alcohol industry in the coming years? I think that the processes will be modified, especially with respect to energy consumption and more controlled processing of the raw materials. One should bear in mind that the potable alcohol industry is a very old one where traditions are very important. It is not possible to change raw materials and processing without years of testing in order to ensure that the quality of the products has not changed.

In the traditional batch processing method, energy consumption was about 17–24 megajoules per litre ethanol (Hagen, 1981). Of this amount, 7–8 megajoules per litre were consumed in the cooking of the raw materials, the rest during the distillation. With modern continuous processing methods, the indications seem to be that the energy consumed can be decreased to 6–9 megajoules per litre (1–2 megajoules per litre during cooking and 5–7 megajoules per litre during distillation) without changing the quality of the product. This saving is obtained by increasing the temperature at which the cooker feed is preheated from 20°C to 60°C using recycled surplus process energy, increasing the mash concentration from 20 to 30 per cent dry substance and reducing the cooking temperature from 150°C to about 100°C (the yield loss is limited to a very few per cent, dependent on grain quality and particle size).

Recently, work in Japan (Suntory) has shown that it is possible to reduce the processing temperature to 35°C when using microbial enzymes and a holding time of up to five days. Presumably, further optimization will be reported within this field in the coming years.

The continuous development which is taking place is also important in that it allows industrial enzymes of microbial origin to replace at least part of the expensive and troublesome use of malt and koji. Furthermore, more and more enzymes are being marketed, enzymes which can perform increasingly more specialized reactions and make the substitution for malt and koji easier so that commercial benefits can be obtained without changing the quality of the potable alcohol.

5. Production size (Schrøder)

When trying to determine the basic production of potable alcohol (*see* Table 4.1.4), one runs into a number of statistical problems in addition to the normal statistical uncertainties involved.

One problem is that, owing to the fact that potable alcohol is taxable, it is tempting for manufacturers to give too low production figures, which may be a significant factor of uncertainty in some countries. A more serious and frequent problem is the fact that industrial alcohol is sometimes included in distilled alcoholic beverages. This category is the most heterogeneous in other ways too, for it may include beverages with an alcohol content varying from a few per cent to 80 per cent. This heterogeneity may become more marked as new types of mixed drinks are introduced based on distilled alcohol but with a very low alcohol content.

Most countries, however, report production in consumption strength which is assumed to be 40 per cent by volume, on an average. This assumption is in accordance with the practice of the Produktschap Voor Gedestilleerde Dranken (Netherlands).

TABLE 4.1.3

Relative proportions of raw materials used for production of potable alcohol

	100% ethanol ($\times 10^6$ hectolitres)	Per cent
Molasses-based	12	35
Grain-based	10	29
Whisky	5	15
Wine-based	3	9
Potato-based	2.5	7
From fruits etc	1.5	5
	34	100

TABLE 4.1.4

Split on areas of the production of potable alcohol

	100% ethanol $(\times 10^6$ hectolitres)	Per cent
Asia and Oceania	4.4	13
Africa	0.2	1
Western Europe	17.2	51
Eastern Europe	4.5	13
North America	6.1	18
Latin America	1.6	5
	34	100

References

Aschengreen, N. H. *Process Biochem.* **8,** 23–25 (1969).

Hagen, H. A. *Meeting on Bio-fuels,* Bologna, Italy, June (1981).

Lützen, N. W. *VI int. Fermentation Symp.*, London, Ontario, Canada, July 20–25 (1981).

Nakano, M. *4th int. Fermentation Symp. Kyoto, Japan,* March 19–25 (1972).

Norman, B. E. & Lützen, N. W. *Int. Symp. Cereals,* Carlsberg Research Center, August 11–14 (1981).

Rosén, K. *Process Biochem.* **5,** 25–26 (1978).

Schrøder, P., Novo Industri A/S, DK-2880 Bagsvaerd (personal communication).

Ueda, S. & Koba, Y. *Int. Fermentation Technol.* **3,** 237–242 (1982).

ALCOHOL – POWER/FUEL
T. P. Lyons

1. Introduction

In little more than a century the industrialized world has consumed material energy resources which accumulated in the Earth over many millions of years. With the now clearly forecast exhaustion of these resources, major alterations in the economic activities of the world may be necessary as the crisis becomes more acute. In an effort to avoid the clearly disastrous consequences alternative energy sources are being investigated throughout the world. Renewable energy sources offer one of the best short-term and long-term solutions, and include among their number feedstocks containing starch, sucrose, cellulose and hemicellulose. These feedstocks are synthesized in plants from water and carbon dioxide, using solar energy. A key role in this area will be played by the emerging biotechnology industry. While biotechnology is regarded as a new science, however, one of the processes it involves is as old as civilization itself – namely alcohol production by fermentation using the enzymes of yeast.

The use of alcohol as an automotive fuel is neither a new technology nor a new concept. Extensive literature on the subject dates back to the 1920s, and alcohol has been used both in times of war and peace. Henry Ford's Model T car was designed so that it could run on alcohol, petrol or any mixture in between. With the easy availability of petrol, however, this technology was not utilized until the late 1970s when, with the doubling of oil prices, and in some cases the total unavailability of oil, alcohol once again received attention. In the USA, alcohol blended with gasoline at a ten per cent inclusion was named gasohol. The federal government

---

EDITORS' NOTE ON UNITS OF MEASUREMENT

Throughout this section the use of several measurement systems of direct relevance to North American operational practice may be a little confusing for readers operating in the metric systems. Rather than alter the text or insert metric equivalents throughout, we suggest the use of a conversion table or chart for temperatures will give direct equivalence. For the other measurements readers are advised that:

$$1 \text{ US gallon} = 3.78 \text{ litres}$$
$$1 \text{ US bushel} = 56 \text{ pounds}$$
$$= 25.5 \text{ kilogrammes}$$

encouraged the production of this alcohol by reducing the excise tax on gasohol by four cents per gallon, and this, in addition to a reduction in local state taxes, led to tremendous interest in its production both on a small and large scale. By the end of 1980 alcohol production was running at an annual rate of some 80–120 million gallons and the federal government set targets to increase production to 500 million gallons by the end of 1981 and to 1.8 billion gallons by the mid-1980s (*see* Figure 4.2.1).

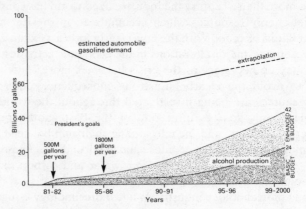

Figure 4.2.1   US goals for alcohol fuels production

On the other hand, Brazil has set a target of 3 billion gallons by the same date (*see* Figure 4.2.2), and with the use of a World Bank loan has recently approved construction of no less than 388 projects in alcohol production (*Chemical Week*, 1981). Other countries,

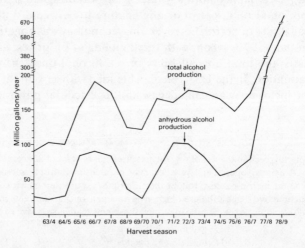

Figure 4.2.2   Ethanol production in Brazil

including Canada, France and the Phillipines, have followed a similar approach, setting up national gasohol commissions.

2. Overall view of alcohol production

Before embarking on a detailed discussion of alcohol production, it is necessary first to consider the overall process. Regardless of the raw materials being used to produce alcohol, there are invariably four major steps involved (*see* Figure 4.2.3).

(*i*) Pretreatment: raw materials rich in carbohydrates must in some way be converted into fermentable sugars.

(*ii*) Fermentation: the fermentable sugar is utilized by yeast (or possibly some other microorganism) to produce alcohol and carbon dioxide.

(*iii*)Concentration: the alcohol produced during fermentation is concentrated by distillation to 100 per cent alcohol.

(*iv*)Byproduct formation: the spent mash left after the alcohol has been removed is processed into a by-product.

In the first two steps a major role is played by a biological catalyst or enzyme. Enzymes are characterized by such features as their capacity to function at moderate pH and temperature, high substrate specificity and extremely efficient conversion rates. Without their use, the raw material cannot be efficiently broken down to low molecular weight sugars such as glucose, maltose and sucrose. Enzymes are widely used in the food processing industry. This chapter examines their role in alcohol productions.

Raw materials. Enzymes relevant to alcohol production can perhaps best be classified by looking at the individual raw materials or feedstocks from which the alcohol is obtained. While it is true that any carbohydrate raw material can thoeretically be considered as a feedstock for alcohol production, inevitably some are more suitable than others. Suitability is dictated by the yield of alcohol

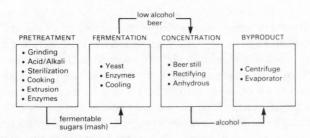

Figure 4.2.3   The four main steps in alcohol production

per ton of the individual feedstock, the availability of the feedstock and the yield of alcohol per acre. In future, consideration of suitability as an alcohol feedstock will generally be based on this overall alcohol yield per acre.

Feedstocks can be classified into three main areas:

(*i*) Those containing starch as their source of carbohydrates.

(*ii*) Those containing pre-formed sugar such as sucrose as a fermentable sugar source.

(*iii*)Those containing cellulose as a carbohydrate source.

3. Feedstocks containing starch

The feedstocks containing starch include most of the cereals, and tuberous roots such as potatoes. Starch is not a homogeneous entity, being composed of two high molecular weight polysaccharides known as amylose and amylopectin. The two polysaccharides are distinctly different in their physical properties (*see* Table 4.2.1), notably their solubility in water, their molecular size, their ability to stain with iodine and their susceptibility to enzymatic degradation.

TABLE 4.2.1

**Comparison of some properties
of amylose and amylopectin**

Property	Amylose	Amylopectin
Basic structure	Linear	Branched
Stability in aqueous solution	Retrogrades	Stable
Degree of polymerization	About 10^3	$10^4 - 10^3$
Average chain length	About 10^3	$20-25$
Iodine complex (λ_{max},nm)	650	550

(Fogarty, 1981)

Amylose has an unbranched linear structure whereas amylopectin is a branched polymer. The differences in the properties of the two structures arise mainly from the presence of an α-1,6 D-glucosidic linkage in amylopectin which is not present in the predominantly α-1,4 D-glucosidic-linked amylose molecule. The structure of amylose makes it unstable in aqueous solution and gives it a tendency to precipitate (retrograde) by the formation of hydrogen bonding between the long chains (*see* Figure 4.2.4).

Although starch is an excellent material for alcohol production, it cannot be fermented directly to alcohol by yeast (*Saccharomyces cerevisiae*), and must therefore first undergo a pretreatment step to

AMYLOSE

AMYLOPECTIN

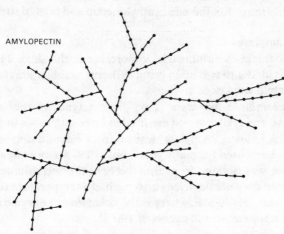

Figure 4.2.4   Structure of the components of starch

convert it to a fermentable sugar or glucose. This conversion is brought about by the combined effect of grinding, heating and the use of microbial enzymes. There are three groups of microbial enzymes of industrial importance: endoamylases, exoamylases and debranching enzymes (*see* Table 4.2.2).

TABLE 4.2.2

Starch-degrading enzymes

Endoamylases	Hydrolyse α-1,4 glucosidic bonds in the interior of the substrate (α-amylases)
Exoamylases	Hydrolyse alternate (β-amylases) or successive (amyloglucosidases) α-1,4 glucosidic bonds from the non-reducing terminal of the substrate
Debranching enzymes	Hydrolyse α-1,6 glucosidic bonds in amylopectin (pullulanases and isoamylases)

(Fogarty, 1981)

Endoamylases are typified by enzymes such as α-amylase which hydrolyse the α-1,4 linkage but not the α-1,6 bond in amylopectins. They degrade the substrate by a random endomechanism of attack which brings about a rapid reduction in viscosity. Exoenzymes, on the other hand, while they also hydrolyse the α-1,4 bond, act in a sequential fashion and effect a slow reduction in viscosity as the number of bonds hydrolysed increases. Debranching enzymes hydrolyse only the α-1,6, linkage in amylopectin and related structures.

4. Endoamylases

The endoamylases of industrial importance in the alcohol industry can be divided into two main groups: thermostable α-amylases and thermolabile amylases.

Thermostable α-amylases. Two distinct types of thermostable α-amylase are available and are used in large quantities in the food processing industry. Amylase derived from *Bacillus amyloliquefaciens* has been used for many years, but in 1973 a more heat-stable α-amylase was isolated from *Bacillus licheniformis* (Madsen *et al.*, 1973). The *amyloliquefaciens* enzyme had a temperature optimum of approximately 175°F, whereas the *licheniformis* enzyme remains stable at temperatures in excess of 190°F.

TABLE 4.2.3

Principal end-products of degradation of starch by thermostable α-amylases

B. amyloliquefaciens enzyme:	Maltohexaose
	With small amounts of
	Maltopentaose
	Maltotriose
	Maltose
B. licheniformis enzyme:	Maltopentaose
	Maltotriose
	Maltose

(Fogarty, 1981)

In addition to their temperature stability, the enzymes from the two sources differ in their principal end products (*see* Table 4.2.3). Thermostable α-amylases tend to be used where batch cooking (*see* p. 186) is the favoured method of starch processing or where the type of degradation product is of importance, such as in fructose syrup production.

Thermolabile α-amylases. The heat-labile α-amylases are obtained mainly from *Bacillus subtilis* and are very similar in activity to the enzyme from *Bacillus amyloliquefaciens*. Normally the *subtilis* type of α-amylase is not only more heat labile (temperature optimum 165–170°F) but also less expensive than the heat-stable enzyme. It is more commonly used with the continuous cooking method, in which a liquefaction stand at 170°F following a hold at a high temperature is used.

5. Exoamylases amyloglucosidase

The exoacting microbial amylases cleave the α-1,4 glucosidic bonds in starch. The most important of these enzymes is amyloglucosidase. This enzyme is produced almost exclusively by fungi and is prepared commercially from either *Aspergillus* or *Rhizopus* species. Amyloglucosidases differ from bacterial amylases in being less thermostable and have different pH optima. They cleave both the α-1,4 and the α-1,6 glucosidic bond, although acting at a much higher rate on the former than the latter. In a commercial operation, the bacterial α-amylase is normally used to liquefy a 30–40 per cent slurry of starch at a pH of 6.0–6.5 and a temperature of 180–320°F. The resulting product is a mixture of dextrins of different chain lengths, and these are then degraded to glucose by the amyloglucosidase. This degradation can be performed at 140–150°F and at a pH of 4.5–5.0 under the action of the *Aspergillus niger* amyloglucosidase. It is a more common practice in the distilling industry, however, to add the *Aspergillus* or *Rhizopus* enzyme directly to the fermenter with the yeast at 90°F, as this enables the glucose-producing enzymes to produce fermentable sugar as the yeast utilizes it. The mode of action of endo and exoenzymes can perhaps best be understood by following a typical process involved in the cooking of maize.

6. Typical cooking process

Maize, or corn, contains 60–68 per cent starch and many distillers use the rule of thumb that one 56 lb bushel of corn will give them 32 lb of starch, which when hydrolysed into sugar gives 36 lb of sugar. This sugar, when fermented, will produce 2.6 US gallons of alcohol. The cereal is first ground to reduce the size of each particle so that the water and enzymes can come into intimate contact with

TABLE 4.2.4
Typical sieve analysis

Screen size	Hole size (inches)	% Corn on screen
12	0.0661	3.0
16	0.0469	8.0
20	0.0331	36.0
30	0.0234	20.0
40	0.0165	14.0
60	0.0098	12.0
>60		7.0

the starch. A variety of equipment is available to bring about this grinding, but hammermills are the most favoured means. The hammermill normally has a screen size of $\frac{1}{8}'' - \frac{1}{16}''$ and Table 4.2.4 shows a typical sieve analysis.

Following milling, the fine particles of ground cereal are normally added to water (100–120°F) in an agitated cooking vessel at a ratio of 3–4 lb of water per 1 lb of corn (expressed in the USA as a cooking ratio of 16–25 gallons per bushel). A small percentage by weight of the endoenzyme or liquefying enzyme is added to assist the breakdown of the starch and the mixture is heated by direct steam injection with constant agitation. As the temperature of the grain rises the cereal starch swells and begins to gelatinize at between 145 and 155°F. This progressive swelling of the granules results in a sudden rise in viscosity (*see* Figure 4.2.5) at about 155–160°F, with a consequent increase in the power required to agitate the mash. With continued heating the viscosity rises to a peak at around 190°F and then begins to fall. Depending on whether batch, atmospheric or pressure batch cooking is being used, the mash is heated to between 212 and 320°F and held there for a period to complete the destruction of the starch. The mash at this stage becomes very loose and is easy to pump, the ease with which the mash can be pumped being a measure of the extent of its liquefaction.

Following the hold period at the elevated temperature, the cooked grain mash is cooled to 150–170°F by vacuum cooling, addition of cold water, or the use of external or internal cooling coils. As the paste begins to thicken again, further α-amylase is added. This liquefying enzyme works very rapidly and the mash becomes extremely thin and loses its sticky, paste-like consistency. A hold time of 30–40 minutes is normally given and the mash is then cooled to either 150 or 90°F.

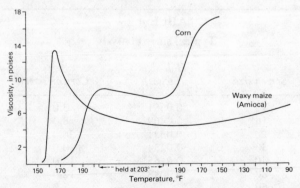

Figure 4.2.5   Viscosity build up in cooking corn

If an *Aspergillus niger* saccharifying enzyme is used, the hold temperature is normally 150 °F, whereas if an amyloglucosidase from *Rhizopus* is employed the mash can be cooled directly to 90 °F and this type of enzyme can be added with the yeast (*see* Figure 4.2.6.).

Depending on the policy of the distillery, pH adjustment may be carried out at both the liquefying stage and the saccharifying stage. The necessity for this adjustment remains a controversial issue, but most distillers work on the premise that as the liquefying enzyme has a pH optimum of 6.5 and a normal mash has a pH of 5.5 – 6.0 no pH adjustment is necessary. On the other hand, the saccharifying enzyme has a pH optimum of 4.5 – 5.5 and the mash is sufficiently close to this to avoid the necessity of adjusting pH at this stage.

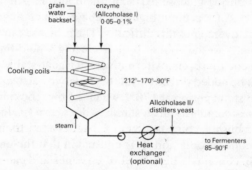

Figure 4.2.6   Batch cooking

7. Continuous cooking
With the continuous cooking method, a system similar to that shown in Figure 4.2.7 is adopted. In the operation stillage, ground corn, water and preliquefying α-amylase are mixed in a slurry tank. After 20 – 30 minutes this is pumped through a mixing chamber where steam is introduced. The mash is then pumped through either

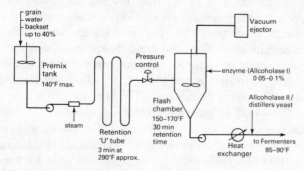

Figure 4.2.7   Continuous 'U' tube

a long pressure tube or a column where it remains for 3 – 30 minutes at temperatures of 250 – 320°F.

Following this hold time the mash is cooled to 180°F and further liquefying enzyme added. The mash is then cooled to either 150°F for amyloglucosidase treatment or to 90°F for simultaneous fermentation and amyloglucosidase treatment.

Although the process described above is basically designed for corn, it can also be used with barley, rye, milo and wheat. Nevertheless, each cereal has individual characteristics which must be mentioned.

Barley. Barley contains a high level of the carbohydrate β-glucan, a β-1,3, β1,4-linked glucose polymer. This glucan is an extremely viscous gum and if not degraded during processing it can cause difficulties during glucose extraction and also in stillage evaporation. The use of a protein digesting enzyme in conjunction with a β-glucanase overcomes this difficulty. The protease can normally be added directly to the cooker by introducing a hold time of 140°F during the cooking cycle. Alternatively, if a heat-stable glucanase is used it can be added directly with the liquefying enzyme during the conversion stand period at 170°F which follows cooking.

Rye. Rye contains as much starch as does corn (yielding 2.4 – 2.5 gallons per bushel, compared with 2.5 – 2.6 for corn) and is widely used in alcohol production. It is unusual in that the ungerminated rye already contains a high level of enzymes and can be liquefied without the addition of microbial enzymes. The processing of rye normally incorporates a preliquefying stand at 150 – 160°F to allow the native enzymes in the cereal to act. The mash is then cooked in the normal fashion with the addition of commercial liquefying enzymes.

Like barley, rye contains high levels of gum and these can cause stillage evaporation problems. As with barley, the use of a glucanase enzyme during the conversion stage helps to overcome this.

Cassava. In many African countries and in South America, cassava or manioc represents a major alcohol feedstock. The cassava root normally contains 25 – 30 per cent starch. After washing the roots are cut into small chips and dried. The dried chips, which have a moisture content of about 12 per cent, are exported in this form for industrial purposes or are cooked, liquefied and saccharified using normal distillery technology. Yields of alcohol are normally between 40 – 44 US gallons per ton (*see* Figure 4.2.8).

The enzymes used for cassava are similar to those used for cereals. Recent work (Menezes, 1978) has indicated that the use of cellulase from *Trichoderma* increases alcohol yield by breaking down cellulose, which acts as a barrier to sugar formation.

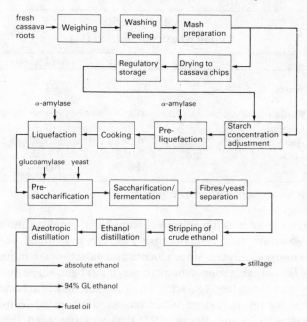

Figure 4.2.8   Production of ethanol from cassava root (Lindeman & Rocchiccioli, 1979)

8. Feedstocks containing sugar

The most important sugar-containing raw materials for alcohol production are sugar cane, sugar beet and their derived extract, molasses. In addition to this, however, various byproducts from the food processing industry contain high levels of fermentable sugar. The waste from the cheese manufacturing industry ranks prominently in this area. These wastes, which are principally whey and washwaters, present the industry with a serious problem since they are both an important effluent of value as food and a potential pollutant. The biochemical oxygen demand level can be from 30,000 to 40,000 milligrammes per litre and can cause major pollution if discharged into the municipal sewage system or into streams. In the USA alone, some 9 million tons of cheese whey are made available each year and this has the potential to produce 90 million gallons of alcohol at a cost per gallon of only 22 cents (SERI, 1980).

The composition of whey varies according to the type of cheese produced, but an approximate estimate is given in Table 4.2.5. Lactose, which consists of glucose and galactose, is the predominant sugar of whey and the alcohol manufacturer, when using whey,

TABLE 4.2.5
Liquid whey composition (%)

	Sweet whey	Acid whey
Water	93	93
Fat	0.3	0.1
Protein	0.8	0.6
Lactose	4.9	4.3
Ash	0.56	0.46
Lactic acid	0.2–0.3	0.7–0.8
	pH 6.1	pH 4.7

(Cunningham and Lyons, 1980)

requires a yeast which is selectively capable of breaking the lactose down into its two components. The yeast involved is called *Kluyveromyces fragilis*. Since glucose and galactose are more universally fermentable sugars than lactose, it has been suggested that a lactose-treated whey (galactosidase) would make a better raw material for fermentation. When lactose was treated in this way (O'Leary, 1977 and Reese, 1975), it was observed that *Saccharomyces cerevisiae*, the normal brewers' or distillers' yeast, could ferment whey containing 30–35 per cent total solids and produce a beer containing six to seven per cent alcohol. Longer fermentation times (120 instead of 72 hours) were required with the *Saccharomyces cerevisiae* fermentation when compared with *Kluyveromyces fragilis*. In a effort to utilize corn and whey together, mixed cultures of *Kluyveromyces fragilis* and *Saccharomyces cerevisiae* have also been used successfully.

Molasses. Molasses is perhaps the simplest and globally the most widely used raw material for alcohol production. The two main sources of molasses are sugar cane and sugar beet, and are the residual products of sucrose separation by crystallization. In the context of alcohol production, fermentation of these raw materials could occur either using whole cane or beet juice or with the residue (molasses) obtained from sucrose crystallization. The method used is dictated by the relative prices of the sugar and alcohol.

As the sugar in molasses is preformed, there is little need for the application of enzymes. However, it is important that the yeast selected for the fermentation has a high invertase activity so that sucrose, the principal constituent of the molasses, can be broken down to glucose or fructose. Invertase is associated with the yeast cell wall and it acts as the sucrose passes into the yeast (Trevelyan, 1958). Supplementation of molasses with invertase is known to improve the yield per ton. Other applications are possible when alternative sugar-containing raw materials such as sweet sorghum are used. These feedstocks often contain high levels of starch which impede the extraction of the sucrose. Degradation of the starch with amylase and cellulase improves extraction.

9. Cellulose-containing raw materials

Although over the next few years the source of alcohol for use as fuel will undoubtedly come from cereals such as corn and wheat and from sugar crops, it will not be long before a virtually untapped source of energy is also utilized - namely agricultural wastes. Cellulose is the most abundant organic material available as a source of food, fuel and chemicals. It has been estimated that the annual worldwide production of cellulose is around 100 billion tons (Spano, 1976) – approximately 70 kilogrammes of cellulose per day for every one of the Earth's 3.9 billion people. In order to use this annually replenishable resource, however, cellulose must first be hydrolysed into glucose. This hydrolysis can be carried out using either acid or enzymes following certain pretreatment steps. The object of the pretreatment is to make agricultural waste susceptible to hydrolysis and methods used for this are normally classified (Dunlap,1976) as physical, chemical or a combination of both. Physical methods include milling and gamma radiation, while sodium hydroxide is the most effective chemical method.

Although many microorganisms are cellulolytic, most of the active developments have been confined to two microorganisms, *Trichoderma* and *Aspergillus*. In one process, the *Trichoderma* is grown on an aqueous solution of cellulose to produce cellulase

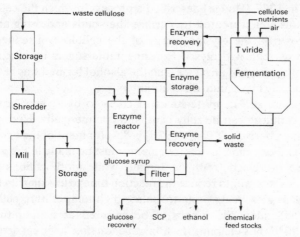

Figure 4.2.9   Natick enzymatic conversion of waste cellulose (Nystrom & Allen, 1976)

enzymes. The exocellular enzymes are then recovered by simple filtration and the enzyme-rich broth is mixed with the pretreated cellulosic waste (*see* Figure 4.2.9). The principal products of the reaction are a crude syrup containing glucose, cellobiose, xylose and a mixture of salts present in the original enzyme broth. The major by-products of the unreacted solids are generally highly crystalline cellulose and unhydrolysable compounds such as lignin. Cellulase enzymes tend to be absorbed onto the cellulose and these can be at least partially recovered and recycled (Nystrom & Allen, 1976).

TABLE 4.2.6
Effect of mixed cellulase enzyme on hydrolysis

Enzyme	Tv	Aw	TV:Aw	Tv:Aw	TV:Aw
			Hydrolysis in first hour produced ml		
Substrate			2:1	1:2	1:1
Cellulose (5%)	3:2	0:9	2:7	1:9	2:5
Bagasse (5%)	1:6	1:6	2:8	2:6	3:4

(Ghose, 1978)

The system has been improved by fortifying the *Trichoderma* cellulase enzyme with enzymes in which it is deficient. One such enzyme is xylanase and Table 4.2.6 shows some typical results.

Most studies involving cellulose are now directed towards developing a one-step process for the production of alcohol wastes. A patent in 1977 (Hodge) described a process in which the cellulosic waste was steam treated to sterilize the solids and bring about a breakdown in the crystal energy of the cellulose. The sterilized waste was then hydrolysed to fermentable sugars by inoculating with both yeast and enzymes, and the alcohol formed was removed continuously by vacuum distillation.

There are still many technical barriers to overcome, however, before cellulose can be fully exploited commercially. High on the list of such barriers (Spano, 1976) is development of the necessary enzymes. For example, improvements must be made in the yield of cellulase produced from the fungus, while modifications to the hydrolysis step would reduce the reaction time to less than 24 hours.

Work is also being done to isolate high temperature cellulases which will allow the fungus to be cultured at temperatures of 150–190°F thus enabling the total reaction time to be carried out at a much higher temperature, resulting in higher reaction rates.

One thing is certain: if the alcohol industry is to survive, alcohol must be produced from this saccharification of cellulose. At present cellulose saccharification is only at the development stage, while the corn/sugar industry is already well established. Nevertheless, it is expected that cellulose saccharification will begin to compete with the corn/sugar industry in the not to distant future. The yield of alcohol per acre will make the base cost of cellulose clearly superior to that of corn and the removal of hemicelluloses and lignin may be equated cost-wise to the physical removal of the whole germ of the corn kernel. Thus starch, while it has the major advantage of solubility and therefore lower cost in terms of extraction, may soon be overtaken.

References

Cunningham, D. & Lyons, T. P. *Am. Dairy Rev.* November (1980).

Dunlap, C. E., Thompson, J. & Chiang, L. C. *Am. Inst. chem. Engng Symp. Ser.* **72,** 58 (1976).

Fogarty, W. L. in *Gasohol, A Step To Energy Independence* (ALLTECH Technical Publications, Lexington, Kentucky, 1981).

Ghose, T. K. & Sahai, V. *Biotechnol. Bioengng Symp.* **21,** 283 (1979).

Hodge, W. H. *U.S. Patent 4009075* (1977).

Lindemann, L. R. & Rocchiccioli, C. *Biotechnol. Bioengng Symp.* No. 2, 1107 (1979).

Lyons, T. P. in *Gasohol, A Step To Energy Independence* (ALLTECH Technical Publications, Lexington, Kentucky, 1981).

Madsen, G. B., Norman, B. E. & Slott, S. *Die Starke* **25,** 304 (1973).

Menezes, T. J. B. *Process Biochem.* **13,** (9) 24 (1978).

Nystrom, J. M. & Allen, A. L. *Biotechnol. Bioengng Symp.* **6,** 55–74 (1976).

O'Leary V. S., Green, R., Sullivan, B. C. & Holsinger, V. H. *Biotechnol. Bioengng Symp.* **19,** 1019 (1977).

Reese, E. T. *Biotechnol. Bioengng Symp.* No. 5 (1975).

SERI *Small Scale Fuel Alcohol Production*, March (Washington, DC, 1980).

Spano, L. A. *Symp. Clean Fuels from Biomass, Sewage, Urban Refuse* (Orlando, Florida, 1976).

Trevelyan, W. E. *Chemistry and Biology of Yeast* (Academic, New York, 1958).

ANALYTICAL APPLICATIONS
T. J. Langley

1. Introduction

The routine analytical use of enzymes began in earnest in the early 1960s. Following research into the significance of serum markers in disease states, the medical diagnostic use of enzymes developed quickly. All of the significant developments in enzymic analytical testing in the past two decades have been in the medical field, and the experience gained has spilled over into other industrial areas, The important developments in analytical enzymology have been:

The important developments in analytical enzymology have

(*i*) Improvements in reagent design and instrumentation

(*ii*) The miniaturization of routine testing, allowing reduction of cost per test

(*iii*) The availability of bulk purified enzymes

(*iv*) Legislative and consumer power to make medical diagnostic testing widely available.

In the medical applications the raw material for enzymic diagnostic testing is most often serum. This liquid, while complex in itself, is simply prepared and easily handled by automatic systems. A consistent approach to automation of testing methods has, therefore, been possible. This is not the case for other industrial applications, where basic components such as sugars and organic acids need to be measured in a variety of situations, but where in many cases sample preparation requires a complex set of manipulations. It has not been possible to treat industrial applications with the same consistent approach as for medical diagnosis. For this reason the routine use of enzymes in industrial analysis has lagged behind the medical diagnostic industry. However, the need to evaluate and control raw materials accurately for cost control and to limit contamination has emphasized the qualities that enzymes can bring to analytical procedures.

2. Principles

Enzymes are protein catalysts and are governed by kinetic and practical considerations (*see* Chapter 2). There are a number of practical advantages in using enzyme mediated tests, the greatest advantage being the specificity of an enzyme for its substrate, as this enables a series of similar compounds, for example monosaccharides, to be simply and quantitatively differentiated. A further advantage of enzymes is that they will function in a complex mixture. Substrate preparation compared with that for other procedures, such as gas and high performance chromatography, is also relatively simple. This is particularly relevant in the reduction of the cost per test performed when operator costs are rising fast. Finally, the reactions are fast and simple, allowing frequent repetition or large numbers of tests to be performed with high accuracy.

Specificity. The specificity of an enzyme for its substrate is determined by the three-dimensional structure of the protein and the steric relationships of functional groups at the 'active site'. A high degree of specificity is exhibited by many enzymes and is utilized to great effect in analytical testing methods.

Whereas chemical methods of analysis identify a functional group within a compound, an enzyme has the capacity to differentiate between two compounds carrying the same functional group. Specificity is not always absolute, however, and low levels of background activity can be obtained even with purified enzymes. In fact, certain enzymes have a wide specificity towards a range of similar substrates, for example sorbitol (polyol) dehydrogenase. Care is thus needed when choosing an enzyme for use in a test procedure.

Purity. Whereas specificity is an in-built characteristic of an enzyme, the purity of an enzyme preparation is determined by the enzyme producer. In terms of the effective use of the enzyme in an analytical system purity is as important as specificity, especially where a complex sample containing a number of similar substrates is being tested.

Ideally, an enzyme preparation for analytical systems should not have any contaminating activities. To achieve this would make enzyme prohibitively expensive, and a compromise is achieved by defining the limits of any contaminating activities compared with the principle activity. Generally, the contaminating levels are below 0.05 per cent of the principle, and at this level have little effect on the accuracy of the assay.

Two points need to be noted when considering purity and contaminating activities; first, whether the assay procedure used to determine the contaminating activity is optimal (a contaminating activity can be significantly understated if a sub-optimal assay is used), and second, most analytical enzymes have been developed for the medical diagnostic field. Contaminants and specified levels have been determined with this in mind.

Where unspecified side reactions take place at a low level, causing 'creep' at the end-point, it is possible to employ a simple extrapolation procedure to determine the true end-point position (*see* Figure 4.3.1).

Enzymic reactions. There are three types of reaction which can occur in enzymic analysis: *(i)* Substrate reaction, *(ii)* indicator reaction and *(iii)* trapping reaction.

(i) A substrate reaction is the primary reaction which takes place between the substrate being tested and the enzyme. For example:

$$\text{Starch} + (n-1)\text{H}_2\text{O} \xrightarrow[\text{(EC 3.2.1.3)}]{\text{Amyloglucosidase}} (n)\text{Glucose}$$

$$\text{Glycerol} + \text{ATP} \xrightarrow[\text{(EC 2.7.1.30)}]{\text{Glycerokinase}} \text{Glycerol-3-phosphate} + \text{ADP}$$

(ii) Indicator reactions are necessary to enable a stoichiometric estimation of the substrate reaction to be made. This can, for example, be carried out by a ultraviolet method utilizing a dehydrogenase enzyme and NAD(P) or NAD(P)H cofactors. An increase or decrease in the optical density of these cofactors in the long-wave ultraviolet region can be determined using a spectrophotometer by reading at the 340 nm absorption maximum (*see* Figure 4.3.2). Measurement can also be made employing spectrum line photometers, equipped with a mercury vapour lamp, using wavelengths of 365 or 334 nm.

In the simplest cases the indicator reaction is also the substrate reaction. For example, the determination of D-isocitrate:

$$\text{Isocitrate} + \text{NADP}^+ \xrightarrow[\text{(EC 1.1.1.41)}]{\text{Isocitrate dehydrogenase}} \alpha\text{-Ketoglutarate}$$

$$+ \text{NADPH} + \text{CO}_2 + \text{H}^+$$

When none of the reactants of the substrate reaction enables an optical measurement to be made, a specific indicator reaction is

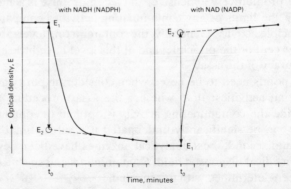

Figure 4.3.1   Extrapolation to the time end-point (E_2) shown by the broken line to time zero (t_0).

introduced. For example, the determination of lactose:

$$\text{Lactose} + H_2O \xrightarrow[\text{(EC 3.2.1.23)}]{\beta\text{-Galactosidase}} \text{Glucose} + \text{Galactose}$$

$$\beta\text{-Galactose} + NAD^+ \xrightarrow[\text{(EC 1.1.1.48)}]{\text{Galactose dehydrogenase}} \text{Galacturonic acid}$$

$$+ NADH + H^+$$

Alternative indicator reactions are based on enzyme-mediated colorimetric reactions. A typical example would involve the use of an oxidase enzyme to produce hydrogen peroxide in the substrate reaction. For example, the determination of cholesterol:

$$\text{Cholesterol} + O_2 \xrightarrow[\text{(EC 1.1.3.6)}]{\text{Cholesterol oxidase}} \Delta^4\text{-Cholestenone} + H_2O_2$$

$$2H_2O_2 + \text{Reduced chromogen} \xrightarrow[\text{(EC 1.11.1.7)}]{\text{Peroxidase}} \text{Oxidized chromogen} + 2H_2O$$

In this reaction the reduced chromogen is a phenol-4-aminophenazone mixture which develops a red colour upon oxidation. The absorption maximum for this colour is at 500 nm. With the emergence of new oxidases this colorimetric indicator reaction

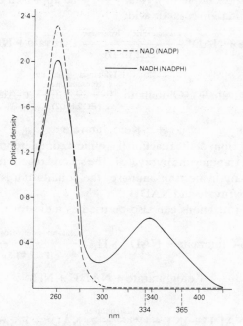

Figure 4.3.2 Absorption curves of NAD(P) and NAD(P)H.

(Trinder reaction) is finding increasing favour in the medical diagnostic field. The primary benefit of colorimetric methods is that less sophisticated and less expensive instrumentation is adequate.

A further colorimetric method enables dehydrogenase mediated substrate reactions to be read in the visible wavelength range. This reaction uses the enzyme diaphorase. For example:

$$\text{NAD(P)H} + \text{H}^+ + \text{Oxidized chromogen} \xrightarrow[\text{(EC 1.6.4.3)}]{\text{Diaphorase}}$$

$$\text{Reduced chromogen} + \text{NADP}^+$$

The chromogens most often used in this reaction are iodinitrotetrazolium chloride and nitrotetrazolium blue.

(iii) Trapping reactions are so called because the equilibrium of certain reactions lies heavily in favour of the substrate so that a 'trapping', or 'disequilibrium', reaction is necessary to displace the reaction in favour of the product. A chemical can be used for this effect, for example in the determination of ethanol:

$$\text{Ethanol} + \text{NAD}^+ \xrightarrow[\text{(EC 1.1.1.1)}]{\text{Alcohol dehydrogenase}} \text{Acetaldehyde} + \text{NADH} + \text{H}^+$$

Semicarbazide is added which reacts with the acetaldehyde and thus displaces the equilibrium which is otherwise heavily in favour of ethanol. An example of a totally enzymic approach to trapping is the determination of lactic acid:

$$\text{L-Lactate} + \text{NAD}^+ \xrightarrow[\text{(EC 1.1.1.27)}]{\text{L-Lactate dehydrogenase}} \text{Pyruvate} + \text{NADH} + \text{H}^+$$

$$\text{Pyruvate} + \text{L-Glutamate} \xrightarrow[\text{(EC 2.6.1.2)}]{\substack{\text{Glutamate–pyruvate} \\ \text{transaminase}}} \text{L-Alanine}$$

$$+ \alpha\text{-Ketoglutarate}$$

The equilibrium of the reaction lies almost completely on the side of lactate. By trapping the pyruvate in the second reaction catalysed by glutamate–pyruvate transaminase, the equilibrium is displaced in favour of pyruvate and NADH.

Trapping reactions can also be used as indicator reactions. For example:

$$\text{L-Glutamate} + \text{NAD}^+ + \text{H}_2\text{O} \xrightarrow[\text{(EC 1.4.1.3)}]{\text{Glutamate dehydrogenase}}$$

$$\text{Ketoglutarate} + \text{NADH} + \text{NH}_4^+$$

$$\text{NADH} + \text{INT} + \text{H}^+ \xrightarrow[\text{(EC 1.6.4.3)}]{\text{Diaphorase}} \text{NAD}^+ + \text{Formazan}$$

A list of enzymes used in the determination of important constituents in the food industry is given in Table 4.3.

Sensitivity. The sensitivity of enzymic assay methods depends upon the detection methods to which they are linked. Ultraviolet assays employing the cofactors NADH or NADPH all have a similar degree of sensitivity that depends upon the extinction coefficient of these compounds at 340 nm. Similarly, the sensitivities of colorimetric methods are governed by the relevant extinction coefficients of the selected chromogens.

The minimum quantity of a substance that can be detected with accuracy in the ultraviolet assay is 0.01 micromoles. For glucose this would be 1.8 microgrammes. The same order of magnitude applies for colorimetric assays. Thus, for the sub-microgramme assay to be accurate an alternative technique must be employed.

The upper limits of detection of the methods outlined are generally 25–50 times higher than the minimum, the practical limits depending consequently upon the accuracy of dilution techniques used for the sample preparation.

Reproducibility. Enzymic assays are reproducible within plus or minus two per cent if all factors are adequately controlled. Variations up to double the value would be considered acceptable in a routine application. The principle factors which affect reproducibility are related to operator techniques, and in particular pipetting error, robustness of the assay itself, temperature control and reagent efficacy. These points are dealt with in more detail below.

3. Methodology

The industrial application of enzymic testing is still evolving, so that modifications to existing tests and the design of new tests will be required. To ensure that these new tests are robust and accurate, it is important to know and understand the properties of the enzymes involved. Particular attention must be paid to their limits of activity and susceptibility to extraneous factors. The next part of this section will highlight some of the areas for consideration.

Reagents and assay conditions. Enzymes function at their optimum when pH, ionic strength and temperature are controlled within specified limits specific to each enzyme. The individual profile for these factors for a particular enzyme can be further influenced by the choice of buffer ion (*see* Figure 4.3.3). Much of the basic information concerning enzymes used in reagents can be obtained from technical information data published by enzyme manufacturers, or from the general literature.

In multi-enzyme assay systems compromises to these factors are required in order to obtain a workable assay. This may result in one particular enzyme working at the limit of its effective range, and care is needed if changes are contemplated with regard to the reaction conditions. Such changes should not result in a less reliable assay system. An understanding of all the factors involved in enzymic activity will reduce this risk.

The sample solution under test is the most likely source for error and variability in the enzyme assay system, and this is particularly the case if a large and concentrated test sample is used. This situation will risk variation in pH and ionic strength and thus the possible reduction in activity of the assay enzyme(s). This is especially relevant when using enzymes with narrow pH optimum bands, where a change of half a pH unit can reduce the enzyme activity by up to 50 per cent (*see* Figure 4.3.3). The ionic strength of the assay buffer should be sufficient to compensate for small

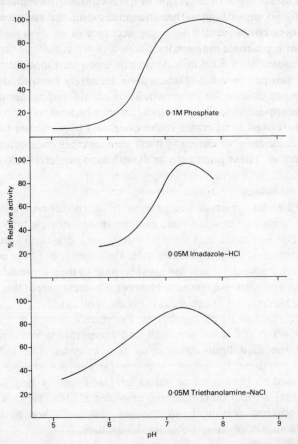

Figure 4.3.3   pH activity profile of glutamate hydrogenase in three buffer systems (data from Whatman Biochemicals Ltd).

changes in sample pH, but high salt concentrations in the sample can often exceed this capacity in the system and result in a shift of assay pH.

The introduction of inhibitors into the system, either from the sample or by generation in the reagents themselves, should also be considered. Heavy metal ions, for example, have an inhibitory action on a large number of enzymes but equally effective inhibition can be observed when chelating compounds or non-aqueous solvents are introduced.

In the reagents themselves, care must be taken to prevent the formation of compounds by NADH and NADPH in solution which will inhibit dehydrogenases, and inhibitory breakdown products formed by the action of some organic buffers (*see* Figure 4.3.4). The simplest way to do this is to ensure that all reagents are fresh on a daily basis and made up in freshly distilled water. Deionized water can contain organic impurities and microorganisms and so is generally less satisfactory.

From a practical viewpoint, where it is not possible to make up the reagents fresh daily, solutions should be stored in a refrigerator ($+4$°C) in stoppered containers to minimize microbial contamination. Where longer storage is intended for dilute enzyme solutions and coenzymes, storage at -20°C may be preferable. This should be done in small lots in order that the cycle of freezing and thawing occurs only once for each aliquot to avoid enzyme denaturation.

Reaction temperature. As enzyme reaction rate is proportional to temperature, and the ideal assay temperature is 30°C, temperature control should be straightforward, since this value is generally

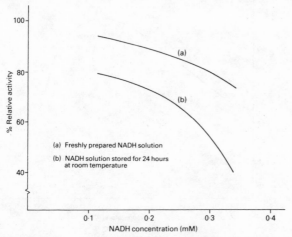

Figure 4.3.4   The effect of inhibitors formed in NADH solution on glutamate dehydrogenase activity (data from Whatman Biochemicals Ltd).

above laboratory ambient temperature. Reaction rates will be high enough to ensure completion within a few minutes, and the risk of heat denaturation is minimal (*see* Figure 4.3.5).

With end-point reactions, the control of temperature is not as critical as for rate reactions. However, it is recommended that reaction temperature is controlled to within $\pm 0.1\,^{\circ}\mathrm{C}$ and that incubation times and time intervals for measuring the results be strictly adhered to. Deviation from projected time curves can then be used as indicators of other errors in the system.

The most frequent source of error relating to temperature is the failure to warm buffers, which may have been stored chilled, prior to use in the assay. Equilibration times in a reaction cuvette are usually not sufficient to ensure that reaction temperature is reached if one of the components is of large volume and not already up to room temperature.

Pipetting. A good pipetting technique is essential and the following recommendations can be made: use top-quality pipettes on which the graduations do not extend to the tip; always wipe traces of liquid from the outside of pipettes before delivering volumes from them; volumes of less than 0.1 millilitre should be pipetted on to a plastic spatula and mixed into the contents of the cuvette.

Pipetting errors are the largest single source of enzyme assay error, and a consistent and careful approach is essential to minimize the effects.

Glassware. All glassware used for enzyme assays, and particular-

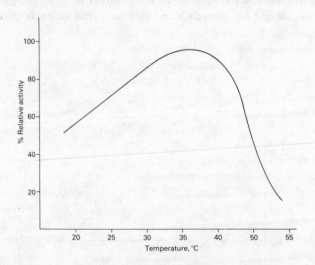

Figure 4.3.5   A typical reaction temperature–activity curve for a dehydrogenase enzyme. Above 45 °C denaturation begins and observed activity falls.

ly the reaction cuvettes, must be absolutely clean. Some enzymes attach easily and firmly to glass surfaces and form a potential source of cross-contamination of tests. Cleaning can be achieved with commercially available cleaning agents, and denatured protein can be removed by a 30 per cent solution of sodium hydroxide used before the cleaning agent treatment. Care must be taken to ensure that all traces of the cleaning agent are thoroughly rinsed away, first with tap and then with distilled water.

Blanks and standards. In the same way as with other assays, the enzyme assay must be run with an appropriate set of blanks and standards for each set of samples. The blank is obtained by replacing the sample solution by water. A set of at least four standards of different strength are required to enable the construction of a 'standard curve' and to provide internal control. They will also be valuable in highlighting any deficiencies in the range of the assay or in the reagents themselves. The estimation of the standards should be performed with each batch of test samples. Batch to batch control can be obtained by running a further control from a standardized substrate solution with each batch of tests.

Equally important in the analysis of a component from a complex starting material is the determination of the presence of inhibiting or interfering compounds. This is done by adding a known quantity of the substrate to the sample prior to preparation. If quantitative recovery is then achieved it can be taken as proof of the absence of interfering compounds. In addition, proportionality between several different weighed sample quantities and the analytical results is further evidence of the absence of interfering compounds, thus confirming correct performance of the assay.

Instrumentation. The instruments used in enzymatic assay are either automatic reagent and sample dispensers or spectrophotometers, or a combination of the two. The medical diagnostic field is well served with instruments that combine both functions and the fully automatic approach is possible because the sample is a liquid – normally blood serum – which is consistent and simply prepared. Analysis of industrial samples is often more complicated and the numbers of samples to be tested often cannot justify the capital outlay for such systems. Generally, only the provision of a spectrophotometer need be considered for routine industrial assays. The market is well served with double-beam, ultraviolet and visible instruments with attachments for automatic sampling and reading. Specialist instruments to cover single, high volume, medical tests (e.g. glucose in serum) have emerged in recent years and can be adapted for routine use in those industrial laboratories with a high frequency of use of a single test method.

4. Other analytical uses of enzymes

Immobilized enzymes. The enzymes themselves represent the highest single component cost in enzyme assay systems. For tests performed frequently in the medical diagnostic area, specific immobilized enzymes have been developed to reduce the actual cost of enzyme per unit test performed. The serum glucose test has received particular attention and an immobilized hexokinase–glucose-6-phosphate dehydrogenase system and a separate glucose oxidase system have been developed by Technicon Instruments Corporation. Both enzyme systems are immobilized in coils for the Technicon Autoanalyser instrument. Glucose oxidase has also been immobilized on a membrane and used successfully by attachment to an oxygen detection system by Yellow Springs Instrument Company Inc. Oxidase systems lend themselves to this latter approach because of the sensitivity and rapid response times of oxygen detectors.

The extension of this approach to a range of industrial applications will depend upon the development of a suitable range of oxidase enzymes, and upon their subsequent performance in immobilized systems. The potential for application in on-line testing is evident, but the associated problems of fouling by crude substrate solutions may limit their introduction.

Sample preparation. Enzymes able to degrade natural polymeric materials are increasingly finding applications in sample preparation, particularly in drug analysis. Drugs bind to serum and body proteins and the analysis of total levels requires their release prior to assay. This can often be simply accomplished by hydrolysis of the binding protein with, for example *Subtilisin A* (Novo Industri A/S), a protease with a wide spectrum of attack on native protein. Since the hydrolysis can be performed at mildly alkaline pH, the risk of forming breakdown products of the substances being tested for is much lower than with acid or alkaline extraction alone. A further example of the specific use of enzymes in sample preparation is the treatment of serum with β-lactamases to remove penicillins and cephalosporins. Without this pretreatment it would not be possible to measure the much lower levels of, for example gentamycin, in the sample by microbiological methods, or to perform bacteriological counts on the serum after incubation, because of the supression of bacterial growth by the high residues of lactam antibiotics.

5. The future

The use of enzymes in industrial analysis will continue to expand because of the increasing need for better quality assurance, backed by legislation, and the advantages conferred by the inherent specificity of these bio-catalysts. In addition to the routine testing of fundamentally basic compounds, it is foreseen that new areas of enzymic testing will emerge as further insight into the biochemistry and enzymology of areas such as flavour formation is developed.

The sensitivity of enzymic testing is dependent upon the detection methods used. At present, microgramme quantities of compounds can be accurately measured. Improvements in sensitivity to match the levels of detection of gas-liquid chromatography or high-pressure liquid chromatography would expand the scope even further. Two such areas are now being developed, namely fluorimetry and luminescence. the latter can be directly applied to existing analytical systems using cofactors such as ATP or NADH, and will detect stoichiometrically sub-nanomolar quantities of these compounds. The development of suitable enzyme systems has only recently begun and further developments to obtain cheap reliable operations will be needed before this methodology finds universal acceptance.

Further reading and background data

Bergmeyer, H. J. *Methods in Enzymatic Analysis* 3rd edn (Verlag Chemie, Weinheim/Academic, New York, 1974).

Methods of Enzymatic Food Analysis (Boehringer Mannheim Gmbh).

Schmidt, E. & Schmidt, F. W. *Fundamentals of Diagnostic Enzymology* (Boehringer Mannheim Gmbh).

TABLE 4.3.1

Enzymes used in analytical tests for industrial applications

| Substrate | Substrate reaction | Enzymes used in | | Assay type | Indicator | Comments |
		Indicator reaction	Trapping reaction			
1. Organic acids						
Acetic	Acetic-CoA synthetase	Citrate synthetase Malate dehydrogenase	Citrate synthetase	Ultraviolet (UV)	NAD$^+$	Indicator reaction driven by trapping reaction, not substrate reaction
Citric	Citrate lyase	Lactate dehydrogenase Malate dehydrogenase	—	UV	NADH	
D-Gluconic	Gluconate kinase	6-Phosphoglucanate dehydrogenase	—	UV	NADP$^+$	
Isocitric	Isocitrate dehydrogenase	—	—	UV	NADP$^+$	
D- and L-Lactic	D-Lactate dehydrogenase L-Lactate dehydrogenase	—	Glutamic-puruvic transaminase	UV	NAD$^+$	Equilibrium in favour of lactate; pyruvate trapped by GPT reaction
L-Malic	L-Malate dehydrogenase	—	Glutamic-oxaloacetic transaminase	UV	NAD$^+$	Equilibrium in favour of malate; oxaloacetate trapped by GOT reaction

Substrate						Notes
	dehydrogenase		—	UV	NADH	
Succinic	Succinyl-CoA synthetase	Pyruvate kinase/lactate dehydrogenase	—	UV	NADH	
2. Sugars						
Fructose	Hexokinase	Phospho glucose isomerase Glucose-6-phosphate dehydrogenase	—	UV	NADP+	PGI used to convert fructose-6-phosphate to glucose-6-phosphate
Glucose	Hexokinase	Glucose-6-phosphate dehydrogenase	—	UV	NADP+	
Galactose	Galactose dehydrogenase	—	—	UV	NAD+	
Sorbitol	Sorbitol dehydrogenase	As for fructose	—	UV	NADP+	Sorbitol dehydrogenase also oxidizes other polyols, therefore fructose formation is measured in mixed samples, rather than NADH
Lactose	β-galactosidase	As for galactose	—	UV	NAD+	
Maltose	α-glucosidase	As for glucose	—	UV	NADP+	
Raffinose	α-galactosidase	As for galactose	—	UV	NAD+	
Starch	Amyloglucosidase	As for glucose	—	UV	NADP+	
Sucrose	β-fructosidase	As for glucose	—	UV	NADP+	Substrate reaction completed at pH 4.6 prior to determination of glucose

Substrate	Enzymes used in			Assay type	Indicator	Comments
	Substrate reaction	Indicator reaction	Trapping reaction			
3. Amino acids						
L-Aspartic	Glutamic-oxaloacetic transaminase	Malate dehydrogenase	—	UV	NADH	
L-Asparagine	Asparaginase	As for aspartate	—	UV	NADH	
L-Glutamic	Glutamate dehydrogenase	Diaphorase	Diaphorase	Colour	Iodophenyl nitrophenyl phenyl tetrazolium chloride	Trapping reaction also indicator reaction; calorimetric assay although NADH formed
4. Others						
Acetaldehyde	Alcohol dehydrogenase	—	—	UV	NADH	
Cholesterol	Cholesterol oxidase	Peroxidase	—	Colour	Phenol-4-aminophenazone	
Cholesterol esters	Cholesterol esterase	As for free cholesterol	—	Colour	As for free cholesterol	
Creatine	Creatine kinase	Pyruvate kinase/lactate dehydrogenase	—	UV	NADH	
Creatinine	Creatininase	As for creatine	—	UV	NADH	

Ethanol	Alcohol dehydrogenase	Semicarbazide	UV	NAD⁺	Equilibrium in favour of ethanol; acetaldehyde trapped by semicarbazide
Glycerol	Glycerol kinase	Pyruvate kinase/ lactate dehydrogenase	UV	NADH	—
Guanosine-5-monophosphate	Guanosine-5-monophosphate kinase	Pyruvate kinase/ lactate dehydrogenase	UV	NADH	Indicator reaction stoichiometric with half substrate reaction
Triglycerides	Kipase	As for glycerol	UV	NADH	Broad spectrum lipases and lipase mixtures can be used to obtain complete hydrolysis

BAKING
J.R. Reichelt

1. Introduction

Wheat and other cereal grains which are used as sources of flour for baking purposes contain a number of naturally-occurring enzymes that are essential for germination and growth. These enzymes are also very important in the baking process, which would not be possible without them.

The most common source of flour for baking purposes comes from wheat, and the major constituent of flour is starch, although most of the unique properties of dough are largely due to the insoluble or gluten protein they contain.

The two major components of starch are amylose and amylopectin. Amylose contains unbranched chains of 1,4 α-glucosidase glucose linked residues, 250–300 units in length. Amylopectin is a branched structure, again with 1,4 α-glucosidase glucose linkages, but with predominantly 1,6 α-glucosidase linkages at the branch points, up to 1000 units in length. Details of the structure and linkages are described in Chapter 4.15, 'Starch'.

The most important enzymes contained in flour for the baking process are the amylases and proteases. The amylases act on any damaged starch from the milling process to produce dextrins and low molecular weight sugars. These sugars are then made available to the yeast during the baking process. The proteases act on the protein material, breaking down the tertiary protein structure and peptide bonds, weakening the gluten. The modification of baking flour by these enzymes is summarized in Table 4.4.1.

TABLE 4.4.1
Enzyme modification of baking flour

Flour component	Starch	Protein
Enzyme type	Amylase	Protease
Products	Dextrins, sugars	Peptides, amino acids
Baking effects	Increased bread volume	Gluten modification, dough viscosity reduction and energy-saving
	Improved crust colour	
	Improved crumb structure	Improved machinability in biscuit production
	Anti-staling effects	

At least two types of amylase, each with different characteristics, occur naturally in cereals, namely α- and β-amylase. β-amylase is found in almost equal amounts in germinated (sprouted) or ungerminated (unsprouted) grain. However, with α-amylase the ungerminated grain contains very small amounts, but increases greatly during germination (Table 4.4.2).

TABLE 4.4.2

Amylase content in germinated and ungerminated cereal grain

Enzyme	Ungerminated grain	Germinated grain
β-amylase[1]		
Wheat	23	22
Barley	28	28
α-amylase (SKB units per gramme)		
Wheat	0.04	200
Barley	0.04	54

[1]According to Kneen & Sandsted (1941).

Consequently, the α-amylase content of the flour is greatly dependant on the conditions of growth and harvesting. In damp climates the tendency will be for high α-amylase activity due to germination, in dry climates, low α-amylase, as there is little germination. This results in large differences in α-amylase content between batches of flour. High amylase content leads to high dextrin production and low water retention within the dough, which results in considerable 'loaf stickiness', open crumb structure with little strength, and high crust colour. High levels of loaf stickiness can 'gum up' commercial bread slicing machines. Low amylase content leads to low dextrin production and results in poor gas production, which results in inferior quality bread of small size and low crust colour.

During the last 20 years or so, millers have supplemented the α-amylase content of flours, originally with malt flour and more recently with fungal α-amylase. This has enabled the miller to offer a bread flour of suitable enzyme content. Where very high α-amylase occurs naturally, this can be diluted by using flours of low amylase activity.

2. Baking process developments

Traditional baking process. Originally, the bread-making process employed a 'bulk fermentation' period of three hours; during this time sufficient sugars were formed to ensure that adequate gas levels were produced to give bread of acceptable quality and volume. The disadvantage to the traditional baking process was the time required for the bulk fermentation.

Chorleywood Bread Process. In 1961 the Chorleywood Bread Process was introduced in the UK and this overcame the problems of the bulk fermentation period. The dough, after mixing, is chopped into 400-gramme or 800-gramme loaves, then proven and baked. This process takes only one hour, compared with the four-and-a-half hours required for the traditional process.

This process also enabled mechanical handling systems to be used and continuous bread production became more widely accepted. Of the total bread sold in the UK today, almost 75 per cent is made by the Chorleywood Bread Process, and over half this amount is in the form of white sliced and wrapped loaves.

The reduced fermentation time used with this process requires a level of α-amylase in the flour with sufficient diastatic activity to produce adequate sugars for yeast gas production.

Activated Dough Method. Recently, another process, called the Activated Dough Method, has been developed, which uses L-cysteine hydrochloride to split the disulphide bonds within the gluten protein matrix. Because of the reduced dough contact time, to ensure sufficient sugar to yield adequate gassing power, this process requires much higher levels of α-amylase supplementation.

3. α-amylase supplementation

Most flour contains only a small amount (one to two per cent) of fermentable sugars, the main supply being produced by the diastatic action of α- and β-amylases on the starch granules damaged in the milling process.

β-amylase is an exo-enzyme, which is normally abundant in flour and hydrolyses the α-1,4 glucosidic bonds from the end of the chains to produce maltose. It is able to break down the amylase (approximately 26 per cent of the starch) completely to maltose, but cannot operate beyond the α-1,6 branching links in amylopectin.

Cereal α-amylase is an endo-enzyme and hydrolyses the α-1,4 glucosidic linkages within the starch producing dextrins of relatively low molecular weight for the β-amylase to act upon. However, the α-amylase content is usually very low, and the miller usually supplements the α-amylase content to a standardized level to ensure adequate baking performance.

In the past, the miller used malt flour for this purpose, but nowadays it is more common for fungal α-amylase to be used. The levels of supplementation used today are the same as those used before the Chorleywood Bread Process was developed. These levels are commonly between 5 and 15 SKB units per 100 grammes flour. The millers today use a variety of analytical test procedures to regulate the α-amylase content for product standardization.

These levels are for the traditional baking process, with bulk fermentation time, and are therefore not sufficient for use in the Chorleywood or Activated Dough processes and must be supplemented at the bakery. This is normally accomplished by the addition of α-amylase directly or, more commonly, by 'bread improvers', which contain fungal α-amylase, ascorbic acid, potassium bromate, emulsifying fats, yeast foods and some soya flour (lipoxidase activity). These are added according to the recipes or formulation given by the enzyme companies or improvers suppliers after extensive baking trials. This latter process is often referred to as bakery technology.

Malt. In the past, malt was added to supplement α-amylase flour content. Usually this was either added as malt flour by the miller or as 'malt extract' by the baker.

However, the addition of malt, although adding α-amylase activity to the baking flour, has certain disadvantages. Commercial malt preparations differ widely in their enzyme activity, and often con-

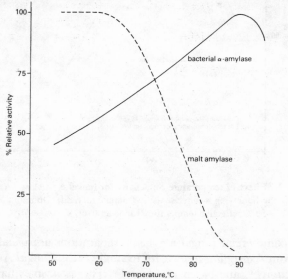

Figure 4.4.1 Effect of temperature on activity for bacterial amylase (thermostable) and malt amylase.

tain high protease levels in addition to amylase, which can adversely affect the bread quality. Malt is also darker than flour and tends to reduce the whiteness of the final flour. One further disadvantage of malt flour supplementation is the comparatively high thermal stability of the malt α-amylase (70–75 °C) and proteases, when compared with that of fungal α-amylase and cereal β-amylase (60–65 °C). The cereal and malt α-amylase continues to act on starch past the gelatinization point of 70 °C and produces excess dextrin material, producing 'sticky crumb' and frailer bread structure (Figure 4.4.1).

Fungal α-amylase. Fungal α-amylase does not give rise to these problems, as baking temperatures inactivate the enzyme at around 60 °C, before any appreciable amounts of starch are gelatinized. Gelatinization temperatures for starches are shown in Chapter 4.15, Table 4.15.2. Cereal β-amylase is also inactivated at around 65 °C in the baking process, and then excess production of dextrins and maltose is avoided. Figures 4.4.1 and 4.4.2 show the effect of temperature on enzyme stability.

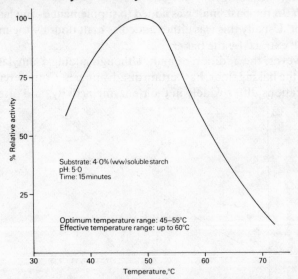

Substrate: 4·0% (w/w) soluble starch
pH: 5·0
Time: 15 minutes

Optimum temperature range: 45–55 °C
Effective temperature range: up to 60 °C

Figure 4.4.2   Effect of temperature on activity for fungal α-amylase (*Aspergillus niger*) using the liquefying α-amylase assay (manual method). Optimum temperature range 45–55 °C; effective temperature range up to 60 °C.

Most commercial fungal amylase preparations are usually obtained from *Aspergillus oryzae*, for example MKC *Fungal Amylase* and *Takamyl®*, and Novo *Fungamyl®*. These enzymes have low protease content, standardized α-amylase activity, usually expressed in SKB units, and are whiter than malt. They are ideally suited for use in the baking process, as their temperature and pH optima are essentially the same as those of cereal amylases. The usual dough range of pH is 4.5–5.5.

Apart from supporting gas production, fungal α-amylases appear also to add to the ability of gas retention within the dough and to increase loaf volume, as well as providing crust colour improvements. Further extensive work is now going on to study these effects, but cannot be discussed further at the time of writing.

Bacterial α-amylase. Bacterial α-amylases are usually obtained from *Bacillus subtilis* sources, and due to the very high thermal stability they exhibit (85–110°C), they are of limited value in the baking process and have little known use. Bacterial α-amylases work right through the baking process producing excessive dextrinization, with resultant stickiness and structural failures, even at very low addition levels.

Baking temperatures. The thermal stabilities of the three types of α-amylase are shown in Figures 4.4.1 and 4.4.2. The range of temperatures normally found within bread during the baking process is shown in Figure 4.4.3.

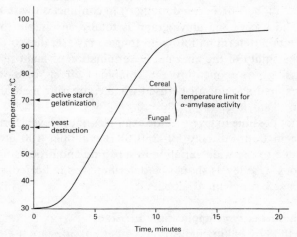

Figure 4.4.3   Interior loaf temperatures during the baking process.

During the baking process, when the dough is placed in the oven, the progressively increasing temperature leads to an increase in enzyme rates of reaction, with resultant sugar and gas production. The volume of the bread increases and the dough starts to rise due to the carbon dioxide gas production by the yeast. At about 60°C, the yeast is inactivated and no further carbon dioxide gas is generated beyond that point. The starch then begins to gelatinize at around 70°C and provides further fresh substrate for any remaining α-amylase to act on, and excess dextrins are produced. The temperature curves clearly show fungal α-amylases to be inactivated before this occurs, and little excess dextrin is produced.

However, any cereal α-amylase present will continue to work on gelatinized starch until inactivated, at 70–75 °C, and produce excess dextrins. The amount of cereal α-amylase present in the flour must also be restricted to minimize these deleterious effects in the final baked product.

When diastatic action is deficient, fungal α-amylase is the addition of choice, particularly for sliced bread. An excess of fungal amylase (more than 20 times the normal level – protease free) can be added without adversely affecting baking results. Loaf volume is increased and there is a slight decrease in crumb firmness (rather more than with malt flour), the crumb resilience is virtually unaffected (malt flour causes a significant reduction).

Determination of amylase supplementation levels. The supplementation levels for fungal α-amylase should only be determined using accepted analytical techniques for α-amylase activity (e.g. Sandstedt, Kneen and Blish method – SKB – or Hagberg 'Falling Number' or Ferrand method) in conjunction with baking trials. The use of the amylograph is totally unsuitable for supplementation determinations with fungal amylase, due to the low thermal stability of the enzyme. Determination of maltose values and gassing power measure the combined effects of α- and β-amylases, and should not be used when comparing malt amylase with fungal amylase.

It is not possible to give specific examples of use levels for fungal amylase in flour (usually 50–250 SKB units per 100 grammes flour), due to the wide variation in substrate and processing plant conditions. Use levels should be determined as indicated, in conjunction with supplement suppliers.

4. Additional enzyme applications in baking

Within the UK baking industry, no other enzymes are used at the present time, excluding lipoxidases, which are added to dough by addition of soya flour (0.3–0.5 per cent) or in improvers to whiten the bread.

In the case of biscuit manufacture, a 'soft flour' is essential to produce a dough with very plastic properties for machining and pumping processes. A soft flour is one with a very weak gluten protein structure. The UK is fortunate in that adequate supplies of soft flour are still readily available and any protein network modification can be achieved (at the present time) by chemical reduction methods. These are by addition of L-cysteine hydrochloride, or more usually by sulphite addition.

In some other European Community countries, notably West Germany, chemical dough modification by sulphite is no longer permitted, and microbial proteases are used instead for protein modification.

The American baking industry has used both microbial amylases and proteases for many years to improve dough handling procedures; this has improved the quality of baked goods and has resulted in energy reduction. Most of the wheat produced in the USA gives 'hard flour', with high gluten protein and low α-amylase content. Enzyme supplementation is used to modify the gluten and increase the level of α-amylase for sugar production.

Proteases. The rheological characteristics of wheat flour dough are dependent upon the gluten protein they contain. This protein forms a network during dough preparation which is responsible for the mechanical loading ability it exhibits. The strength of these networks is partly governed by the 'sulphur bridges' formed between the peptide units within the protein matrix as they become folded when the dough is mixed.

For production of dough for bread, the level of protease must be carefully controlled to avoid the production of a sticky loaf with poor volume and crumb characteristics.

Since cereal proteases are normally found at very low levels, supplementation is essential. Fungal proteases are normally used for bread production, due to the low thermal stability and their amylase content. Usually protease and amylase are added in a controlled manner determined by baking trials.

Fungal Proteases. The fungal proteases used are obtained from *Aspergillus oryzae* and contain both endopeptidase and exopeptidase activity. The endopeptidase hydrolyses the interior peptide bonds of the gluten, thus modifying the viscoelastic properties of the dough. Used at the correct level, fungal protease can improve the handling and machining properties of dough, together with the elasticity and texture of the gluten. Doughs with improved extensibility have better machining characteristics and require shorter mixing time (10–30 per cent) with lower energy input. The doughs obtained from these processes yield much improved bread loaves with high symmetry and uniformity, improved grain and texture with added crumb softness.

The exopeptidase liberates amino acids by its action on gluten, and these free amino acids can intereact with the sugars in the baking process to contribute flavour and crust colour through Maillard reactions. The amino acids can also contribute to yeast growth and gassing power.

There is often a need for fungal protease enzymes to be standardized with varying ratios of protease and amylase (diastase) activity, depending on the flour, other dough ingredients, variety of baked goods and process employed.

Use levels. To determine the extent of desired conditioning, it is recommended to carry out extensibility testing on the dough, observing both physical properties of the dough and test baking results. Overdosing must be avoided, as this produces slack, porous, sticky dough which is very difficult to handle.

Normal use levels for fungal protease are 2.2–4.4 grammes per 100 kilogrammes flour for MKC *Fungal Protease* 31,000 Haemoglobin Units per gramme, or 1.7–4.4 grammes per 100 kilogrammes flour for MKC *Fungal Protease* 61,000 Haemoglobin Units per gramme. An example of *Fungal Protease* 31,000 at a use level of 2.2 grammes per 100 kilogrammes flour is given (0.0022 per cent).

	No enzyme	With MKC Fungal Protease 31,000*
Absorption %	68.0	68.0
Mixing time (mins)	10.0	6.5
Fermentation time (hours)	4.5	4.5
Loaf volume (ml)	2895	2910
Crust characteristic	smooth	smooth
Crumb colour	100 bright	100 bright
Grain and texture	silky	very silky

*0.0022 per cent of flour

This example indicates that a very small quantity of Fungal Protease can reduce dough mixing time significantly and produces a finished loaf with greater volume and improved grain and texture over the control loaf.

Bacterial proteases. The use of bacterial proteases in the manufacture of cracker, biscuit and cookie dough has become widespread as the amylase contained in them presents no problems as the baking temperature rises very rapidly and reaches higher levels than those in the bread baking process. The amylase is quickly inactivated and the 'sticky crumb' problem is eliminated.

The use of bacterial neutral proteases for cracker doughs significantly reduces the tearing problems with the very thin machining necessary for their production, with a reduction of bubbling and buckling during baking as well. They give an improved dough with decreased mixing and proofing times.

Bacterial proteases are usually obtained from *Bacillus subtilis* and standardized to consistent activity levels with maltodextrins.

These enzymes are highly specific endopeptidases which are active on the inner peptide linkages of the gluten protein with resulting reduced elasticity and improved extensibility. Bacterial proteases are ideal for use on high protein flours for cracker, cookie, pizza and biscuit dough production.

Use levels. Use levels will vary with the type and protein content of the flour, and the desired degree of protein modification required for the product. Extensibility trials should be carried out on the dough in conjunction with test baking trials. General use levels are 0.8–1.6 grammes per 100 kilogrammes flour in soda crackers and 10–18 grammes per 100 kilogrammes flour in biscuit doughs, when using MKC or Miles *HT Proteolytic* 200 (200 Northrop Units per gramme).

5. Other enzymes

Recently work has been done on the action of pentosanases on pentosans in wheat (six per cent) and rye (nine per cent) flours. Pentosans are important because they bind water and are responsible for post-bake stiffening (staling of bread) and pentosanases are capable of clearing these materials and reducing the staling characteristics. Hemicellulase enzymes are a rich source of pentosanases and have shown good anti-staling characteristics in long-life bread trials carried out in West Germany.

6. Future trends

In the past there have always been sufficient quantities of 'soft flours' for biscuit and cookie production. However, nowadays more and more high glutten wheat is being grown. These flours are now starting to be used for baking processes and in some countries future legislative constraints on the use of chemical modifications will mean that more enzyme processing will be required to produce acceptable products. High gluten wheat flour will also probably require more pentosanase activity and the current work shows good results in reducing water binding and anti-staling in bread. The new work being carried out on the effects of fungal alpha-amylase in dough is extremely interesting and may result in its increased use in baked goods in the future.

References

Chamberlain, N. & Collins, T. H. 'The Chorleywood Bread Process', *Bakers Digest* (February, 1979).

Chamberlain, N., Collins, T. H. & McDermott, E. E. *J. Fd. Technol.* **16**, 127–152 (1981).

Farrand, E. A. *Cereal Chem.* **41**, 98 (1964).

Fish, A. R. & French F. D. 'High Speed Mechanical Dough Development', *Bakers Digest* (October, 1981).

Kneen, L. & Sandstead, R. M. *Cereal Chem.* **18,** 237–252 (1941).

McDermott, E. E. *J. Fd. Technol.* **9,** 185 (1974).

Underkofler, L. A. 'Enzyme Supplementation in Baking', *Bakers Digest* (October, 1961).

BREWING
T. Godfrey

1. Introduction

For many centuries the production of traditional fermented beverages by the extraction of cereals has been a mainstay of populations and governments. The technical skills of the brewer have never been in doubt, and the developments in the processes have been largely those of the engineering and packaging specialists. The traditional enzyme source for the conversions has been a variety of malted grains, but this chapter is not intended to provide an account of malting technology. If it is accepted that malted grain represents the mobilization of seed resources, via the synthesis and consequent action of hydrolytic enzymes resulting upon germination, then it can be readily appreciated that a number of enzymes will be present that have at least some contribution to make to the brewer's intentions. The germination process is arrested by heat and water removal, and provides a partly modified substrate together with dormant enzymes that can be activated, by grinding and mashing with warm water, to produce a typical brewer's mash. The main features of the conversion achieved with malted cereals can be described through the two main process methods of brewing.

2. Dominant malt enzymes

There are four enzymes in malts that are of primary benefit to the brewer; these are carboxypeptidases, α- and β-amylases, and β-glucanases. The upper limits of their thermal activity are set out in Table 4.5.1. It is clear that the next most important feature of malt brewing is the creation of conditions that utilize the enzymes to the maximum.

TABLE 4.5.1
Upper limits of the thermal activity of malt enzymes

Malt enzyme	Thermal limit (°C)*
α-amylase	68
β-amylase	64
β-glucanase	62
Carboxypeptidase	58

*Reduced to 20 per cent of original activity after 60 minutes.

221

This is largely achieved by attention to two parameters. First, enzymes are almost invariably stabilized towards adverse conditions of temperature and pH by being surrounded by an abundance of their substrate. The brewer therefore makes a mash of the highest concentration of ground grain that is compatible with the later separation of dissolved material from the mash; this is generally in the range of 28–35 per cent as dry solids. Second, a temperature for the mash is chosen that will either give a good compromise on the differing stabilities of the various enzymes, or a series of rising temperatures are selected which are optimal for the majority. These two quite different mashing methods are usually known as 'isothermal infusion' and 'programmed infusion' mashes.

3. Infusion mashing

Ground malt from barley, wheat, rye and sorghum will be converted to a fermentable material when mashed with hot water to give a mixed (strike) temperature of 63–65°C and a typical pH of 5.4. None of the enzymes released from the malt will be at its optimum temperature, but all will be protected to some extent by the starch and proteins released into the mash liquor. The key to the choice of temperature is that almost all of the starch should be fully gelatinized within 15–20 minutes, thus providing a substrate for the amylases. There is little evidence of proteolysis during mashing and it has been suggested that most of the observed protein solubilization has already occurred during malting. This can be confirmed by the thermolabile nature of the carboxypeptidases of malts, which will be almost completely inactive at the mash temperature.

Hydrolysis of β-glucans. This will similarly be limited by the temperature, and glucans are frequently present if the malt is poorly modified and low in β-glucanases.

Amylolysis. This is almost optimal in the infusion mash and both α- and β-amylases are active in converting the starch to a mixture of maltose and glucose, giving about 80 per cent fermentability of the original starch content. The remainder of the soluble starch products consist of higher dextrins and oligosaccharides that, whilst not being fermented, provide much of the characteristic body and mouthfeel of traditional beers. Figure 4.5.1 illustrates the action of these enzymes on starch components.

Malt α-amylase is an endo-enzyme with a random action that is limited to the outer linear sections of amylopectin. It is unable to degrade closer than a few glucose units from the α-1,6 branch points. Malt β-amylase, a sulphydryl exo-enzyme, also acts upon the linear amylose and linear portions of amylopectin to generate maltose units. It acts from the non-reducing ends of the chains.

The rate of these reactions is strongly dependent on temperature and it is very slow when acting upon ungelatinized starch. The infusion temperature is thus selected with reference to the gelatinization range of the total package of cereals in the mash (the combined materials being termed the grist). The ultimate fermentation character of the wort liquor will depend on the way these various factors of gelatinization and enzyme activity are combined.

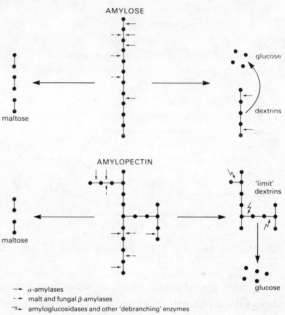

AMYLOSE

glucose

maltose

dextrins

AMYLOPECTIN

'limit' dextrins

maltose

glucose

→ α-amylases
- -→ malt and fungal β-amylases
⌐ᴢ► amyloglucosidases and other 'debranching' enzymes

Figure 4.5.1 Schematic representation of the hydrolysis of starch components.

4. Programmed infused mashing

Basically, this series of stepwise rises in temperature has been adopted to reduce the total process time for wort production, and to permit the adequate utilization of poorer quality malts. Typically, steps are created at 52, 65, 68 and 75°C for varying times (*see* Figure 4.5.2). Up to 25 per cent of unmalted barley or cereal starches may be included in grists for this method of mashing, although extended times at some steps may be necessary to achieve satisfactory conversions in the absence of additional enzymes (*see* page 243, also Figure 4.5.3). Reference to Tables 4.5.1 and 4.5.2 shows that enzyme action, heat inactivation and gelatinization points will not occur in an ideal sequence, but a compromise is achieved that yields a satisfactory extract.

At the initial stage (52°C) a small amount of further proteolysis occurs, resulting in an increase in the soluble nitrogen levels to

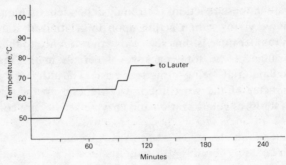

Figure 4.5.2 Mashing diagram: all malt infusion programme.

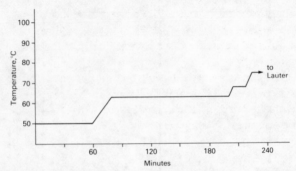

Figure 4.5.3 Mashing diagram: 30 per cent adjunct: 70 per cent malt infusion programme.

match the figures found when mashing with 100 per cent good quality malt. Glucanase attack upon the various gums of the grain begins here and continues throughout the heating up to stage 2. By the time mash reaches 65°C, gelatinization of starch has commenced and the amylases are active. This process is completed during the rise and hold at 68°C, and accelerates rapidly before being stopped by the rise to 75°C.

TABLE 4.5.2
Gelatinization temperatures of various starches

Starch type	Gelatinization temperature (°C)
Maize (high amylose)	68–105
Maize	63–74
Rice	68–75
Sorghum	68–75
Barley malt	63–66
Rye	58–70
Wheat	55–65
Barley	53–58
Potato	53–60

5. Decoction mashing

This method of achieving conversion is characteristic of the production of lager beers, with a typical high percentage of unmalted cereals, and has also been used to facilitate the inclusion of low grade malts.

In addition to the basic infusion mashing procedures, a separate high temperature stage is included to facilitate the liquefaction of the starch adjuncts (*see* Figures 4.5.4 and 4.5.5). Traditionally, a portion of the malt is mixed with the adjunct in a separate cooker and then heated to boiling before being mixed back with the main mash. This hot material raises the temperature of the mash, which is then converted to fermentable products as already described. In some operations this process is repeated by removing a portion of the mash, adding more malt and boiling again before mixing back. In the USA and Canada it is common to use a 'double mash' system (*see* Figure 4.5.6).

The grist consists of a relatively small amount of malt, often as little as 35 per cent. The starchy adjuncts consisting of maize grits or rice are mashed in the 'cereal cooker' with an initial hold at 35 °C for up to 60 minutes, to activate the small amount of malt enzymes.

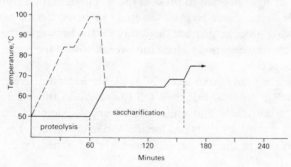

Figure 4.5.4 Mashing diagram: single decoction adjunct cooking with malt mash.

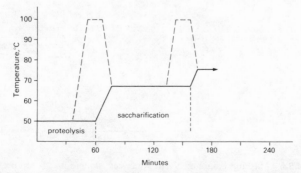

Figure 4.5.5 Mashing diagram: double decoction adjunct cooking with malt mash.

This adjunct mashing is designated the 'first mash' and is heated to 70 °C and held for 20 minutes at this temperature for saccharification. It is then further heated to boiling and maintained there for 45 minutes. When this last heating stage has begun, the 'second mash', consisting entirely of malt, is initiated at 35 °C. Finally, the boiling first mash is combined with the malt mash to give a strike temperature of 65–68 °C. Saccharification now proceeds on all the released starch materials, and about two hours after the start of the process the mash is heated to 73–75 °C and pumped to the Lauter.

The introduction of microbial enzymes as a replacement for the malt that is lost in the decoctions has made a positive contribution to the economics of this method of brewing (*see* page 234).

6. Economic brewing with additional enzymes

The rising costs of malt production and the variations in malting quality of the barley, due to choice of variety and the weather during the growing season, have encouraged the brewer to include a number of additional enzymes in his process. Adjuncts provide a cheaper extract but lack the necessary enzymes for their conversion, and without additional enzymes their use rates are limited by the amount of malt needed to process them. These extra enzymes are available as standardized products that can give accurate control of the brewing process, provide flexibility for the brewer in the choice of grist components and reduce the overall cost of the materials for beer production.

There are many commercially available industrial enzymes for the brewing process, but they fall broadly into three categories: proteases, amylases and glucanases. Fungal, bacterial and plant sources are used for their production and many different aspects of the processes in the brewery are involved. A general guide to the

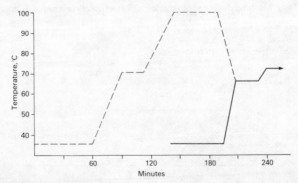

Figure 4.5.6 Mashing diagram: North American 'double mash' adjunct cooking with malt.

types of enzymes and their points of influence is given in Table 4.5.3. In order to follow the basis for the selection of enzymes to aid the brewing process, each stage in the natural sequence converting raw materials to finished beer will be considered in turn, according to the very simple chart set down in Fig. 4.5.7.

TÁBLE 4.5.3
Typical exogenous enzymes applied to brewing

Enzyme type	Beneficial action	Point of application
Bacterial α-amylases	Adjunct liquefaction	Decoction vessel (cereal cooker)
	Adjunct liquefaction Malt improvement	Mash vessel
	Set mashes	Mash vessel
	Starch positive worts	Lauter or mash filter
Fungal α-amylases (maltogenic action)	Improved fermentability	Fermentation
	Low calorie and 'diet'	Fermentation
	Set mashes	Mash vessel
	Starch positive worts	Lauter or mash filter
Fungal amyloglucosidases	Low calorie and 'diet'	Fermentation
	Maximum fermentability	Fermentation
	Priming replacement	2° fermentation or post-pasteurization
Bacterial debranching enzymes	Maximum fermentability	Fermentation
Bacterial glucanases	Increased extract	Mash vessel
	Improved wort separation	Mash vessel
	Improved filtration	Mash vessel/fermentation/ conditioning tank
Fungal glucanases (including cellulases)	Improved extraction	Mash vessel
	Improved wort separation	Mash vessel
	Improved filtration	Mash vessel/fermentation/ conditioning tank
	Increased adjunct (especially sorghum)	Mash/decoction vessel
	Haze prevention	Mash vessel
	Haze removal	Fermentation/ conditioning tank
Bacterial neutral protease	Increased adjunct	Mash vessel
	Nitrogen regulation	Mash vessel/fermentation
Plant proteases (papain)	Chillproofing against protein hazes	Conditioning tank
Fungal pentosanases	Prevention/removal of specific haze components	Mash vessel/fermentation/ conditioning tank
	Improved extract (especially wheat and sorghum)	Mash vessel

In the initial mashing stages, the selected grains and malts are steeped in warm water, either separately or together, and ultimately

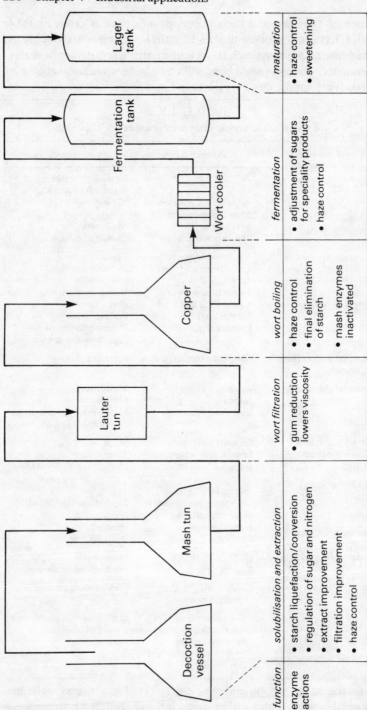

function	solubilisation and extraction	wort filtration	wort boiling	fermentation	maturation
enzyme actions	• starch liquefaction/conversion • regulation of sugar and nitrogen • extract improvement • filtration improvement • haze control	• gum reduction lowers viscosity	• haze control • final elimination of starch • mash enzymes inactivated	• adjustment of sugars for speciality products • haze control	• haze control • sweetening

Figure 4.5.7 Operational sequences of the brewing process indicating the principal influences of enzyme action.

combined in a vessel designed to facilitate the separation of soluble from insoluble material. Modern methods for this separation include the traditional Lauter, mash filter systems and also centrifugation. The resulting fluid containing the extracted soluble material from the grains is termed the 'wort', and is destined to become the fermentation medium for the beer production.

The second main stage is the boiling of the wort, with or without hops or their extracts. The chemical changes characteristic of the flavour of hopped beers are active during boiling. In addition, this heat treatment is used to concentrate the wort to a level that will give sufficient fermentable sugars to yield the desired ethanol content for the final beer. A third function of the boiling stage is the sterilization of the wort to provide a clean medium for the yeast fermentation. This last effect includes the destruction, by thermal denaturation, of the malt and any added enzymes. This is to some extent a positive control step, ensuring that no further changes in the chemical composition of the wort will occur unless directly as a result of yeast fermentation or subsequently added enzymes. For many breweries utilizing a number of additional enzymes in the mashing stages, this is also confirmation that the enzymes are no longer active and not therefore likely to be subject to complicated labelling regulations. However, this does not remove responsibility for the selection of 'approved' additives at all stages in the process. These factors have been of great significance in the increasing drive to restrict adjustments to the brewing process to the mash vessels. There are certain specific cases, however, such as the production of fully fermented 'diet' products or the treatment of hazes appearing at later stages, where it can be beneficial to make further enzyme additions so as to reintroduce specific activities (*see* pp. 252, 255).

The cooled wort is then fermented with selected yeasts under controlled conditions to produce the required alcohol levels, and some beers then undergo a period of maturation, usually after the removal of the bulk of the yeast, to complete the development of specific product character.

The following sections will illustrate the various enzymes applied to each main step in this brewing process. There are many instances where more than one stage can be the point for a particular target change to be accomplished, and in many cases the same activity can be used at different stages for slightly differing targets.

7. Malt mashing improvement

Although many different factors influence the performance of malted cereals, there are three very general factors for the brewer to consider: (*i*) the type of barley (or other malting cereal), and the manner of its cultivation, together with the vagaries of seasonal weather; (*ii*) the method of malting and any additional chemical treatments at the maltings; (*iii*) the mashing system selected by the brewer.

For the first two of these factors, economic considerations coupled with market forces may often combine to produce a grain of much reduced quality. The professional maltster can improve on the quality by making various adjustments to the malting process, perhaps by altering the steeping times, or adding plant hormones to stimulate the germination, and monitoring the influence of his kilning programme on the quality of the extract obtained.

In this last factor, the brewer chooses the malt to suit his method or evaluates the available malt in terms of its economy together with the use of exogenous enzymes. These enzymes are added to the mash to supplement malt enzymes of insufficient potency, and in many cases to provide additional activities not inherently present in even the best quality malts.

Undermodified malts. These are characterized by their lower amounts of total extractable soluble material and lower glucanase and amylase activities than their high quality counterparts. Worts produced unaided from them often have a higher viscosity, with slower run-off and often smaller wort volumes, as a result of the larger amounts of unhydrolysed glucans present, together with lower soluble nitrogen components and poorer fermentability. These various shortcomings can be tackled by the selection of appropriate mashing enzymes. Microscopic examination of the mash often shows that considerable amounts of starch remain trapped in the fibres of the cell debris. These starch particles can be released by the addition of fungal cellulases, which hydrolyse β-1,4 bonds of both cellulose and hemicellulose fibres of cereals. Extract improvements of up to 15 per cent can be obtained by the addition of as little as 0.2 kilogrammes *Celluclast*® 2.0L (Novo) per tonne of malt. Furthermore, the glucanase action of the cellulases generally results in improved filtration of the wort. However, the depressed levels of malt glucanases so characteristic of poor malts are aggravated by the usually elevated levels of glucans in these malts. The most economic improvement in this imbalance is usually achieved by the addition of combinations of bacterial and fungal glucanases other than the cellulases. Both types hydrolyse β-1,3 and -1,4 bonds of glucans but with differing endproducts. The bacterial enzymes generate small oligosaccharides of three to seven glucose units, but cause a dramatic reduction in wort viscosity, while the fungal types yield glucose with a wide range of residual oligosaccharides and useful reduction in viscosity. Practical experience shows that the use of one of each type of these enzymes is far more effective than an increased dose rate of either alone. There is little to choose between the various bacterial glucanases, but quite marked differences are seen in the action of the same activity dose of different fungal glucanases. An appropriate selection may easily be made on the basis of small-scale laboratory mashes with the chosen grist materials. Generally, it is found that the bacterial glucanases are used at twice the rate of the fungal glucanases, assuming they are all on the same activity scale.

The fermentability of the wort is determined after its separation from the grains. The value is nevertheless determined by the degree of saccharification of the mash, and this in turn depends upon the activity of the glucogenic and maltogenic enzymes present. Whilst these are generally not found to be limiting in even very undermodified malts, the restricted release of starches from the cell matrix, described above, can slow down the overall development of fermentable material to a point where the mashing cycle is disrupted. The dextrinizing activity of the mash is thus frequently supplemented with bacterial α-amylase. Rapid hydrolysis of α-1,4 bonds in amylose and amylopectin by the addition, for example, of up to 1 kilogramme BAN 120L (Novo) per tonne malt will provide an improved substrate for the malt enzymes and re-establish the fermentability of the extract.

To produce worts equivalent to high quality malts, the amount and distribution of nitrogen compounds may need adjustment. The ratio of peptide to α-amino nitrogen should be maintained when increased protein hydrolysis is introduced. Although there are many proteases apparently suited to this application, many of them are susceptible to inhibitors produced by raw barley and often present in mashes of undermodified malts. The low levels of protein in the mash, coupled with generally adverse temperatures and often slightly sub-optimal pH values, tend to generate unstable conditions for even those proteases that can function otherwise adequately in the mash. In programmed mashes the stand at $52-55\,^{\circ}\text{C}$ helps to give a greater dose–time response for proteolytic action. In isothermal mashing, it should be assumed that the proteases will only remain at above 50 per cent of their added activity for about 30 minutes. The enzyme $Neutrase^{\circledR}$ 0.5L (Novo), used at $0.3-1.5$ kilogrammes per tonne malt, can raise the soluble nitrogen levels very effectively and maintain the required ratios, as demonstrated in Figure 4.5.8. It is not normal to raise the nitrogen levels by more than about 20 per cent over the untreated values to correct for undermodified malts. Some typical malt improvement enzymes and doses are given in Table 4.5.4.

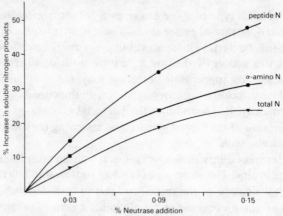

Figure 4.5.8 Increase of soluble nitrogen compounds by the application of *Neutrase®* 0.5L to a malt mash.

TABLE 4.5.4
Enzymes for malt mash improvement

Enzyme type	Example product (Novo)	Use rate (kg per tonne malt)
Bacterial α-amylase	*Bacterial Amylase Novo* 120L	0.5–1.0
Bacterial β-glucanase	*Cereflo®* 200L	0.5–1.0
Fungal β-glucanase	*Finizym®* 200L	0.2–0.5
Fungal cellulase	*Celluclast®* 2.0L	0.2–0.4
Bacterial proteinase (neutral)	*Neutrase®* 0.5L	0.3–1.5

8. Adjunct processing

The considerable economic advantage of the extract obtained by adding unmalted raw materials has encouraged their inclusion in the mash bills of most breweries. In addition to the lower cost of the extract, they contribute to the regulation of several other features of beer production. These include provision of fermentable compounds without adding tannins or proteins, reduction in beer colour whilst maintaining fermentable gravity, and improvement in beer stability and foam character.

The choice of adjunct depends as much on the country, availability and price as upon the particular brewing change intended, but would generally be made from the following: maize grits, wheat flour, rice, sorghum, starches from these and other sources, and of course, sugar syrups from starch conversions and sucrose. As these last two groups are fully processed products, there is no need to consider them here except to refer the reader to Chapter 4.15, 'Starch', for a discussion of the production of enzyme-converted starch sugars.

Some commonly used brewing adjuncts are given in Table 4.5.5, with additional indications of their protein content and benefit from enzymic processing. Although the first four adjuncts have a significant protein content, this is not generally extracted at the brewery (but *see* page 231 for the adjustment of protein). Adjuncts 5, 6 and 7 are substantially free of proteins so that the first 7 can all be used to provide fermentable products without adding either proteins or tannins to the wort. The last two adjuncts are used to replace malt but also to retain the wort composition. Barley will be discussed more fully on page 240.

Regulation of the levels of proteins and tannins in beer is very important in preventing the formation of hazes, as well as for the effects of chilling of the finished beers. Elevated levels of a group of substances described as glycoproteins are claimed to enhance the foaming character of beer and it is often beneficial to add wheat flour or barley, which readily contribute such substances.

Enzymic processing of adjuncts. In all cases, starchy adjuncts require gelatinization, liquefaction, dextrinization and saccharification to convert them to fermentable products. Table 4.5.2 illustrates the wide variation in gelatinization temperatures needed for the different starch sources, but it should be noted here that 'flaked' cereals are processed prior to sale to breweries. The treatments include heating to temperatures that are generally in excess of the gelatinization minimum, which permits them to be processed directly in the mash tun together with the malt grist. However, if the rate of adjunct use is to exceed the conversion power of the malt content, then even flaked cereals may require either separate cooking or the addition of the same range and doses of enzymes as

TABLE 4.5.5
Protein content and enzyme processing benefit of adjuncts

Adjunct	Protein content	Process benefit when enzymes used
1 Maize	+	+
2 Wheat flour	+	+
3 Rice	+	+
4 Sorghum	+	+
5 Starch (various)	±	+
6 Starch syrups	−	−
7 Sucrose	−	−
8 Malt extract	+	−
9 Barley	+	+

required for malt improvement (*see* page 229).

Adjunct liquefaction and dextrinization. Separate cooking in the decoction (cereals) cooker is used to raise the adjunct to gelatinization temperature without heating the whole malt mash to too high a temperature for the survival of the malt enzymes.

The traditional source of liquefying amylases is malt, but the proportion used for this stage is lost by the boiling. To avoid this loss and also to improve the overall liquefaction rate, it is common practice to use bacterial amylases for the cooking. The conventional use of amylases derived from *Bacillus subtilis* has provided economic improvements over malt for three decades, but recently the introduction of the more heat-stable amylases from *Bacillus licheniformis* has produced still further improvements.

Table 4.5.6 shows the essential features of these two types of amylase and the ways in which their performance differs. Improved yields from adjuncts with small starch granules are the result of more complete gelatinization at the higher temperatures made possible by these amylases.

Conventional amylases. These require a hold of 20–30 minutes at around 70–75°C in the heat-up cycle of the cooker. Typically, doses of two to three kilogrammes of, for example, *Bacterial Amylase Novo* 120L per tonne starch content would be added at the charging of the cooker. These amylases also show a marked requirement for the presence of free calcium for their stable action. Calcium hydroxide is frequently added to the mash before cooking to raise the pH to around 6.5 and raise the calcium levels to the 150–200 parts per million required. The boiling stage of the cooking also inactivates the amylase so that, although full gelatin-

TABLE 4.5.6

Comparative operating characteristics of conventional and thermostable bacterial α-amylases for adjunct liquefaction

Amylase	Conventional	Thermostable
Source	B. subtilis	B. licheniformis
Optimum operating temperature	80–85°C	95–105°C
Optimum operating pH	6–6.5	6–6.5
Minimum calcium for stability	150 ppm	70 ppm
Upper starch concentration for practical thinning	30%	45%
Typical dose level	2–3 kg per tonne	1–1.5 kg per tonne

ization may be achieved, there is always the possibility of incomplete dextrinization of the starch. This is completed in the subsequent malt mashing with the cooked adjunct, but can be a limitation on the upper levels of adjunct that can be used overall.

Thermostable amylases. These may be used at much smaller dose levels, for example 1.0–1.5 kilogrammes *Termamyl®* 60L (Novo) per tonne starch content, and without the addition of any calcium salts except where the starch and water supply are both very low in calcium, so that less than 70 parts per million would be present in the mash. Such a combination of circumstances is very uncommon.

To reduce cooking times, it is also possible to mash the adjunct directly with hot water at, say, 90°C. In all cases the cooker may be rapidly heated directly to the boil and maintained for 30 minutes to complete the liquefaction. At the end of the cooking cycle the mash viscosity is generally down to a pumpable 400 centipoise and the mash is being rapidly dextrinized under the action of the amylase that will have survived to a large extent. Figures 4.5.9, 4.5.10 and 4.5.11 show examples of the mashing programmes and viscosity characteristics for the different liquefying enzymes.

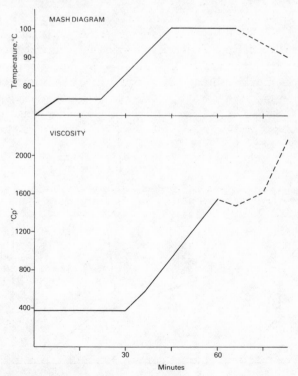

Figure 4.5.9 Liquefaction of maize grits with malt.

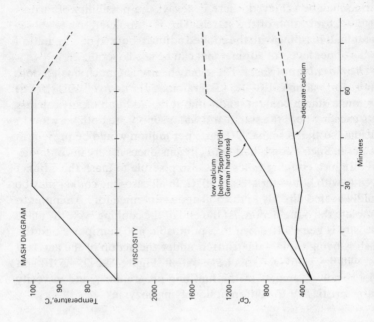

Figure 4.5.10 Liquefaction of maize grits with conventional amylases.

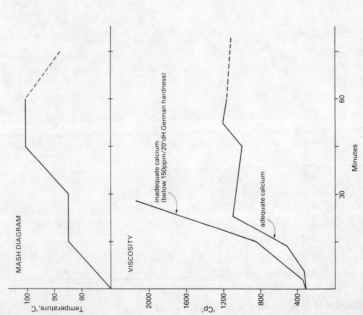

Figure 4.5.11 Liquefaction of maize grits with thermostable amylases.

Adjunct mash concentrations. The thermostability scale for these enzymes approximates to their liquefying performances, which can be translated into increasing cooker capacity potential. The typical malt enzyme adjunct cook will be operated at 25 per cent solids in the cooker; conventional bacterial amylases (with adequate calcium) will act at up to 32 per cent solids, while the thermostable amylases will run up to 40 per cent solids, thus making it practical to increase the capacity of the cooker by up to 60 per cent.

Continuous cooking of adjuncts. A further property of the thermostable amylases is the survival of almost 100 per cent of their activity when they are briefly exposed to high temperatures in the presence of stabilizing levels of starch substrates. Whilst primarily a technique for the starch syrup industry (*see* Chapter 4.15), the application of the continuous cooking of starch adjuncts in brewing has been increasing.

The process can utilize a simple stirred tank cooker, but is at maximum efficiency when a pressurized jet cooker (such as those supplied by the Hydrothermal Corporation of Milwaukee, USA) is used. The jet system consists of separate feeds of starch slurry and clean steam to a venturi device that creates a combination of instant heating and shearing turbulence. The slurry is usually maintained at elevated pressure for 5–10 minutes in a holding tube before being flashed down to atmospheric pressure into a further holding tank for 30–45 minutes to ensure dextrinization of the starch. Energy savings on the cooker system are reported to be up to 25 per cent compared with batch cooking. The main criterion for the adjunct for this method of cooking is that it be finely ground as a flour so that it can pass freely through the jet system. Any cereal flour may be jet cooked with advantage using equipment based on the system shown in Figure 4.5.12.

In this system, a slurry of the adjunct in water to the chosen solids concentration, typically 35–40 per cent, is fed to the jet from a stirred holding tank into which the enzyme is metered. Using *Termamyl®* 60L, the jet cooker is operated at a pressure sufficient

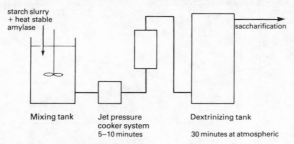

Figure 4.5.12 Schematic layout of a typical jet cooking conversion plant.

to generate a temperature of 105–110°C in the holding cell where the slurry is liquefied for 5–10 minutes. A dose of 1.5 kilogrammes *Termamyl®* 60L per tonne starch solids will result in a fully liquefied and iodine-negative dextrin slurry after a final hold of 30 minutes at atmospheric pressure and 95°C.

Special considerations for processing rice. In batch cooking of rice, as flour or as broken rice grits, gelatinization and liquefaction can be achieved at very much reduced enzyme levels. Typically, only 50 per cent of the levels needed for maize will produce excellent results; for example, *Bacterial Amylase Novo* 120L at 1–1.5 kilogrammes per tonne starch content, and *Termamyl®* 60L (Novo) at 0.5 kilogrammes per tonne starch content (*see* Figure 4.5.13).

Special considerations for processing sorghum and wheat. The presence of fibres and cellular debris, including in the case of wheat a considerable amount of pentosan polymers, can prevent the cooking systems from acting at optimal efficiency. A programmed batch cooking of these adjuncts, to include hydrolysis by cellulases prior to the high temperature starch liquefaction, has been found to enhance the processing characteristics of the subsequent combined adjunct and malt mashes. Increased levels of these adjuncts can be easily introduced by this process. Figure 4.5.14 illustrates the system for sorghum, but it is equally applicable to wheat. An initial dosing of the mash with, for example, *Termamyl®* 60L (Novo) at 1 kilogramme per tonne of adjunct starch, together with *Celluclast®* 2.0L (Novo), also at 1 kilogramme per tonne, is followed by a 20-minute hold at 60°C. During this period, the ther-

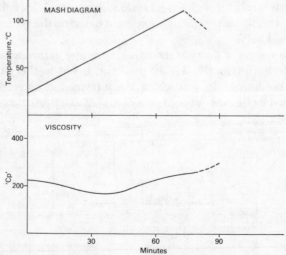

Figure 4.5.13   Liquefaction of cracked rice with *Termamyl®*.

molabile cellulase is acting. The temperature of the cooker is then raised to boiling and held for 30 minutes for the starch to liquefy as in the previous examples. To transfer this system to continuous cooking simply utilizes the jet feed tank to give the cellulase treatment at as high a temperature as possible without inducing gelatinization of the selected adjunct (*see* Table 4.5.2).

Saccharification of dextrinized adjunct mashes. At modest levels of adjunct utilization, it is normal practice to combine the decoction mash with the malt mash and to achieve saccharification to fermentable compounds by the action of the malt enzymes. When higher adjunct levels are being used for the production of lighter coloured beers and 'diet' products, it may be practical to perform the adjunct saccharification separately.

The mash in the decoction vessel, or dextrinizing tank of the jet system, is cooled to a temperature suited to the saccharifying enzyme to be used, and the pH is adjusted downwards to the optimum for the chosen enzyme or enzymes, which are then added. The choice of enzymes depends on the spectrum of sugars one wishes to obtain, which range from very high levels of glucose, through a mixture of glucose and maltose to a low glucose–high maltose pattern. Fungal amyloglucosidases will degrade most of the dextrins resulting from the cooking stage down to glucose and also degrade many of the branched dextrins having α-1,6 branch bonds. Fungal β-amylases, on the other hand, have a much more limited degradation of these branched dextrins, but are highly specific for

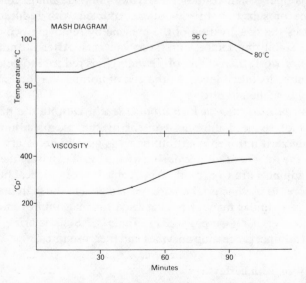

Figure 4.5.14   Liquefaction of sorghum starch with *Termamyl*® and *Celluclast*®.

the production of maltose as the predominant fermentable sac-charide.

By selecting one of these enzymes, or a combination of the two, it is possible to generate a variety of sugar spectra with different characteristics of fermentation. Some of these possibilities are illustrated in Table 4.5.7, using the enzymes *Amyloglucosidase Novo* 150L and *Fungamyl*® 800L (Novo) at the levels indicated. These levels are far from rigid, and the dose/time/proportion of enzyme used should be determined by practical observation with the adjunct of choice.

After the desired saccharification stage has been reached, the adjunct is either filtered independently and passed to the copper, or added to the finished malt mash for separation of wort using the method preferred by the brewery.

Enzyme inactivation considerations. Of the enzymes mentioned in this section, the commercially available cellulases, conventional amylases and fungal β-amylases are all thermolabile and are inacti-vated at the cooker stage, or when the mash is heated before separation. The commercial amyloglucosidases so far investigated have a stability that requires heating to at least 82 °C for 30 seconds to inactivate them. Thus the enzymes used for malt improvement (*see* page 232) and those for adjunct processing will be inactivated during the next stage, that of wort boiling in the copper. Detailed studies of the deactivation of even the thermostable amylases such as *Termamyl*® have confirmed that they, too, are rapidly inactivated under copper boiling conditions. It is likely that the almost complete absence of substrate by this stage, coupled with reduced pH, enhances the deactivation of the enzyme by removing two of the critical stabilizing factors for these enzymes. After 30 minutes' boiling of wort, the activity of *Termamyl*® is reduced to below 20 per cent of its initial level. At the end of normal wort boiling, the activity has fallen to zero.

Nitrogen extraction at high adjunct levels. Despite the primary benefit of reduced nitrogen content by the use of adjuncts of virtually zero nitrogen contribution, it is usually necessary to in-crease the release of soluble nitrogen from the malt part of the mash when adjuncts are to exceed 30 per cent. It is of practical benefit, therefore, to use bacterial neutral proteinases in the malt mashing stage in a similar manner to that used for malt improvement as described earlier (*see* page 231). Table 4.5.8 summarizes some typical adjunct processing enzymes and their use rates.

9. Brewing with barley
Building on the earlier work of various brewing researchers, several

TABLE 4.5.7
Variations of sugar spectrum from adjunct saccharification with enzymes and subsequent mashing with malt (30% adjunct)

Amyloglucosidase (e.g. AMG 150L (Novo), % v/w of adjunct)	Fungal β-amylase (e.g. Fungamyl® 800L, % w/w of adjunct)	Temperature (°C)	Time (hours)	% Sugar distribution			
				Glucose	Maltose	Maltotriose	Dextrins
0.15	—	60	1.0	36	30	12	22
0.10	0.075	60	1.5	23	41	13	23
—	0.25	55	1.0	4	52	20	24
—	0.10	55	1.5	3	54	21	22

TABLE 4.5.8
Enzymes for adjunct processing

Enzyme type	Example product (Novo)	Use rate per tonne malt		Point of addition and function
		Maize	Other	
Conventional bacterial α-amylase	Bacterial Amylase Novo 120L	2–3 kg	1–1.5 kg	Cooker, liquefaction
Thermostable bacterial α-amylase	Termamyl® 60L	1–1.5 kg	0.5–1.0 kg	Cooker, liquefaction
Fungal cellulase	Celluclast® 2.0L	—	0.8–1.2 kg	Cooker, cellular breakdown
Fungal amyloglucosidase	Amyloglucosidase 150L	1.0–1.5 litres		Converter, saccharification
Fungal β-amylase	Fungamyl® 800L	0.75–2.5 kg		Converter, saccharification
Bacterial neutral proteinase	Neutrase® 0.5L	0.3–1.5 kg		Malt mash, nitrogen control

reports of enzyme-assisted barley brewing appeared in the early 1970s. The European Brewery Convention Congress at Estoril in 1971 provided a forum for much of the discussion and exchange of data that led, by 1973, to the accumulation of considerable practical evidence for the economic advantages of using raw barley as a major source of extract. Developmental studies and large-scale trials were reported frequently. By the time that the European harvest was to be so severely distressed by the drought of 1976, a considerable body of data existed to allow a confident brewer to embark on this route to economic brewing. In that and the subsequent year raw barley was to provide some 20 per cent of the extract in UK beers. In some subsequent years the overall proportion declined slowly to a steady level of approximately 15 per cent of brewer's extract, with a parallel gradual introduction of the technology into other European countries and recently into North America.

The most generally accepted method uses an infusion with an upward temperature gradient in agitated mash vessels. Interestingly, some of the main manufacturers of brewery equipment have designed plant especially suited to the conditions required for barley brewing and have installed them in many major breweries.

Barley forming up to 80 per cent of the grist can be used, but the most common levels are from 30 to 60 per cent, and some consideration to the milling is needed, as barley is much harder than malt. Various schemes have been adopted to make economic milling practicable, and these include the adjustment of malt mills to a more compromising tolerance and the insertion of harder steel rolls. The most successful method seems to be to arrange for a premilling steeping of the combined malt and barley grist followed by wet-milling.

Mashing conversion is required to produce a final wort that gives a finished beer with the character and quality of that from grists with a much larger malt content. Whilst the starch conversion chemistry of microbial enzymes is readily appreciated by the brewer familiar with adjunct processing, the adequate nutrition of the yeast and the elimination of filtration limiting and haze-forming potentials from the process are not so well known. The use of a range of enzymes in the barley mashing system can stabilize these various factors and provide substantial economic benefits for this form of extract. The processing of raw barley, with a more varied range of industrial enzymes than generally required for the other adjuncts, enables the brewer to produce a wort very closely related to that from an all-malt grist.

Enzymes for mashing barley. The addition of the microbial

equivalents for the malt proteinases, amylases and glucanases has proven a practical scheme since, in particular, barley has high levels of β-amylase and a protein content that is a good substrate for yielding soluble nitrogen precisely like that of malt. Barley starch also gelatinizes at mashing temperatures and therefore does not need to be cooked separately from the malt. A typical barley–malt mashing programme is illustrated in Figure 4.5.15. Proteolytic action is developed in the first part of the programme at 50°C. The enzymes that are entirely of the neutral and uninhibited type are essential here if the maximum benefit is to be obtained, since many of the mixed proteases have undeclared amounts of alkaline optimum activity that is inhibited by barley. An example of the preferred type is *Neutrase®* 0.5L from Novo, which is used at levels of 1.2–1.5 kilogrammes per tonne barley. At this level there will be sufficient for both the release of nitrogenous soluble compounds and the activation of the β-amylase. It has been established that more than half of the β-amylase in barley is latent and in an inactive form which can be activated by the action of this type of protease; the effect is an almost trebling in the level of this carbohydrase in the mash.

Carbohydrase action reaches its maximum when the temperature of the mash is raised first to 65°C and then to 68°C, during which the starch becomes gelatinized and is converted by the combined action of the β-amylase from the barley, the added α-amylases and the contribution from the malt component of the mash. It has been rapidly established that, in addition to β-amylase, an amount of β-glucanase is required for the efficient control of the glucan gums that are extracted from the barley in the mash process. Many industrial fermentations for α-amylase production yield significant levels of a suitable β-glucanase for this purpose, and appropriate strains of *Bacillus subtilis* are cultivated for the preparation of a

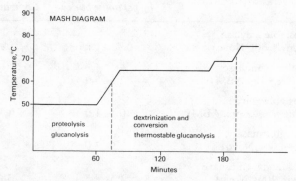

Figure 4.5.15 Mashing programme for barley brewing.

mixture of these two enzymes in appropriate proportions. Such enzyme preparations are added at the start of mashing, together with the proteinase, and it is considered that glucanase activity begins right away. Different producer strains of bacteria yield enzymes with small but useful differences of thermal stability and the more thermostable examples allow the maximum attack on glucan gums throughout the rising temperature of the programme. An example of this combined enzyme would be *Cereflo*® 200L from Novo, which would typically be used at a rate of 2.0 kilogrammes per tonne barley.

When α-amylase and β-glucanase are taken from separate sources, it is possible to adjust their proportions more precisely to the actual grist being used, and also to take advantage of both the alternative thermostable amylases, previously described in the section on adjunct cooking (*see* page 234), and the various β-glucanases. Selection of thermostable glucanases such as those from *Penicillium emersonii* in combination with thermostable amylases offers the opportunity of heating the mash more rapidly and so shortening the mashing cycle. There are additional benefits from the selection of more than one β-glucanase for the mash, as recent practice has demonstrated that there are small differences in the specificity of the various enzymes which combine to give a far higher rate of degradation of the most viscous materials. Although it is essential that the correct enzyme doses be established for the grist in question, some typical doses are indicated in Table 4.5.9. The dextrinization of the starch is then completed at the higher temperatures of the mash programme and the mash is finally raised to 78°C before lautering.

TABLE 4.5.9

Suggested combinations of amylase and glucanase for barley brewing

	kg per tonne barley	Combination
Thermostable amylase (*Termamyl*® 60L Novo)	0.75–1.25	A
Conventional amylase (*Nervanase*® 180 ABM)	2.5–5.0	B
Thermostable glucanase (*Penicillium emersonii* ABM)	0.5–1.0	A/C
Standard glucanase (*Cereflo*® 200L Novo)	2.0–3.0	C
Standard fungal glucanase (*Finizym*® 200L Novo)	1.5–2.5	B/C

All-in-one enzyme systems. All-in-one enzyme systems are available in liquid form and allow a simple single addition of enzyme at the beginning of the mash. Since the internal ratios of the three declared enzyme activities – amylase, glucanase and proteinase – are fixed, one should not assume that they are necessarily optimally suited for all grists. In general, they contain a substantial excess of carbohydrases when used at the rates suggested and are therefore dosed on the basis of a correct protease level. This is not unreasonable, since there is little evidence of problems arising from the overuse of carbohydrases, while protein balance is far more critical for good beer production. It is also of importance from a practical monitoring aspect, since analysis of the nitrogen components of the mash (or the wort) is comparatively lengthy, whilst a simple iodine test will establish the satisfactory degradation of starch and an elementary filtration test for the removal of hindering gums takes only a few minutes. An example of the products available is *Ceremix®* from Novo, which is recommended to be dosed at 2.5–3.0 kilogrammes per tonne barley.

Extended saccharification. Only quite recently has extended saccharification in the mash tun vessels become a major target. This was previously carried out in the fermenter if an alteration in the amount of fermentable sugars became necessary. Growing concern regarding the labelling of finished beers has begun to alter the practice, however, despite the very severe economic penalties in the use rates of the selected enzymes. The combination of short contact times and adverse pH coupled with inactivation at the copper boiling mean that dramatically higher enzyme rates are required than would be the case at the fermenter (*see* page 249). However, where it is decided to increase fermentability at this stage it becomes a matter of choosing the enzymes according to the spectrum of fermentable sugars to be produced. Conventional amyloglucosidases and fungal amylases with maltogenic characteristics can be used by virtue of the same considerations already discussed for the saccharification of adjuncts (*see* pages 239 and 241). With a maximum of an hour in which to function, the dose rates must be much higher than those shown in Table 4.5.7 and can be as much as three times greater. As will be seen in the discussion on fermentation, these levels are anything up to 100 times greater than needed for the same conversion under more optimal conditions. Table 4.5.10 summarizes some typical barley brewing enzymes and their use rates.

Selection of enzyme rates to increase barley content. This can be a considerable task, and a foolproof rule has yet to emerge to aid the brewing chemist in making an accurate prediction. However, by

TABLE 4.5.10
Enzymes for barley brewing

Enzyme type	Example product	Dose rate (kg per tonne barley)
Bacterial neutral protease	Neutrase® 0.5L	1.2–1.5
Bacterial α-amylase	BAN 120L	2.0–3.0
Bacterial β-glucanase (with α activity)	Cereflo® 200L	2.0–3.0
Fungal β-glucanase	Finizym® 200L	1.5–2.5
Thermostable α-amylase	Termamyl® 60L	0.75–1.25
Thermostable fungal β-glucanase	β-Glucanase ABM	1.5–1.0
Combination products amylase, glucanase, protease	Ceremix®	2.5–3.0
Saccarification enhancing amyloglucosidase	Amyloglucosidase 150L	2.0–5.0 litres
Fungal amylase (maltogenic)	Fungamyl®	1.0–3.0

pooling the considerable practical experience gained in the past seven years, it is possible to draw some informative but tentative conclusions.

Laboratory extract measurements on the actual grist in the ratios at which it is to be used, yield data on the unaided performance of the intrinsic cereal enzymes present. If the results are plotted against typical doses of microbial enzymes that have been found suitable to achieve worts of high and reliable quality, a series of wedge-shaped plots of dose level emerge. Furthermore, the point at which the various enzymes become relevant can be observed. These data have been set out in Figure 4.5.16, which expresses the traditional extract in both Brewer's pounds per quarter and the Congress extract percentages. The approximate values for typical levels of barley at 25, 50 and 80 per cent of grist are indicated for reference.

It has been assumed that the most economical stage at which to add saccharification enzymes to improve fermentability will be at the fermenter, although by raising the base line dose as discussed earlier, these can be added to the mash. To establish the starting dose values for the left hand side of each 'wedge', it is practical to adopt the enzyme dose recommended (or experimentally determined) to give the maximal improvement of an all-malt mash using poor quality malt; this would be between 93 and 96 Brewer's pounds per quarter, or 75–77 per cent Congress extract.

The increasing use rates of each enzyme type then follow according to the figure, but on different gradients, so that amylases and glucanases increase eightfold, proteinase by one and a half times,

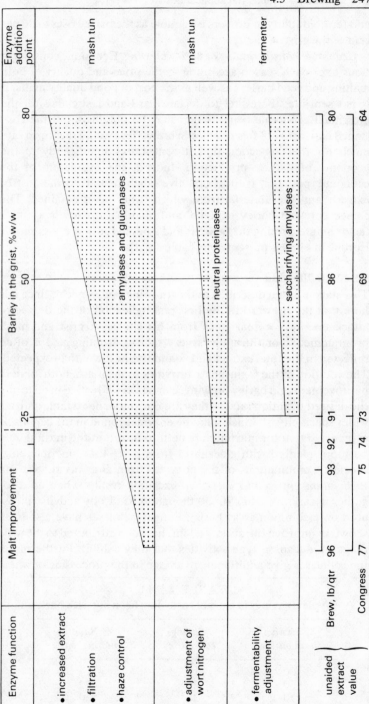

Figure 4.5.16 Selection of enzyme dose for increasing barley content.

and fermentability improvers by double as the barley rises to 80 per cent of the grist.

   Economic considerations of barley brewing. Economic considerations alter each season according to the price and quality of both malting and feed barleys as well as the cost of good quality malts. It is not sensible, therefore, to declare hard and fast claims for the individual economic benefits that may be obtained, but the continued expansion of this practice in breweries around the world says much for the conviction that it contributes significantly to the economy of extract production. Indeed, the evaluation of the economic potential is most effective when related to the cost of extract, and the material from which it has been obtained. The choice between barley of 'feed' and 'malting' quality is a major factor in interpreting the benefits and a pooled survey of figures for Europe in 1980 is presented in Table 4.5.11.

10. Mash filtration

This stage is often described as the 'brewing bottle-neck', since it is here that the brewer has the first real indication of the degree of attack upon the cereal gums from his choice of grist and mash programme. Poor filtration is very time-consuming and is often reflected in the total extract and volume being lower than expected. The addition of the β-glucanase enzymes to the mash, both for malt improvement and barley brewing, can be very effective in restoring the desired filtration rate, as they are both more heat stable and of a wider specificity of attack than the enzymes found in the cereals. It is reported that for maximum benefit to filtration the mash should be supplemented with glucanases from more than one microbial species. Combinations of the enzymes from *Bacillus subtilis* and *Penicillium emersonii* have given excellent results when used together in ratios of from 70:30 though to 85:15 by activity (measured on a common assay basis). Fungal cellulases have also been shown to improve filtration, and this has been attributed to the wide range of glucanase-type activities that they exhibit. Furthermore, the cellulases give additional advantage in the processing of wheat

TABLE 4.5.11
Typical savings in extract cost when brewing with barley

% Extract from barley	% Savings 'feed' barley	% Savings 'malting' barley
30	18	14
50	27	22
70	36	30

in the mash because they often also contain hemicellulase (pentosanase) activity.

Examples of the effects of the presence of β-glucanase in the mash upon the filtration of worts are illustrated in Figure 4.5.17. From an economic standpoint a rate of enzyme is generally chosen that gives a 35–50 per cent improvement in filtration rate, with a viscosity reduction of between 15 and 20 per cent over the unaided mash, as determined in the brewery laboratory. As an example, this would indicate a rate of around 2.0 kilogrammes Cereflo® 200L (Novo) per tonne barley in the grist, and about 0.2 kilogrammes per tonne of grist for malt improvement. Table 4.5.12 compares the characteristics of wort produced by the application of the appropriate enzymes to a 50/50 barley malt mash in relation to that from a good malt and the unaided grist.

11. Enzymic treatments at the fermenter

Having produced a fermentation medium from a selection of raw materials processed by an appropriate method, some opportunities for adjustment still remain by treatment at the fermenter stage.

In the context of beer fermentation, several ways of expressing the proportion of the soluble wort solids actually converted to alcohol have been developed. For the purposes of this section the term 'real attenuation' will be used; this expresses the percentage soluble material converted to alcohol. The values vary depending

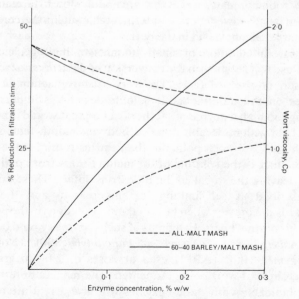

Figure 4.5.17 Effect of β-glucanase on mash filtration (Cereflo® 200L).

TABLE 4.5.12
Typical characteristics of barley brewing worts and beers

Raw materials	Viscosity at 10 °B Cp at 25 °C	Extract °B	Reducing Sugar (g glucose per 100 ml)	Soluble N (mg per 100 ml)
100% Quality malt	1.45	11.8	4.74	114
50/50 Barley/malt	1.71	11.3	4.12	67
50/50 Barley/malt + enzymes	1.38	11.7	4.63	108

on the composition of the grist, the manner of mashing and the yeast selected for the fermentation, together with the actual conditions of fermentation. For the production of most standard beers, real attenuation values lie in the range 58–68 per cent.

Adjustment of fermentation in standard beers. Only in exceptional circumstances is it necessary to make enzyme additions to the fermentations for these products, but there are situations in which it very significantly assists in recovery procedures. If heating at the mashing stage is inadequately controlled and too rapid or too high heating occurs, it is likely that a large part of the malt enzymes and any added enzymes will be destroyed before they have completed the conversion. Another situation is in the use of high syrup, or adjunct levels inadequately processed with saccharifying enzymes from microbial or malt sources that are unduly slow to ferment, and may even have a lower content of fermentable sugars than required for the target alcohol level in the beer.

The careful addition of small amounts of maltogenic fungal 'β'-amylase will reestablish the conversion of dextrins and facilitate correction of these defects. The very selective action of these enzymes on the unfermentable soluble dextrins is important in avoiding too high a degree of saccharification that would otherwise leave a beer with noticeably reduced 'body and mouth-feel'. Attack upon the α-1,6 branch points in the dextrins is minimal, and the main product is the splitting off of maltose units from peripheral chains, leaving the so-called 'limit dextrins' intact. These enzymes are described as self-limiting for this reason, and their use safeguards the brewer when he is making corrections in the production of his normal beers. Examples of such enzymes would be those from *Aspergillus oryzae* such as *Fungamyl*® 800L (Novo) or *Amylozyme*® 100L (ABM) used at doses of 2.0–20 grammes per hectolitre of wort being fermented. The maximum fermentability obtainable with these enzymes will give a real attenuation in the range 80–85 per cent as illustrated in Figure 4.5.18.

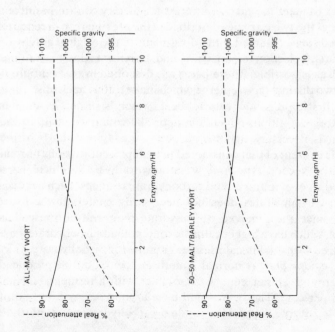

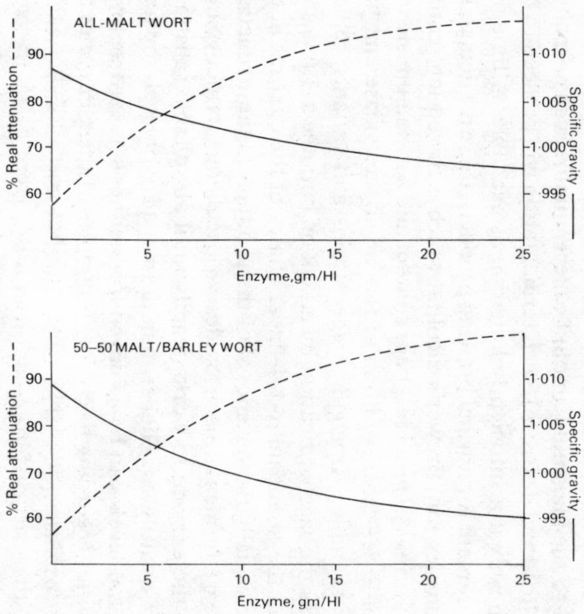

Figure 4.5.19 Increased fermentability by the addition of amyloglucosidase (*AMG* 150L; Novo).

Figure 4.5.18 Increased fermentability by the addition of fungal 'β'-amylase (*Fungamyl*® 800L; Novo).

Maximum fermentability of worts. If the wort carbohydrates are degraded to the maximum amount of fermentable sugars (maltotriose, maltose and dextrose), such fermentation will proceed to the maximum only if a sufficiently long time is available. Although a rare requirement for normal brewing products, there is a demand in some countries for 'Brewer's alcohol' which is mixed into carbonated soft drinks to provide a beer content and a maximum alcohol content of 2 per cent. Addition of the amyloglucosidase enzymes from, for example, *Aspergillus niger* strains and variants, rapidly accelerates the saccharification by attack on both the α-1,6 branch points and the abundant α-1,4-linear chains of the dextrins in the wort. The equally rapid removal of the products of these reactions by the yeast fermentation provides an excellent thermodynamic drive for the enzyme to work steadily until all suitable substrates have been hydrolysed. Higher than normal alcohol levels, or standard alcohol levels can be achieved from either standard or lower wort gravities according to choice. Examples of these enzymes are *Amyloglucosidase* 150L (Novo) or *Ambazyme*® LE90 (ABM) added at the start of fermentation in doses from 5–25 grammes per hectolitre depending upon enzyme selection and the fermentation target. Some characteristics of this type of fermentation with all-malt and 50/50 malt/barley worts are given in Figure 4.5.19.

Production of 'diet' or 'lite' beers. The comparatively recent development of beer products with quite dramatically altered composition to meet the market interest for dietary control has taken the use of the two enzymes described in the previous two sections to a refined stage. Coupled with adjustments to the original gravity of the worts, selection of enzymes and enzyme ratios to suit the objectives is possible. These beers are described in many different ways, two distinct types emerge which make different claims.

The first type is 'low calorie' beer, which is subject to various regulatory definitions depending upon the country of sale. In the USA it is necessary that such a beer has only two-thirds of the calorific content of that considered to be representative of a normal beer. The second type is 'low carbohydrate' beer, which has a reduced level of unfermented carbohydrate residues. Such a reduction in carbohydrates does not necessarily confer 'low calorie' status, since the removed carbohydrate may well be present as alcohol, which has a higher calorific content than carbohydrate – 6.9 compared with 4.0 kilocalories per gramme for carbohydrate.

Low calorie beer. A normal American beer would be produced from around 11 per cent wort solubles with a fermentability of 64–68 per cent. The beer would have around 3.5 to 3.8 per cent alcohol, 2.9 to 3.1 per cent residual carbohydrates and 39 to 42 kilocalories per 100 grammes. For this wort to become a 'low calorie' beer, the energy content would have to fall to some 26 kilocalories per 100 grammes, which could not be achieved by simply raising the fermentability with amyloglucosidase enzymes, because at 85 per cent fermentability the beer would have 4.6 per cent alcohol, less than 1 per cent residual carbohydrates, but still around 38 kilocalories per 100 grammes. This problem has been solved by reducing the gravity of the wort to 7.5 per cent solids and fermenting to 85 per cent real attenuation. The target 26 kilocalories per 100 grammes can be reached at an alcohol value of 3.2 per cent. It will be appreciated that the use of high levels of unmalted adjunct in the mashing, or large amounts of syrups in the copper for these beers would seriously dilute the flavour and texture characteristics when fermenting from such low solids levels. These beers are usually produced from all-malt mashing, or from barley/ malt mashes with barley substitution no greater than 50 per cent.

To obtain the desired fermentability, it has been found beneficial to use both amyloglucosidases and fungal 'β'-amylases together in proportions that will depend on several factors: (*i*) the intrinsic fermentability of the wort; (*ii*) the target fermentability; (*iii*) the desired retention of texture and 'body' in the beer; (*iv*) the desired alcoholic content; (*v*) consideration of the carry-through of enzymes in the beers. Some examples of the influence of different proportions of these two enzyme types are found in Figure 4.5.20, and the alcohol, carbohydrate and calorific values from the foregoing discussion are set out in Table 4.5.13.

TABLE 4.5.13
Examples of product characteristics for 'lite' beers

	% Wort solids	% Real attenuation	% Beer alcohol (w/w)	% Residual carbohydrate	Kilocalories per 100 g
Normal beer	11	65	3.6	3.0	40
Low calorie	7.5	65	2.4	1.95	26
	7.4	85	3.2	0.55	26
Low carbohydrate	7.0	85	3.0	0.5	25
	8.0	87	4.0	0.5	28
	8.0	74	3.0	1.5	29
	9.1	76	3.5	1.5	32

The thermal stability of the amyloglucosidases becomes relevant when carry-through to packaged beer is considered, for it is considered likely that at least 20 per cent of added enzyme will survive beer pasteurization. Conversely, the fungal 'β'-amylases are thermolabile and do not survive. With this in mind, there is a developing interest in the use of higher proportions of the latter enzymes, despite the fact that they need longer fermentation times. The resulting reduction in the level of amyloglucosidase is reflected in lower levels in the finished 'lite' beer product.

Low carbohydrate beer. As there are no clear cut normal values for the actual degree of reduction in carbohydrate values for this type of 'lite' beer, the demands on the production technique are lower. More significant constraints come in relation to the flavour, texture and alcoholic content, so that the adjustments are made by altering fermentability on different levels of wort solids. The mechanisms are much the same as for 'low calorie' beers. General experience shows that only when the residual carbohydrate is below one per cent does the 'low carbohydrate' beer also conform to 'low calorie' status. Slightly higher than normal alcohol levels are often found in this type of beer, which indicates that the wort gravities are maintained near to normal, probably to conserve some of the organoleptic qualities of a standard beer.

All the enzyme considerations mentioned in the production of 'low calorie' beers apply equally here, and in addition, there is some concern in many European breweries for the retention of good beer

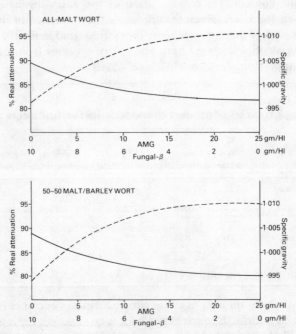

Figure 4.5.20   Increased fermentability by the addition of amyloglucosidase (*AMG* 150L; Novo) and fungal 'β'-amylase (*Fungamyl*® 800L; Novo).

foam (head) character. A number of amyloglucosidases particularly low in proteolytic side activities have been offered to meet the criticisms levelled against the standard enzymes. It remains to be seen if they will be entirely satisfactory, as detailed trials at some breweries are revealing several minor differences in the specificity of these proteases. These small differences can become large in regard to selective effects on head retention.

A third enzyme type for the conversion of wort carbohydrates has recently appeared; this has a potent debranching action on residual dextrins. Usually called 'pullulanase', but sometimes referred to as 'isoamylase', these enzymes may be of slightly differing action and become separately considered in the future. A combined tolerance towards fermentation pH levels, and a thermolability that prevents any carry-through risk, could make them very attractive. The cost may well delay rapid acceptance, although the heightened action in combination with the fungal 'β'-amylases keeps the use rates down to the range of 0.5–1.0 grammes per hectolitre of each.

Sweetening of finished beers (primings replacement). The addition of small amounts of sugar to sweeten certain dark beers is practised in many regions. The degradation of some of the residual unfermented dextrins in the beer after pasteurization can be used to generate the sweetness. Amyloglucosidase is generally used for this, but legislative considerations may mean that it will not be used so much in future. Addition before pasteurization requires that an allowance for the 70–80 per cent loss in the treatment is made in calculating the use rate of around 5 grammes per hectolitre, whereas only 1 to 2 grammes per hectolitre would be needed if the enzyme was added post-pasteurization.

Treatment of glucan hazes at the fermenter. The addition of fungal β-glucanases at this stage represents the last opportunity for the degradation of glucan gums that may give problems either at the beer filter or in the packaged product. As discussed on page 244, the choice of enzymes from *Aspergillus niger* or *Penicilium emersonii,* or a combination of both, provides a variety of substrate specificities that increases the potential for complete elimination of these haze-inducing polymers. The presence of glucans can be established on evidence from the filters or accelerated shelf tests on packaged beers and treated in following worts during fermentation. Typically, very low doses are required, 0.5 to 1.5 grammes per hectolitre being adequate. In some breweries it is the practice to delay the treatment until the wort reaches the maturation tanks, so that no further glucans can be contributed from dying yeast cells and the enzymes can provide a final solution to the problem.

12. Enzymatic chillproofing

This method of limiting the risk of haze in stored beers has been used for many years. It is becoming apparent that these hazes are complexes of several other substances including polyphenols and possibly carbohydrate fragments in addition to proteins, but also that the successful treatment of one component severely reduces the potential for the hazes to form. Papain, a protease extracted from the plant *Carica papaya*, is the preferred enzyme and is usually supplied to the brewing industry in a relatively pure liquid form. The mechanism of action of this enzyme in this situation is still not clear, but it is thought that it may initiate a coprecipitation of chill-haze proteins by a priming hydrolysis, in a manner similar to that of cheese rennets (*see* Chapter 4.6, 'Dairy').

Opinion varies as to the best stage for the addition of papain at the end of the beer production line, but it is agreed that it should be no earlier than the cold maturation tanks. This would provide the longest action time before pasteurization reduced the enzyme activity to very low levels. It is claimed for some preparations that the enzyme has been stabilized (it requires cysteine and a reducing environment for maximum stability) so that it will survive pasteurization. Such preparations are therefore recommended for addition immediately prior to pasteurization. Use rates vary widely with the activity and degree of stabilization of each preparation, but generally lie in the range 1.0 to 5.0 grammes of liquid products per hectolitre of beer treated.

Immobilized enzymic chillproofing. Recent interest in the application of enzymes immobilized into fixed packed columns of reactors has reached the brewing industry, and early trials with microbial enzymes prepared in this way have been encouraging. The mechanistic understanding for a satisfactory chillproofing action in only a short contact time as the beer flows through the reactors, and the absence of the enzyme protein to form a coprecipitate leave many questions to be answered. Perhaps the immobilized systems are operating on a totally different mechanism of haze formation. The single most attractive point of the use of immobilized enzymes is the total absence of the enzyme in the packaged product, but a further advantage is in the possible reduction in costs by having a system available for multiple re-use.

13. Related industries producing syrups, flavours and vinegar

Large areas of the process schemes and methods described in this chapter have been adopted, or independently arrived at, for the production of cereal extracts for other industrial applications. Syrups are prepared from both malt and barley grists to provide the flavour in the baking and confectionary industries as well as for the breakfast cereal market. Critical selection of the saccharification enzymes for the specific sugar spectrum and reaction with protein products forms the foundation of the use of enzymes to control flavouring (*see* Chapter 4.9, 'Flavouring and colouring'). Highly active enzyme syrups are also prepared as additional support to the malt complement, and emphasis is usually on the release of β-amylase by proteolytic addition as described on page 243. The preparation of barley syrups is largely for use by breweries with limited mashing capacity and uses the methods described on page 243, although generally higher enzyme use rates are applied to accelerate the processing. The use of thermostable enzymes for these syrups is not always recommended and must be related to knowledge of the market for the syrup. Where a diverse range of customer outlets is envisaged, it can be assumed that at least some of them will be for baked goods and other starch-containing products. If thermostable enzymes are to be used, it will be necessary to test the product from the syrup evaporators to confirm enzyme inactivation, so that starch degradation will not be a risk in the customer's product.

Non-alcohlic beer-like beverages are also prepared by enzymic conversion of a mash similar to that used for brewing standard beers, but with extra emphasis on the extraction of flavour and body-giving materials. The use of the maltogenic amylases is usually increased, together with additional proteolytic enzymes. High percentages of malt together with speciality malt-derived syrups form the basis of many of these products which are then either simply diluted and stabilized at low solids levels to be compatible with market acceptance or fermented as for the traditional product, the alcohol subsequently being stripped off under vacuum.

Where the regulations permit the wider use of cereals for the production of fermentation vinegars, most of the practices described for the cooking and saccharifying of adjuncts (*see* page 232) and barley brewing (*see* page 242) are utilized quite fully. Whilst some of the flavour character of the grain is required in all the brewed but not all distilled vinegars, the objective is largely to obtain a very high efficiency of alcohol production in the beer which is to form the feedstock for the acetification fermentation. Thus, for the vinegar brewer the use of saccharifying enzymes and their contribution both at mashing and during fermentation form an important part of his overall processing economy considerations. It should be noted that another factor is the use of the glucanases as regulators of haze development, a very important consideration for the vinegar producer, who markets a very clear product.

Where limitations on the grist put malt use at a premium, the use of enzymes for malt improvement is usually adopted for the best extraction at the mash tun, but enzymes are added less frequently to the saccharifying and fermentation systems.

14. Future developments in enzymes for brewing

Mashing and filtration enzymes. The development of a wider range of thermostable enzymes, in particular proteases and gum hydrolysing glucanases, can be expected. This will bring further reductions in the mashing times by reducing the need for a hold period at the initial lower temperature. This would also increase the possibility for the development of a downward temperature gradient programme, which is generally accepted as being a likely source of significant fuel economy at the brewery.

Further developments among the cellulases and pentosanases will increase the levels of wheat starch that can be processed successfully in single mashing systems. Fermentability adjustment at the mashing stage suggests a growing use for the saccharifying enzymes, and these should have higher thermostability than at present and must include the newer 'debranching' activities.

In the case of mashing capacity limitations, or the expansion of capacity without increased capital investment in mashing equipment, a rising grist to liquor ratio in the direction of the much-discussed 'high gravity brewing' can be expected. To simplify dosing and to provide economically acceptable extracts under these conditions, it is anticipated that higher potency enzyme preparations will be produced, and of higher priority still could be the development of more effective multi-enzyme products to enable one-shot additions to be made. Increasing automation in modern brewhouses should be part of the planning of enzyme treatments, too, and single dose systems will fit most readily into automated programming.

Adjunct processing enzymes. An increased use of saccharifying enzymes to prepare highly fermentable worts to add at the copper, rather than at the mashing stage, can be expected to reduce extract costs. By this means, the equivalent of starch syrups can be provided at the brewery in areas where they are not readily available from starch processing industry.

New processing systems for reduced energy at adjunct cooking have high priority. It has been calculated that, by using the continuous cooking systems of the starch industry together with a downward temperature programme for conversion and saccharification, a further reduction in fuel of 20–25 per cent can be established (*see* Chapter 4.15, 'Starch').

Fermentation enzymes. In the efforts to limit the carry-through of enzymes to finished products, it is anticipated that saccharifying and debranching enzymes will be prepared in thermolabile forms, but otherwise retain the same activities as used for precopper saccharification.

Probably the most glamorous developments will come from brewing research in the genetic engineering of yeasts. One target is the production of brewing yeasts with their own complement of saccharifying enzyme activity, and early results show that the concept is entirely feasible. The distilling industry, in particular the fuel alcohol side, has an equally large interest for similar yeasts, although opinion differs about the temperature of fermentation considered optimal, and through this route we can see more chance of covering research costs. Between them, the two industries will almost certainly fund the development of such yeasts and then incorporate them into regular production applications once they are regulatorily approved.

Chillproofing enzymes. As more detailed knowledge of the nature of the chill-haze compounds and the mechanism of the haze dissolution from the beer is gained, it is likely that more specific enzymes can be selected to attack one or more of the components to prevent haze formation. The survival of papain into the beer is considered part of its chillproofing function and future work will be directed to establishing if this is the case. If not, then the immobilization of papain together with its stabilizing chemistry will be a commercial target. Similarly, the immobilization of other proteases, and possibly tanninase and polyphenolases from the research into cellulose utilization (which has a high priority for removal of lignin), will be used to develop alternative chillproofing systems.

Further reading

Proc. Congr. European Brewing Convention: 13th Congress, Estoril (1971); 14th Congress, Salzburg (1973); 15th Congress, Nice (1975); 16th Congress, Amsterdam (1977); 17th Congress, Berlin (1979); 18th Congress, Copenhagen (1981) (Elsevier, Amsterdam).

Proc. a. Meet. Am. Soc. Brewing Chemists.

Hough, J. S., Briggs, D. E. & Stevens, R. *Malting and Brewing Science* (Chapman & Hall, London, 1971).

Harborne, J. B. & van Sumere, C. F. *The Chemistry and Biochemistry of Plant Proteins* (Academic, New York, 1975).

Rose, A. H. *Economic Microbiology* Vol. 1 (Academic, New York, 1977).

DAIRY
K. Burgess and M. Shaw

1. Introduction

Milk is a multicomponent secretion of the mammary gland containing important sources of nutrition: protein, fat and lactose, together
with small amounts of minerals and vitamins. Milk is also a source of
a large number of enzymes (approximately 40), albeit in very small
quantities, some of which have important implications with regard
to milk processing. For example, naturally occurring proteases are
important in determining the flavour characteristics of maturing
cheese, while the presence of naturally occurring lipase may result
in the development of rancid flavours in products containing milk
fat. A review of the major properties and effects of naturally
occurring milk enzymes has been given by Brunner (1977).

The application of enzymes in milk processing is well-established
through the use of rennet which probably represents the earliest use
of enzyme technology in the food industry. More recently β-
galactosidase has found commercial use in the hydrolysis of lactose
in milk products. These two applications represent the major use of
enzyme technology in the dairy industry although a number of
smaller applications exist for the production of highly specialized
milk products.

2. Milk clotting enzymes

Historically the use of milk clotting enzymes for the transformation
of milk into products such as cheese can be traced back to around
5000 BC. Nowadays, the most prevalent rennet in use is that derived
from the abomasum or fourth stomach or vell of the suckling calf.
The active, milk coagulating enzyme present in the calf rennet
extract is an acid protease, designated as rennin or chymosin
(EC 3.4.23.4).

Acidification of milk results in precipitation of casein which
encloses the other major milk components. The reaction is complete at the casein isoelectric point of pH 4.6. Curd produced as a
result of acidification tends to be of a granular and inelastic nature.

Enzymic coagulation, on the other hand, is carried out at a higher
pH in the range 5.8–6.5. The resultant curd is altogether smoother,
more elastic and capable of shrinkage allowing whey drainage.

Enzymic milk coagulation. Milk coagulation catalysed by proteases takes place in two distinct, but normally overlapping, stages.

The first stage is enzymic and involves the cleavage of the k-casein fraction resulting in destabilisation of the casein micelles

$$\text{k-casein} \xrightarrow{\text{enzyme}} \text{para-k-casein} + \text{macropeptide}$$

Chymosin cleaves the k-casein at phenylalanine–methionine bonds. Other enzymes catalyse a similar reaction but are generally not as specific.

Concurrently, the second stage proceeds in the presence of available calcium and aggregation of the casein micelles occurs

$$\text{para-k-casein} \xrightarrow[\text{pH } 6.0-6.4]{Ca^{++}} \text{dicalcium para-k-casein}$$

There is a third stage where continuing proteolysis of the casein occurs during cheese maturation resulting in flavour development. Some 6 per cent of the chymosin activity is retained in cheese curd.

Factors affecting milk coagulation. The following factors have been shown to influence enzymic activity and hence curd tension or firmness.

(a) Curd firmness increases as rennet addition is increased from 5 to 30 millilitres per 100 litres of milk. Further addition will have no affect.

(b) Curd firmness increases with temperature up to 40°C and then declines.

(c) Decreased pH (i.e. more acid) increases curd firmness down to pH 5.8 when the firmness begins to decrease. Most milks are renneted between pH 6.3–6.5.

(d) Decreased curd firmness and longer coagulation times result if milk is subjected to extended cold storage before renneting. Curd firmness increases with the addition of calcium chloride.

(e) The ratio of fat to solids not fat in milk can influence curd firmness. High fat milk results in soft curds.

(f) Reaction of k-casein with protein or free fatty acids as a result of proteolysis and lipolysis will affect the completeness of the coagulation process.

(g) Non-specific proteolysis of the k-casein around the phenylalanine (105) and methionine (106) bond may result in incomplete cleavage and hence soft curd formation.

(h) Incorporation of whey proteins into the curd by various methods may affect the curd firmness.

(i) Different coagulants may result in different curd strengths.

(j) Mastitic milk affects curd formation.

Curd firmness determination. It is particularly useful to measure curd firmness in order to assess the progress of milk coagulation, to compare strengths of different enzymes and to determine optimum firmness for cutting. A number of methods are available and these have been reviewed by the International Dairy Federation (1977).

Rennet activity determination. The coagulating power of coagulants can vary and since rennets of different strengths will produce different curds it is important that this strength be measured. Manufacturers of natural rennets and alternative coagulants use a variety of methods to determine the coagulating strength or power, for standardization of their products. The strength is often determined by a modified Berridge method where reconstituted skim milk powder, dissolved in 0.01 M calcium chloride is used as a substrate (Berridge, 1952). The strength is expressed as Rennet (or soxhlet) Units per millilitre, and can be defined as the number of millilitres of milk which can be clotted in 40 minutes at 35 °C by 1 millilitre of rennet.

There is a British Standards Institution method for measuring clotting strength (BS 3624, 1963) which uses a freeze-dried purified rennet as a standard. These methods give the coagulation strength of the rennet extracts but are not a measurement of the degree of proteolytic activity. This is becoming an increasingly important area following the introduction of alternative coagulants, some of which may show different proteolytic specificities when compared with standard rennets. The clotting strength to proeolytic activity ratio (CPR) is a useful index for coagulants. A number of methods have been developed to assess the activity spectrum of rennets and alternative coagulants.

Standard rennets. Presently there are no accepted standards of identity for rennet and alternative coagulants. The UK cheese Regulations SI 1970 No:94 as amended by SI 1974 No:1122 do not define rennet. These regulations are presently being revised and representations have been made for the incorporation of a revised definition of cheese which will allow the use of a range of 'permitted coagulating agents' to include rennet, rennet/pepsin mixtures, microbial rennet substitutes and various acids such as lactic, acetic, hydrochloric and orthophosphoric.

Calf rennet is still the most prevalent coagulant in the cheesemaking industry and is regarded as the standard rennet against which alternative coagulants are measured. The rennet extracted from the fourth stomach of the suckling calf contains 88–94 per cent chymosin and 6–12 per cent pepsin while extracts from the older bovine animal contain 90–94 per cent pepsin and only 6–10 per cent chymosin. Thus variations in the ratio of chymosin to pepsin can be expected depending on the age of the calf at slaughter.

Standard rennet is marketed to a standardized strength in the UK of 15,000 Rennet Units per millilitre. The ratio of chymosin to pepsin is not declared. Analysis of the relative proportions of these two enzyme components in commercially available standard rennets show that the chymosin content is presently in the range 75–80 per cent while mixtures of standard and bovine rennets contain 40–50 per cent chymosin and bovine rennet 25 per cent chymosin.

Standard calf rennet is supplied by manufacturers as a pure solution of rennet enzymes with added salt, propylene glycol and sodium benzoate as preservatives, together with permitted colour and flavour. In warm countries where liquid rennet will not keep powdered or paste rennets are available.

Rennet extract is introduced into the milk in a diluted form, usually 1:40 volumes, with cold tap water, to disperse it more uniformly. The milk must then remain quiescent until coagulated, for even slight vibrations result in hindrance of the development of a homogeneous, compact curd. It is usual practice to continue stirring milk for approximately five minutes after addition of the rennet and this gives sufficient time for good mixing.

The amount of rennet added to cheese milk will be governed by the type of cheese being manufactured. The specific recipe will also dictate the temperature and acidity of the milk necessary to produce a coagulum ready for cutting within a specified time.

Temperature is an important factor: between 20–25°C the resultant coagulum will tend to be soft, between 30–32°C the curds will be firm and cut well without disintegrating, while between 33–35°C the curds will be rather tough and rubbery with a tendency to retain whey. The acidity of milk at renneting is crucial: if too high, subsequent curd drainage will be reduced and the curd will retain excess moisture. If too low the curd will be hard and appear dry. A typical pH range for milk coagulation would be 6.5–6.7.

The quantity of rennet used depends on the cheese being manufactured and other factors such as seasonal variation in the chemical quality of milk and variations in processing conditions. The range of addition is in the order of 10–45 millilitres per 100 litres of milk when using a standard rennet (15,000 Rennet Units per millilitre). Typically, the addition rate for cheddar, for example, would be 0.022–0.025 per cent v/v (22–25 millilitres per 100 litres milk), while in the manufacture of some cheeses produced by the acid precipitation of curd, such as cottage cheese, a smaller quantity of rennet is added (3–5 millilitres per 100 litres milk), in order to strengthen the subsequently produced curd. The use of calcium chloride is often recommended up to a maximum addition rate of 200 parts per million anhydrous calcium chloride.

Rennet rapidly loses activity if stored diluted for long periods.

Light too causes deterioration and thus rennet must be stored in light-proof containers. During normal storage calf rennets will lose between 1.0–1.5 per cent of coagulating activity per month.

The actual time from rennet addition to optimal coagulum cutting varies depending on the recipe. The time range is generally between 30 minutes and 1 hour 30 minutes, with cheddar, for example, being typically 45 minutes. The actual point of cutting is related to curd firmness and measurement of the optimal value is presently a subject of considerable interest within the industry.

Alternative coagulants. Recently there has been a great deal of discussion concerning the availability of standard calf rennet for cheese manufacture. Over the years there has been a worldwide increase in milk production, a subsequent increase in cheese production, with a decline in the actual numbers of cows involved due to individual milk yield increases. This, together with other factors has resulted in a reduction in the number of calves for slaughter thus affecting the availability of calf rennet. If calf rennet were to be used as the sole milk coagulant there would be insufficient to supply the manufacturers in all countries.

The supply and demand picture for rennet has been brought into focus in the minds of cheesemakers by the rapidly increasing price, the reduced availability of true calf rennet on world markets and the increasing availability and acceptance of alternative milk coagulating enzymes. Table 4.6.1 illustrates some of the alternatives which have been investigated as possible coagulants for cheesemaking.

TABLE 4.6.1

Sources of alternative coagulants

Group	Source of enzyme	Comments
Animal	Calf	High chymosin content
	Ox	Low chymosin, high pepsin
	Pig	Pepsin
	Chicken	Pepsin
Plant	Pawpaw	Papain, too proteolytic
	Pineapple	Bromelain, too proteolytic
Bacteria	*Bacillus polymyxa*	Generally too proteolytic
	Bacillus mesentericus	for cheesemaking
	Bacillus subtilis	
Fungi	*Mucor miehei*	Microbial rennets used
	Mucor pusillus	in cheesemaking
	Endothia parasitica	

Milk clotting enzymes are derived from a diverse range of organisms and considerable work has been carried out on the assessment of these materials for suitability in cheesemaking. The plant derived enzymes, papain and bromelain, for example, have been shown to be strongly proteolytic resulting in the development of bitter off-flavours during cheese ripening.

Work with porcine and bovine pepsins and mixtures of these with standard calf rennet has produced encouraging results in cheesemaking trials. Pepsins coagulate milk more slowly than standard rennet and they exhibit greater proteolytic activity at the lower pH of maturing cheese. Some texture defects have been reported but these have not been universally experienced and cheese produced is generally of high quality and essentially the same as that produced using standard rennet.

Bacterial derived coagulants tend towards being too proteolytic for cheesemaking. Most work on microbial rennets has been con-

TABLE 4.6.2
Specificity of enzyme attack upon insulin (oxidized B chain)*

Enzyme	No. bonds vigorously broken	Main bonds attacked
Mucor miehei coagulant	2	Glu–Ala; Leu–Val
Calf rennet (Chymosin)	2	Glu–Ala; Leu–Val
Trypsin	2	Lys–Ala
Ficin	4	Glu–Ala; Tyr–Leu; Phe–Tyr
Pepsin	5	Leu–Val; Phe–Tyr
Fungal (alkaline)	5	Leu–Tyr; Phe–Tyr
Bacterial (neutral)	6	His–Leu; Ser–His; Ala–Leu; Gly–Phe; Arg–Gly
Bacterial (alkaline)	7	Gln–His; Ser–His; Leu–Tyr
Fungal (acid)	9	His–Leu; Gly–Phe; Phe–Phe
Papain	9	Asn–Gln; Glu–Ala; Leu–Val; Phe–Tyr

*This table is also discussed in Chapter 5.

centrated in the area of fungal derived enzymes. Many fungi have been screened for milk clotting enzymes with a few being considered worthy of commercial production. Those currently available include *Mucor miehei* derived enzymes, for example *Rennilase®*, *Hannilase®*, and *Marzyme®*, and *Mucor pusillus lindt* derived enzymes for example, *Emporase®*. Considerable success has been met with the use of these microbial coagulants in the USA, Eire, Australia and Continental Europe for the manufacture of a number of different cheese varieties without major changes having to be made to cheesemaking procedures.

Table 4.6.2 illustrates the degree of specificity of bond cleavage for various enzymes and it can be clearly seen that the plant (papain) and bacterial derived enzymes are nonspecific in action while the *Mucor miehei* enzymes show the same specificity as calf rennet (chymosin).

It is important when evaluating alternative coagulants for cheese manufacture to ensure close monitoring of the make. The firmness of curd at cutting is all important and the use of a curd tension meter is invaluable. Careful monitoring of whey losses in respect of fat and casein fines is essential as is the measurement of cheese yield per unit volume of milk. The characteristics of cheese during maturation should also be carefully monitored to ensure no defects occur such as textural breakdown and development of off flavours.

A further problem which has been highlighted recently is that of residual coagulant activity in whey which can represent a problem

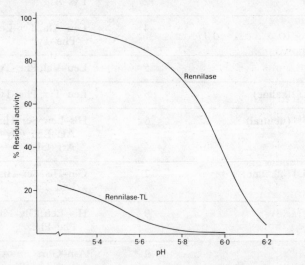

Figure 4.6.1   Influence of pH on the residual activity in whey after pasteurization at 72–74°C for 15 seconds (modified Berridge method; data from Novo Industri).

when this is used in formulating dairy products, baby foods and dietary aids. The presence of residual coagulants may cause proteolysis during whey processing, or coagulation on addition of whey to milk-containing foods. Several test methods are now available to monitor this residual whey activity. A number of thermolabile or second generation, microbial rennets such as *Marzyme 11®*, *Rennilase 50 TL®* and *Modilase®*, are now available, which show good susceptibility to heat treatment during whey processing. This is a pH dependent operation and Figure 4.6.1 illustrates the influence of pH on the residual clotting activity in whey after pasteurization at 72–74°C for 15 seconds for *Rennilase®*, a first generation microbial rennet, and *Rennilase TL®*, for example.

Table 4.6.3 lists the residual activities in whey at pH 6.0 of *Rennilase TL®* after 15 seconds pasteurization at different temperatures (data from Novo Industri).

TABLE 4.6.3
Inactivation in whey of Rennilase TL® at different temperatures

Temperature °C	Per cent residual activity after 15 seconds pasteurization at pH 6.0
68	18
70	7
72	2
74	<1

Rennilase TL® (supplied by Novo Industri) is standardized according to a modified Berridge method using reconstituted skim milk powder as a substrate. The strength is expressed as Kilo Rennet Units (KRU) per millilitre and a preparation containing 50 Kilo Rennet Units per millilitre is comparable in coagulating strength to a calf rennet of 50,000 Rennet Units or soxhlet units per millilitre. Thus when substituting *Rennilase TL®* for standard rennet of 1:15000 strength it is only necessary to use 30 per cent of the dose rate to achieve comparable coagulation. In terms of pH and temperature dependence *Rennilase TL®* resembles that of calf rennet. Mode of use for the microbial coagulants is similar in all respects to standard rennets. Minor changes in processing conditions may be necessary in order to produce optimum curd firmness within the required time for particular recipes. Microbial coagulants can be readily adopted for the manufacture of most varieties of cheese. Like all coagulants they should be stored at temperatures below 5°C. When stored at 4°C enzymes will maintain declared activity for at least three months.

TABLE 4.6.4

Coagulant prices (UK, October 1981)

Name	Source	Manufacturer	Cost/Litre
Standard rennet	Calf	Hansens	*£4.05*
Cabo	Calf/ox	Hansens	£3.35
Stabo	Ox	Hansens	£3.00
50:50	Ox/pig	Hansens	£2.90
Hannilase	Microbial	Hansens	£2.85
Modilase	Microbial	Hansens	£8.10*
Standard rennet	Calf	Marschalls	*£4.02*
Marzyme I	Microbial	Marschalls	£8.01*
Marzyme II	Microbial	Marschalls	£8.16*
Rennilase 14	Microbial	Novo	£1.69
Rennilase 50 TL	Microbial	Novo	£7.60*

*Triple strength products.

Table 4.6.4 lists October 1981 UK prices for various coagulants. Table 4.6.5 summarizes the economic consequences of using a microbial coagulant for cheddar cheesemaking in substitution for standard calf rennet. It can be seen that a saving of £1.68 on rennet cost per 5000 litres of milk processed can be achieved which is equivalent to 0.91 kilogrammes cheese or 0.18 per cent of the cheese yield. Thus, it can be concluded that any loss of cheese yield in excess of 0.18 per cent would negate any cost saving from using a microbial coagulant.

TABLE 4.6.5

Financial implications of substituting microbial coagulant for standard rennet (October 1981)

Assumptions:
1. Milk cost   : 13.4p/1
2. Standard rennet cost   : £4.05/1 (Hansens)
3. Microbial rennet cost : £2.70/1 (Modilase)
4. Cheese yield   : 10 kg/100 l milk
5. Cheese value   : £1850/tonne
6. Rennet usage level   : 0.025% v/v milk

Taking 5000 litres of milk for cheddar making:

Milk cost	: £670
Standard rennet cost	: £5.06 (0.76% of milk cost)
Microbial rennet cost	: £3.38 (0.50% of milk cost)
Saving on rennet cost	: £1.68 (33%)
Cheese yield	: 500 kg
Cheese value	: £925

It is extremely difficult to measure cheese losses to this degree of accuracy and this, together with the inherent risk of development of off flavours or other defects in the cheese, would lead one to conclude that there would seem to be little incentive to evaluate alternative coagulants at present in a creamery situation.

Considering the low percentage cost proportion of rennet as a raw material (0.76 per cent of milk cost), standard rennet prices would need to rise quite considerably in the future before it becomes financially justifiable to change.

Editors' note: Despite the considerations raised above, it is interesting to note that in 1981 more than one third of all cheese produced worldwide utilized microbial coagulants.

3. Lactose hydrolysis

Lactose is the major constituent of milk, skim milk and whey, where it accounts for 40, 50 and 75 per cent of the solids respectively. The problem of whey utilization in particular is therefore very much dominated by the opportunities which exist for the utilization of lactose, although most developments to date have centred on the recovery and utilization of the protein fraction of whey.

More recently, however, a number of technologies have been applied to the utilization of lactose and the two most important of these have been fermentation processes (to produce alcohol, methane etc) and lactose hydrolysis techniques. The added value gained by the hydrolysis of lactose, to its constituent monosaccharides glucose and galactose, lies in the increased usefulness of hydrolysed lactose as a food carbohydrate. Lactose itself has limited use in this respect because of its relatively low sweetness, solubility and digestibility, but the hydrolysis products of lactose, glucose and galactose, are superior in all of these respects. Increased sweetness and solubility improve the technical usefulness of whey products while the increased digestibility of hydrolysed lactose also offers the opportunity of supplying milk solids to populations which have hitherto been unable to consume milk products because of their inability to hydrolyse lactose in the digestive tract.

In common with other sugars, lactose can be hydrolysed using either acids or enzymes. However, recent experience has shown that acid hydrolysis of lactose, using ion exchange resins, gives rise to excessive colour formation, and is also relatively inflexible in that it is only possible to process demineralized, deproteinized streams using this technique. Enzymic hydrolysis using β-galactosidase offers much more flexibility in terms of product quality and range of raw materials which may be processed.

Applications of lactose hydrolysis to dairy products. Lactose hydrolysis in milk and whey products is already carried out commercially in a number of countries. The major applications for lactose hydrolysis are listed below.

(a) *Liquid milk.* Lactose hydrolysis in liquid milk improves digestibility for lactose intolerant consumers. In flavoured milks, lactose hydrolysis increases sweetness and enhances flavours.

(b) *Milk powders.* Lactose hydrolysed milk powders for dietetic uses, especially for infants with temporary β-galactosidase deficiency.

(c) *Fermented milk products.* In some cases, lactose hydrolysis in milk used for the manufacture of cheese and yoghurt can increase the rate of acid development and thus reduce processing time. Cheese maturation time may also be reduced, possibly through the presence of small amounts of proteases in some β-galactosidase preparations.

(d) *Concentrated milk products.* Lactose hydrolysis in concentrated milk products (e.g. sweetened condensed milk, ice cream), prevents crystallization of lactose.

(e) *Whey for animal feed.* Lactose hydrolysis in whey enables more whey solids to be fed to pigs and cattle and also prevents crystallization in whey concentrate.

(f) *Whey.* Lactose hydrolysed whey is concentrated to produce a syrup containing 70–75 per cent solids. This syrup provides a source of functional whey protein and sweet carbohydrate and is used as a food ingredient in ice cream, bakery and confectionery products.

(g) *Deproteinized whey (permeate).* After demineralization, lactose hydrolysed permeate is evaporated to a syrup containing 60–65 per cent solids. This has properties very similar to those of a medium dextrose equivalent glucose syrup.

Characteristics of β-galactosidase. β-galactosidase (EC 3.2.1.23) catalyses the hydrolysis of the milk sugar lactose into its constituent monosaccharides, glucose and galactose. In addition to these two major reaction products, small amounts of di- and trisaccharides are produced as a result of transgalactosidation reactions, particularly at high substrate concentrations. However, these byproducts are eventually hydrolysed after prolonged treatment.

β-galactosidase has been isolated from a wide range of microorganisms but commercially available enzymes are usually derived from yeasts (*Kluyveromyces fragilis, Kluyveromyces lactis*) and fungi (*Aspergillus niger, Aspergillus oryzae*).

The major difference between the yeast and fungal enzymes is the optimum pH for hydrolysis; yeast enzymes have optimum activity in the range pH 6–7, while the optimum for fungal enzymes is pH 4–5. The neutral β-galactosidases derived from yeast are therefore best used for the hydrolysis of lactose in milk (pH 6.8) and sweet cheese whey (pH 6.0–6.8), while the fungal enzymes are best used in the acid wheys derived from the manufacture of acid casein or cottage cheese (pH 4.6).

Both yeast and fungal enzymes are inhibited by one of the reaction products, galactose. Complete hydrolysis is therefore very difficult to achieve unless high concentrations of enzyme are used.

A number of minerals also have a profound effect on enzyme activity. Both fungal and yeast enzymes are inactivated by heavy metals (e.g. copper, zinc, mercury) and the yeast enzymes in particular are also inhibited by sodium and calcium ions. The effect of the latter can be reduced through the addition of phosphates, to complex the calcium, and through the heat treatment of milk or whey to precipitate calcium as calcium phosphate.

Other mineral ions are responsible for increasing the activity of β-galactosidase. The most important activators are potassium, magnesium and manganese ions. Optimum concentrations of these cations lie in the range 10^{-2} to 10^{-1} M for potassium, 10^{-3} to 10^{-4} M for magnesium and 10^{-4} to 10^{-5} M for manganese.

Most of these activators and inhibitors are in fact present in milk and whey and in such complex systems the quantity of enzyme necessary to produce a specified degree of hydrolysis must be determined by experiment using the actual substrate. Addition of activating cations to milk and whey is generally unnecessary except in the case of demineralized whey, where the use of yeast derived enzymes requires the addition of 10^{-2} to 10^{-1} M potassium. The rate of lactose hydrolysis in demineralized whey is then approximately twice that in whole whey because of the reduced levels of sodium and calcium ions in the former.

Lactose hydrolysis techniques.

Assessment of enzyme performance. Although β-galactosidase enzymes are sold with a guaranteed level of activity, the definition of enzyme activity units varies considerably among different suppliers. With lactose as substrate, the unit of activity is defined as the amount of enzyme which releases one μmole glucose in one minute under standard reaction conditions (temperature, pH etc). Another commonly used substrate is ortho-nitrophenyl-β-D galacto-pyranoside (ONPG) and in this case the unit of activity is defined as the amount of enzyme which hydrolyses one μmole of ONPG in one minute under standard reaction conditions.

In view of these differences in definition of enzyme activity, the only practical assessment of an enzyme is the determination of the rate of hydrolysis of lactose in the particular substrate (e.g. milk, whey) at the optimum conditions of pH and temperature specified by the enzyme manufacturer.

The degree of hydrolysis, defined as the percentage of lactose molecules cleaved, is most simply measured by determination of the amount of glucose released, or by following changes in the physical properties of the hydrolysed lactose solution. Solution properties such as freezing point depression, optical rotation and osmotic pressure, change as the disaccharide lactose is converted into the lower molecular weight monosaccharides glucose and galactose. Freezing point depression in particular can be measured very accurately and can therefore be used as a simple measure of the degree of hydrolysis with which it is linearly related.

Batch hydrolysis. In the batch hydrolysis process, pasteurized milk or whey is incubated with β-galactosidase until the desired degree of hydrolysis has been achieved. The reaction is best carried out in an enclosed tank with gentle agitation.

Most experience to date has been with the treatment of whole milk and sweet whey with the neutral yeast enzymes. Enzyme preparations are available in powder and liquid forms with a range of purities. General purpose, or technical grades are supplied for use with whey while high purity grades are for use with milk.

Optimum reaction conditions for the yeast enzymes are pH 6.3–6.7 and temperature at 30–40°C. Milk does not require pH adjustment but whey generally requires the addition of a small quantity of alkali. This should be in the form of potassium hydroxide, since the enzymes are inhibited by sodium ions.

At the optimum reaction temperature, bacterial growth rates are high; therefore the hydrolysis period must be relatively short in order to maintain microbiological stability. This is approximately four hours at 40°C. A preferred approach is to carry out the hydrolysis reaction in the cold (i.e. 5–10°C), and extend the hydrolysis period to 16–24 hours. Under these conditions, typical rates of enzyme addition to achieve 70 to 80 per cent lactose hydrolysis in milk would be as follows:

Lactozym 1500 HP(Novo Industri)	1.3 ml/l
Maxilact LX 5000(Gist Brocades)	0.5 g/l
Hydrolact L50(Sturge Enzymes)	1.6 ml/l

After the required degree of hydrolysis has been achieved, the enzyme can be inactivated by conventional pasteurization (e.g. at 72°C for 15 seconds).

Using this type of batch hydrolysis technique, the enzyme remains in the hydrolysed product and is therefore lost. However, in the processing of deproteinized streams (e.g. permeate), improved utilization of the enzyme can be achieved through the use of ultrafiltration equipment to recover the enzyme after hydrolysis. After a sterilizing filtration treatment, the recovered enzyme is used to treat a further batch of permeate. Small losses in enzyme activity are made up through the addition of fresh enzyme.

Immobilized enzyme hydrolysis. Several companies now supply immobilized forms of β-galactosidase for the treatment of milk and whey products. To the authors' knowledge, only one immobilized enzyme process is used commercially for the hydrolysis of lactose in milk. The process uses a yeast derived β-galactosidase immobilized within cellulose acetate fibres and was developed in Italy by SNAM Progetti. The hydrolysis reaction is carried out in the cold in a batch reactor and the lactose hydrolysed milk is UHT processed for liquid consumption.

The development of immobilized enzyme systems for the hydrolysis of lactose in milk has been severely limited by microbiological problems arising from continuous operation at neutral pH at 30–40°C. This has not been the case in whey processing, where a number of immobilized enzyme systems have been developed for operation at acid pH. These comprise a fungal β-galactosidase immobilized onto a range of supports. The most successful of these systems have been controlled pore silica beads (developed by Corning Glass) and adsorbant resins (developed by Valio Laboratory).

All of the immobilized enzymes developed for whey processing are designed for continuous use in a fixed bed with downward flow. Most experience to date has been gained with the Corning Glass system and since this system incorporates most of the features of the various immobilized enzyme processes, it serves as a good example of this method of lactose hydrolysis.

The raw material for hydrolysis was originally demineralized permeate, but a second generation enzyme was then developed to enable the processing of whole whey. In the case of lactose hydrolysis in permeate, demineralized permeate at pH 3.5 is pumped downwards through a column of the immobilized enzyme (IME) at a rate of approximately 10 litres per kilogramme of immobilized enzyme per hour. Initially the hydrolysis temperature is 32°C, but as enzyme activity falls, the temperature is gradually raised to 50°C in order to maintain the same degree of hydrolysis. The estimated enzyme life is in excess of 4000 hours.

An integral part of the continuous hydrolysis process is the cleaning and sanitation of the immobilized enzyme reactor. In the case of lactose hydrolysis in permeate this is simply achieved by backflushing with dilute acetic acid.

Comparison of hydrolysis techniques. The relative advantages and disadvantages of the free and immobilized enzyme techniques are listed in Table 4.6.6. The obvious advantage of immobilized enzyme systems is the reduced usage of enzyme. However, as with most whey processing operations, hydrolysis of lactose in whey or permeate by immobilized enzymes involves a substantial capital investment and is therefore justified only for large scale operations.

TABLE 4.6.6

Comparison of lactose hydrolysis processes

Mode of hydrolysis	Advantages	Disadvantages
Free enzyme (batch)	Flexibility – may be used with any substrate. Low capital cost	Higher enzyme cost (unless recovered by ultra filtration) Enzyme remains in product
Immobilized enzyme (continuous)	Reduced enzyme cost. No enzyme in product Simple, continuous operation	Less suitable for lactose hydrolysis in milk Substantial capital cost

To date, only one immobilized process is available for lactose hydrolysis in milk; batch hydrolysis is therefore necessary in most cases. Also, in milk processing the ratio of hydrolysis cost to product value is much lower than in the case of whey hydrolysis, and the higher enzyme cost incurred in batch hydrolysis is therefore less significant.

Finally, ultrafiltration for enzyme recovery in permeate hydrolysis is unlikely to be used in future because continuous immobilized enzyme processes are now available which are simpler, and require less labour in operation.

Process schemes for the manufacture of lactose hydrolysed milk and whey. The two most important applications for lactose hydrolysis listed above are the manufacture of lactose hydrolysed milk for liquid consumption and lactose hydrolysed whey for use as animal feed and as a functional food ingredient. The manufacture of these two products will now be considered in more detail.

Process schemes for the production of lactose hydrolysed milk are given in Table 4.6.7. In the simplest process, milk is first pasteurized and then hydrolysed as previously described. When the desired level of hydrolysis has been achieved (usually 80 per cent), the lactose hydrolysed milk is pasteurized or UHT processed prior to bottling or packing. Table 4.6.7 also shows an alternative process for the production of lactose hydrolysed UHT milk. In this case a sterile β-galactosidase solution (prepared by microfiltration) is introduced into UHT processed milk immediately before aseptic packing. Lactose hydrolysis then proceeds during ambient storage of the sterile product. The extended hydrolysis period (approximately two weeks) means that a very small quantity of β-galactosidase is required in comparison with that in the batch hydrolysis process (e.g. less than ten per cent). The enzyme used in this process must be very pure (i.e. protease-free) in order that the milk does not deteriorate in storage.

TABLE 4.6.7

Process scheme for hydrolysis of lactose in milk

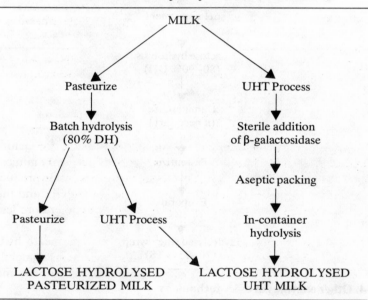

A process scheme for the production of lactose hydrolysed whey is given in Table 4.6.8. Separated whey is first pasteurized prior to pH adjustment to the optimum pH for hydrolysis. The hydrolysis step itself is carried out using the free or immobilized enzyme techniques described earlier, to achieve a degree of hydrolysis of 75–90 per cent. The whey is then repasteurized and evaporated under vacuum to a syrup containing 70–75 per cent solids. In many instances the mineral content of this lactose hydrolysed whey syrup gives rise to adverse flavours in the food product in which it is used.

This can be overcome by partial or complete demineralization of the hydrolysed whey prior to evaporation. Whey demineralization is carried out using either ion exchange resins or electrodialysis and the demineralization step can be performed either before or after hydrolysis. Hydrolysed whey syrup is quickly cooled and stored at 5–15°C in order to prevent colour formation through the Maillard reaction.

TABLE 4.6.8
Process scheme for hydrolysis of lactose in whey

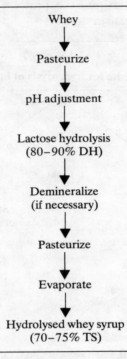

Whey
↓
Pasteurize
↓
pH adjustment
↓
Lactose hydrolysis
(80–90% DH)
↓
Demineralize
(if necessary)
↓
Pasteurize
↓
Evaporate
↓
Hydrolysed whey syrup
(70–75% TS)

---

4. Other enzymes in dairy technology

Protease. Enzyme modified cheeses (EMC) are a unique source of the major flavour components of matured cheese. They are produced by the modification of natural cheeses with multiple enzyme systems (proteases and lipases) to accelerate many of the biochemical changes which occur during traditional cheese ripening. This controlled enzyme curing process ensures consistent flavour character and balance in the final products. Several enzyme modified highly flavoured cheese products are available which are readily incorporated into many food formulations. The technology of manufacture consists of the addition of specific proteases and lipases to a slurry of a natural cheese variety and incubation at a temperature of 37 °C (typically) for a period of up to two weeks. At this stage the enzyme modified slurry is retorted to prevent further enzyme action and the product is packaged in the form of a paste.

Enzyme modified forms of Cheddar, Swiss, Provolone, Romano, Parmesan, Mozzarella, Cream, Blue and Edam are available from manufacturers such as Miles Laboratories, Dairyland Food Laboratories and International Flavours and Fragrances.

The degree of enzyme modification can be controlled to develop a variety of flavour profiles, product compositions and product forms. For example, there are five types of Miles *Marstar® Enzyme Modified Cheddar Cheeses* available with different flavour profiles. *Marstar® EMC 130* and *EMC 245* have a more intense cheddar cheese flavour typical of a sharp cheddar. Based on this typical sharp flavour, *EMC 130* is a 5× flavour concentrate, while *EMC 245* is a 20× flavour concentrate. A very sharp cheddar cheese flavour is associated with *EMC 138* and *EMC 139*, which are both 5× flavour concentrates. They have different flavour profiles due to the use of different enzyme systems. Finally *EMC 131*, a 5× concentrate, has a mild cheddar cheese flavour. It has been shown that the soluble nitrogen and free fatty acids show a progressive increase from the mild to the highly flavoured products as a result of controlled increases in proteolysis and lipolysis.

Typical applications for enzyme modified cheese products include processed cheese, cheese spreads, cheese dips, cheese substitutes and other custom product formulations. They are used either to increase the cheese flavour intensity of a formulation without increasing the total cheese solids, or to maintain cheese flavour character when total cheese solids are reduced. Up to one half of the cheese solids in a formula can be replaced depending upon the type and intensity of cheese flavour required and other ingredients present. As a flavour enhancer in formulated products, enzyme modified cheese can increase the quality and quantity of cheese flavour at a typical level of from one to three per cent.

When used to replace cheese in a formulation, the initial use level for an enzyme modified cheese can be established using the formula

$$\frac{\text{Present cheese in formulation} \times \text{Desired per cent cheese replacement}}{EMC \text{ flavour intensity factor}} = \text{per cent } EMC$$

For example

$$\frac{50\% \text{ cheese} \times 5.0\% \text{ replacement}}{5 \times \text{intensity}} = 0.5 \text{ per cent } EMC$$

Enzyme modified cheese products are stable for a minimum of six months when stored at less than 10 °C.

Apart from the use of proteolytic enzymes in cheesemaking, several other applications exist for the hydrolysis of milk proteins.

Alkaline proteases have been used in the cleaning of ultrafiltration and reverse osmosis membranes used for the concentration of skim milk and whey. However, the development of more resistant membranes has meant that membrane cleaning can now be carried out using acids and alkalis rather than the more expensive enzyme cleaners. Proteases also find limited use in the modification of the functional properties of milk proteins. Neutral and alkaline proteases have been used to partially hydrolyse casein and whey protein in order to increase isoelectric solubility and foaming characteristics.

Lipases. In addition to their use in the manufacture of enzyme modified cheese, lipases have an important role in the manufacture of certain Italian cheese varieties. A variety of lipase powders is commercially available producing specific and reproducible ratios of free fatty acids as a result of lipolysis of milk fat. This specific enzyme action on butterfat results in a characteristic flavour for each animal species from which the lipase is extracted (e.g. Marschall Division of Miles Laboratories produce three lipase powders extracted from kid, goat, lamb and calf glandular sources respectively).

The optimum temperature range for the action of all lipase powders is between 28° and 37°C, while the optimum pH is 6.2.

For cheesemaking, lipase powders are added to milk in the cheese vat prior to the addition of the coagulant. The recommended dosage rate for the manufacture of Italian cheeses such as Romano and Provolone is in the range of 3–18 grammes lipase powder per 100 litres of milk, depending on the degree of flavour development required. Some use has been made of lipases in the acceleration of flavour production in blue cheeses.

Lipase modified butterfat products are a source of major flavour components of butter. The manufacture of enzyme modified butterfat products utilizes high grade butterfat which is modified by controlled enzymic action to release volatile, flavourful fatty acids (butyric, caproic, caprylic and capric), together with other non-volatile long chain fatty acids and end products. The modification process is standardized to produce consistent flavour profiles and performance.

At present lipase modified butterfat products are available in two forms. One series contains flavour profiles developed exclusively by enzymic action on butterfat and qualify as natural butter flavours. The other series incorporates enzyme developed flavour profiles, together with other flavour adjuncts such as culture related flavour components, and these are generally considered to be artificial flavours. The latter enzyme modified butterfats are primarily designed to impart dairy flavour characteristics in specific applications. The enzyme modified butterfats are produced in emulsion form and, as with the enzyme modified cheeses, the product is heat treated to inactivate residual enzymes. Enzyme modified butterfat flavour performance is related to fatty acid profile and other flavour adjuncts present in some products supplement and modify the fatty acid profile. The function of modified butterfats in a food varies with addition level. Added at 0.01–0.05 per cent (of total formula weight), the products impart richness and fullness of flavour without significantly changing essential flavour characteristics. At 0.05–0.2 per cent, modification of certain types of flavour characteristics can be attained, and at levels greater than 0.2 per cent, products will impart selected dairy flavour characteristics. It is essential to allow formulated products to temper for a period of at least 24–48 hours to allow flavour equilibrium.

Product applications are carried out in confectionery, for example, to enhance the buttery character of toffees and caramels etc, and to reduce excessive sweetness. In coffee whiteners, they assist in imparting a rich creamy flavour, while in margarines certain of the enzyme modified products will impart a full, rich, buttery character. They may also be used in bakery products, and dairy products such as fat filled and synthetic milks, and as a flavour enhancer in a variety of cheese products such as dips, sauces and cheesecakes.

Catalase. Hydrogen peroxide (H_2O_2) is an effective chemical sterilant and is used as a milk or whey preservative in some countries lacking refrigeration and perhaps pasteurization equipment. The treatment of raw milk with hydrogen peroxide destroys harmful microorganisms, but does not significantly affect naturally occurring enzymes and advantageous bacteria. The excess hydrogen peroxide remaining in the milk after treatment is destroyed by the action of the enzyme catalase. Commercially available catalase (e.g. *Catalase L*, Miles Laboratories), is a standardized liquid enzyme system extracted from beef liver or from *Aspergillus niger*. Catalase specifically catalyses the decomposition of hydrogen peroxide to water and molecular oxygen

$$H_2O_2 \xrightarrow{\text{Catalase}} H_2O + \tfrac{1}{2}O_2$$

There are two basic steps in the peroxide–catalase treatment of milk or whey.

(a) Addition of five per cent hydrogen peroxide so that amount of peroxide, milk/whey temperature and length of contact time are standardized and controlled.

(b) Catalase solution is added to the peroxide treated milk/whey after step one to allow decomposition of the peroxide to water and oxygen.

A catalase stock solution can be prepared by mixing one volume of Miles *Catalase L* (100 Keil Units per millilitre) with six volumes of water. 5 per cent hydrogen peroxide solution is added to milk at the rate of 0.4 per cent (corresponds to 0.02 per cent addition of 100 per cent peroxide; the addition of up to 0.05 per cent peroxide to cheese milk is permissible in the USA). The peroxide treated milk is held at 30 °C for 20 minutes before addition of 63 grammes of *Catalase L* stock solution per 1000 litres of cheese milk. The milk is slowly agitated for 20 minutes when a test for residual peroxide is carried out.

For *Catalase L* activity, pH 6.5–7.5 is optimum, while optimum temperature range is 5 °C–45 °C. Temperatures in excess of 60 °C inactivate *Catalase L*. For example, at pH 5.0, dilute solutions of *Catalase L* are completely inactivated after 30 minutes at 60 °C.

Treatment of milk with hydrogen peroxide can produce certain defects in cheese, oxidize certain amino acids, particularly methionine, and reduce the biological value of the proteins. The use of hydrogen peroxide–catalase treatment has not been widely accepted for cheese milk treatment in the UK, but has been used for preserving whey for extended periods prior to further processing.

5. Future developments
With the notable exceptions of rennet and β-galactosidase, the dairy industry has been slow to capitalize on enzyme technology. However, more recently, several areas are emerging where considerable potential would seem to exist for the use of industrial enzymes. The development of DNA-recombinant technology will undoubtedly serve to encourage the economic production of a number of potentially useful enzymes.

As the thermolabile fungal proteases become more accepted, with the production of good quality cheeses and increased confidence in yield considerations, there will undoubtedly be a wider use made of these coagulants.

The use of immobilized enzymes attached to inert bases such as glass beads, is a distinct possibility, not only to bring about milk coagulation but for the formation of specific cheese flavour precursors. The first stage of milk coagulation would be carried out by passing cold milk (5 °C) through immobilized enzyme columns followed by induction of the second, aggregation stage, by heating the milk to the normal coagulation temperature of 30–32 °C. However, problems may be experienced with enzyme 'blinding' due to presence of butterfat in the whole milk streams. A further disadvantage is that no residual coagulant activity, which has a role in cheese flavour development during maturation, would remain unless other specific enzymes were immobilized to carry out this function. It would seem more probable that the industry will continue to use added enzymes as long as the coagulant manufacturers can assure the continued supply of good quality products at an economical price.

The manufacture of cheese on an industrial scale is a capital intensive operation and the running costs and interest charges for cheese storage during maturation represent a significant proportion of milk to cheese conversion costs. There are, then, considerable incentives to investigate possible methods for accelerating the cheese ripening process in order to minimize storage costs. Several methods for accelerating the ripening of cheese are presently being investigated.

It is protein, and to a limited extent, fat breakdown in maturing cheese, catalysed by coagulants and endogenous microbial enzymes, which play a fundamental role in flavour development. Consequently, one of the methods being actively pursued is the addition of exogenous enzyme preparations. Some work using commercially available food grade proteases and lipases shows encouraging results. Some proteases, however, while producing strong flavours in cheese in a relatively short time, can produce flavour defects such as bitter notes and imbalance.

TABLE 4.6.9

Exogenous enzymes used to accelerate cheese ripening

Type of cheese	Enzyme type	Source
Cheddar	Acid proteases Neutral proteases Peptidases Lipases Decarboxylases	Various commercial enzymes
Cheddar, Romano, Parmesan	Lipase Protease	Lamb gastric extract
Gouda	Protease	*Aspergillus oryzae*
Blue	Lipase	*Aspergillus* species
Mozzarella	Lipases Esterases	Calf pregastric secretions

Table 4.6.9 illustrates some of the exogenous enzymes which have been used to accelerate cheese ripening. Cheesemaking trials using a range of food grade enzymes have shown that neutral proteases at an additive rate of 0.00125 per cent (w/w) (Novo *Neutrase® 1.5S*) significantly enhance the flavour intensity of cheddar cheese without producing flavour defects. Table 4.6.10 illustrates the statistical taste panel data resulting from cheddar cheese trials carried out at the National Institute for Research in Dairying using a range of commercial proteases. The use of alkaline and acid proteases resulted in bitter off flavours developing in the cheese.

TABLE 4.6.10

Effect of commercial proteinases on flavour development in cheddar cheeses after two months' maturation

Proteinase	Concentration (units g^{-1})[1]	Cheddar intensity (0–8 scale)	Bitter intensity (0–4 scale)	Other off flavours (0–4 scale)
Acid	50	1.8[2]	3.3	1.1
(*Aspergillus oryzae*)	10	2.6	2.8	0.7
	20	2.7	2.6	0.4
	0.4	2.1	2.2	0.3
	0.08	2.2	0.7	0.2
Neutral	50	3.1[2]	1.6	0.9
(*Bacillus subtilis*)	10	3.4[2]	0.7	0.4
	2	3.2[2]	0.1	0.2
	0.4	2.7	0.1	0.2
Alkaline	2	2.9	3.5	0.0
(*Bacillus licheniformis*)				
Pronase	2	4.1[2]	0.5	0.6
(*Streptomyces griseus*)				
Untreated	—	2.4	0.1	0.2

[1] 1 unit = amount of enzyme required to bring about an increase in A_{595} of 0.5 in 15 minutes using the Hide Powder Azure (HPA) assay of Cliffe & Law (1982). 1 HPA unit = 10^{-5} Anson units.
[2] Significantly different from untreated cheese at $P < 0.05$.
(Law & Wigmore, 1982)

Further work using *Neutrase®* protease has shown that if the treatment is combined with temperature control, flavour acceleration is improved. For example, if *Neutrase®* treated cheese is held at 18°C for one month, a cheese with an equivalent flavour intensity of a four month cheese results.

Further work is presently being carried out using other enzymes in an attempt to produce cheeses having well-balanced flavour profiles without textural defects.

One of the technical problems associated with enzyme treatment is their addition to the cheese curd. Powdered enzymes are mixed with the cheese salt and thoroughly mixed before addition to the curd. This does represent a handling problem and research has been initiated to investigate the suitability of various encapsulation methods. Thus cheese ripening enzymes could be enclosed in fat capsules, for example, and these would be added to cheese milk, where the capsules become entrapped within the curd matrix followed by gradual breakdown and release of the enzymes.

The selection and use of enzymes for accelerated ripening in the future will be aided by the development of screening methods. For example, electrophoretic zymogram techniques can be used to screen the protease components of cheese starter cultures which represent a good balance of cheese ripening enzymes. Comparison of commercially available enzyme zymograms with those of the starter components may aid selection of enzymes for accelerated ripening.

The main development in the use of β-galactosidase is expected to come in the form of immobilized enzyme systems suitable for the continuous hydrolysis of lactose in milk.

Finally, two enzymes which occur naturally in milk may have further potential in the improvement of milk quality. Lactoperoxidase is an important component of the lactoperoxidase-thiocyanate-hydrogen peroxide system which possesses important antimicrobial properties. The isolation and immobilization of this enzyme may provide an effective means for the cold sterilization of milk. Another naturally occurring milk enzyme, sulphydryl oxidase, has also been implicated in the processing characteristics of milk through its effect on disulphide bond formation. An immobilized sulphydryl oxidase would have considerable potential as a means of reducing the cooked flavour in UHT milk.

References

Berridge, N. J. *Analyst* **77,** 57 (1952).

Brunner, J. R. in *Food Proteins* (eds Whitaker, J. R. & Tannenbaum, S. R.) 175 (AVI, Westport, 1977).

International Dairy Federation *Methods for the Determination of the Firmness of Milk Coagulum* Doc. no. 99 (1977).

Law, B. A. & Wigmore, A. S. *J. Soc. Dairy Tech.* **35,** (2) 75 (1982).

DETERGENTS
H. C. Barfoed

1. Introduction

The story of enzyme detergents dates from 1913, when the German chemist, Otto Röhm, obtained a patent for a pre-soaking product which contained enzymes from animal pancreas glands. The product, called *Burnus®*, consisted primarily of sodium carbonate to which was added a crude extract of pancreas (pancreatin). The enzyme content was quite low and, furthermore, the enzyme was not very active at the high pH value provided by the soda. Nevertheless, *Burnus®* remained on the European market until several years after World War II.

A new development started in 1959 when the first detergent containing a bacterial proteinase, *Bio 40®*, was launched by the Swiss company Gebrüder Schnyder. The real breakthrough, however, came a few years later when the alkaline proteinase, *Alcalase®*, was developed by Novo Industri in Denmark. *Alcalase®* was incorporated into *Bio 40®* and shortly after into *Bio-Tex®*, manufactured by the Dutch firm Kortman & Schulte in collaboration with Gebrüder Schnyder.

The tremendous success of this product stimulated a rapid growth in enzyme detergents which in certain countries captured up to 70 per cent of the market. A temporary setback occurred early in the 1970s when allergies developed in detergent workers, but such problems were soon overcome by the introduction of dust-free enzyme preparations. Today, enzyme detergents have a significant market share, ranging from 30 to 60 per cent in most industrialized countries.

2. Applications of enzyme detergents

The first enzyme detergents were intended primarily for cleaning heavily soiled working clothes such as those from the fish industry, slaughterhouses, bakeries and hospitals. Stains on this type of laundry contain a high percentage of protein, so the use of protein-splitting enzymes to remove them is logical. It was subsequently realized, however, that protein-containing stains also occur frequently in ordinary household laundry. This is because even at a low concentration, a protein may act as a binder, fixing other soil components to the fabric and making complete removal of the stain difficult.

The catalytic (i.e. time-dependent) nature of the enzyme reaction might suggest that enzymes would be most effective in a soaking process, allowing ample time for the enzyme to act. However, experience has shown that in most cases slight digestion of the protein is sufficient to make the stain soluble, and the modern, efficient laundry enzymes available are able to produce this effect within the time available in a normal machine washing cycle. It is, therefore, rational to use enzymes in ordinary household detergents, as well as in soaking agents.

A fairly recent technique which has shown promising results is so-called pre-spotting. Before pre-soaking (or washing), particularly dirty patches on the laundry, such as shirt collars and cuffs, are moistened with an undiluted liquid enzymatic washing agent. The high local concentration of enzyme and washing agent achieved exerts such a strong effect on the stains that they can then usually be removed by ordinary washing procedures.

3. Dishwashing

A major problem in machine dishwashing is caused by food residues containing starch, for instance from mashed potatoes, oatmeal or spaghetti. This may be overcome by using a detergent containing a heat and alkali stable amylase. However, a reformulation of the detergent is usually necessary, because enzymes currently available are destroyed by active chlorine, a normal component of machine dishwashing powders. Furthermore, the alkalinity of the detergent may have to be somewhat reduced.

4. Energy considerations

In recent years the steep rise in energy prices has provoked renewed interest in pre-soaking as a means of saving energy. At the same time the trend towards low-temperature washing, a result of the increasing use of man-made fibres, has been emphasized by energy considerations. In both instances enzymes have proved very useful and have in fact made it possible to achieve equally good results to those obtained with normal high-temperature washing.

5. Detergent enzymes

Most commercial detergent enzymes are alkaline bacterial proteinases of the serine type (i.e. they contain the amino acid serine in their catalytic centre), and are derived from the common soil bacteria *Bacillus subtilis* or *Bacillus licheniformis*. Figures 4.7.1 and 4.7.2 show temperature and pH profiles of *Alcalase*®, a typical member of this group.

It can be seen that although the activity increases with increasing

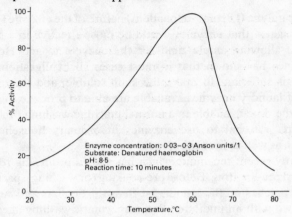

Figure 4.7.1   Activity of *Alcalase*® at different temperatures

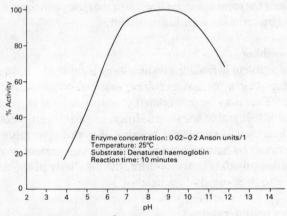

Figure 4.7.2   Activity of *Alcalase*® at different pH values

temperature up to about 60 °C, the slope of the curve is not very steep. This means that the enzyme works well even at the low temperature found during soaking and in the initial steps of machine washing. The pH/activity curve shows an optimum at moderately alkaline conditions (pH 8.5–9.0) but the useful pH range is quite broad and extends up to about pH 10. Consequently, *Alcalase*® may be used in most ordinary household detergent formulations.

For heavy duty formulations and liquid detergents enzymes with a greater alkali tolerance are preferable. Such enzymes, derived from alkalophilic *Bacillus* species, are commercially available. Temperature and pH activity profiles of a highly alkaline proteinase, *Esperase*® (Novo), are shown in Figures 4.7.3 and 4.7.4.

The most striking feature is the very flat pH/activity curve, which shows that the enzyme works well even at pH 12.

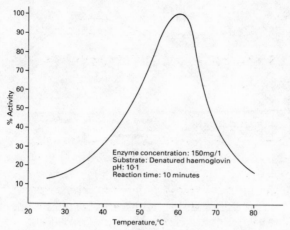

Figure 4.7.3 Activity of *Esperase® M 4.0* at different temperatures

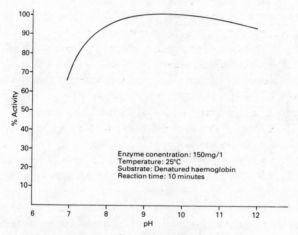

Figure 4.7.4 Activity of *Esperase® M 4.0* at different pH values

In enyzmatic machine dishwashing detergents an alkali-resistant and moderately heat-resistant (up to about 60°C) α-amylase is needed. Ideally, the enzyme should also be able to withstand the bleaching agent (usually a chlorine-containing compound) found in most dishwashing detergents. Unfortunately, the ideal enzyme is not available. The nearest to date is the α-amylase from *Bacillus licheniformis* (e.g. *Termamyl®*, Novo Industri A/S). The temperature and pH activity profiles of this enzyme are shown in Figures 4.7.5 and 4.7.6.

It is seen that while the enzyme tolerates heat very well, the alkali resistance is only moderate. It may be added that active chlorine will destroy the enzyme activity almost instantly.

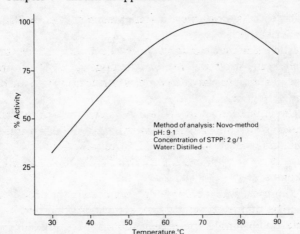

Figure 4.7.5   Activity of *Termamyl*® at different temparatures and in the presence of STPP

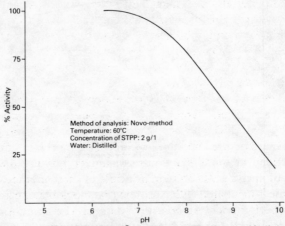

Figure 4.7.6   Activity of *Termamyl*® at different pH values and in the presence of STPP

6. Physical form

Being proteins, enzymes may provoke allergic symptoms in susceptible persons. Because early enzyme preparations were marketed as fine powders, which could form dust, allergy problems occurred in detergent plant workers exposed to high enzyme concentrations. In order to prevent further problems of this nature, the development of dust-free preparations was started. The technology has now been developed through a number of stages to a point where it is fair to say that the enzyme preparations, when properly handled, do not form any dust at all. This has been achieved by shaping the enzyme with fillers and binding agents into small beads or granules, often surrounded by a layer of inert material, such as wax or a soluble cellulose derivative.

The particle size of the enzyme granulate must be similar to that of the detergent itself to ensure a homogeneous blend and to prevent segregation of enzyme particles during transport and handling of the detergent.

For use in liquid detergents the enzymes are supplied in the form of slurries, in which the enzyme powder is suspended in a non-ionic surfactant, and recently as true liquids of suitable potency.

7. Regulations

In most countries, a new enzyme product used in the processing of food must be approved under laws governing food additives or processing. So far similar approval has not been required for the use of enzymes in other industries. However, laws recently passed in both the USA and the EEC require that new industrial chemicals and other materials, including industrial enzymes, comply with certain standards. This means that detergent manufacturers in these areas will have to notify the relevant authorities in their country of their intended use of enzymes.

8. Specifications

Technical enzyme preparations contain only a small percentage of active enzyme protein (typically one to ten per cent) and are sold on an activity basis. This means that in product descriptions the analytical method used must be specified. Although internationally recommended methods exist, most manufacturers use their own methods of analysis. A common principle of these methods is that the enzyme is incubated with a protein (haemoglobin or casein) for a certain period under standard conditions. The amount of hydrolysis products is determined spectrophotometrically, either directly using ultraviolet light, or after coupling with a colour reagent (e.g. Folin-Ciocalteu's Reagent or trinitrobenzene sulphonic acid). Sometimes the addition of typical detergent components, such as sodium tripolyphosphate, to the mixture is specified in order to achieve a so-called functional analysis. In view of the large variations in detergent formulations, however, this practice is of limited value.

Usually the manufacturer will guarantee a certain storage stability under specified conditions, for example, an activity loss of less than ten per cent during the first year of storage. The detergent manufacturer must establish stability data of the enzyme in his own formulations in order to ensure a satisfactory shelf life of the finished product.

Also, the enzymes must meet certain specifications in regard to safety and ease of handling, including absence of toxic materials and pathogenic bacteria, as well as a suitable particle size and bulk density. With respect to microbiological quality (total viable count, total mould count etc), the standards valid for food enzymes will often be applied.

9. Mixing of enzymes into detergent powders

In the manufacture of enzyme detergents in powder form the sensitivity of enzymes to heat and humidity as well as the safety aspects should be considered. Usually the enzyme granulate is added to the otherwise finished detergent together with other heat labile components (e.g. sodium perborate and perfume) immediately before packing.

Unless a microdosing system is available, the more concentrated products (i.e. those used at a level of less than one per cent by weight of the detergents) should be mixed into the detergent by a stepwise procedure. In the first step a pre-mix of enzyme and, for example, sodium tripolyphosphate in a ratio of about 1:10 is made in a double-cone mixer, Lödige mixer or other suitable mixing equipment. The second step of the mixing process (i.e. the addition of the pre-mix to the detergent powder) may be carried out by means of apparatus similar to that shown in Figure 4.7.7.

The pre-mix is fed by vibrating hopper to a short belt conveyor. The feed rate is adjusted by means of an automatic continuous weighing machine controlling the hopper slide gate. From the belt the pre-mix is transferred via a slide to the conveyor carrying the detergent powder. The two components now need only a short final mixing (e.g. in a continuous screw-type blender) before transfer to

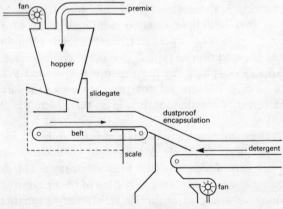

Figure 4.7.7 Typical equipment layout for industrial detergent blending.

the packing machines.

Alternatively, the mixing could take place in a continuous blender directly ahead of the conveyor carrying the detergent to the packing machines. In order to ensure a constant feed to the mixer, it should be fitted with constant head hoppers for detergent and enzyme pre-mix.

When designing plant for the mixing of enzymes into detergents it should be borne in mind that detergent enzymes are active protein-decomposing agents which may cause local irritation to moist skin, mucous membranes and respiratory organs. Furthermore, inhalation of enzyme dust may cause allergy in susceptible persons. Therefore it is essential that all equipment is dustproof and that effective ventilation is provided. The exhaust air from the ventilation system should be passed through effective filters in order to prevent release of dust to the environment.

Detailed recommendations for the lay-out and construction of enzyme detergent plants, as well as for the safety of the workers employed, may be obtained from enzyme manufacturers.

10. Enzyme detergent formulations

The following table provides some guidelines for the formulation of various types of detergents:

Soaking (Pre-wash) detergent

Anionics (LAS)	6–15%
Nonionics	1–3%
Sodium tripolyphosphate	25–45%
Sodium silicate	0–5%
Sodium carboxymethyl cellulose	0.6–1.0%
Optical brighteners, perfume	0.1–0.2%
Alcalase® 1.5 T	0.5–1.0%
Sodium sulphate	to 100%
pH	8.0–9.5

Liquid detergent

Anionics	5–10%
Nonionics	10–40%
Solubilizer	5–15%
Sequestering agent	0–15%
Water	max. 45%
Optical brightener, perfume	0.1–0.5%
Esperase® 8.0 Slurry	0.4–0.8%
pH	7.0–9.5

Heavy duty (General purpose) detergent

Anionics (LAS)	7–15%
Soap	2–4%
Nonionics	1–5%
Sodium tripolyphosphate	20–40%
Sodium perborate	15–30%

Sodium silicate	4–8%
Sodium carboxymethyl cellulose	0.5–1.0%
Optical brighteners, perfume	0.4–0.8%
Alcalase® *1.5 T/Esperase*® *4.0 T*	0.4–0.8%
Sodium sulphate, water etc	to 100%
pH	9.5–10.5

Dishwashing detergent

Sodium tripolyphosphate	20–50%
Sodium metasilicate	10–30%
Sodium bicarbonate	40–60%
Surfactant	3–10%
Termamyl® *60 G*	1–3%
pH	9–9.5

11. Evaluation of enzyme detergents

Generally the washing efficiency of a detergent is evaluated in three steps: (*i*) screening tests in the laboratory; (*ii*) bundle tests in the laboratory field; (*iii*) practical tests in the field. The screening tests in the laboratory are carried out in model washing systems like the *Terg-O-Tometer* and *LaunderOmeter*. These models simulate the washing processes as performed in an agitator and drum household washing machine, respectively. The advantage of using the model apparatus is the much higher testing capacity compared with the household machine.

In these laboratory tests the introductory evaluation of different test parameters like enzyme type and concentration, temperature, water hardness, type of soil etc, is made. Later on in the development process the findings may be confirmed through the bundle and practical tests.

The bundle tests and practical tests are performed in household washing machines with garments and other items soiled under actual use conditions. In such tests the increased efficiency of an enzymatic detergent on certain stains can be demonstrated by comparing results using the same detergent without enzyme.

12. Types of stain

For the laboratory evaluations different types of test soilings (e.g. blood, milk, cocoa, egg, spinach, grass etc) are used. These substances are applied to different textile types found in practice, such as cotton, cotton/polyester blends and polyacrylics. Some of these test materials are commercially available (e.g. the well-known test fabric *EMPA 116* soiled with blood, milk and India ink), whereas others are developed and produced in the laboratory.

13. Future prospects

In response to the improvements in domestic machine systems it is anticipated that increased attention will be paid to liquid detergent formulations. To a lesser extent there will be national encouragement to alter the composition of detergents, especially towards lower phosphate content and increased biodegradability. These attitudes will probably further stimulate the liquid detergent concept.

Energy saving reductions in washing temperatures can also be seen as a challenge and opportunity for the selection and presentation of enzymes that will give high catalytic performance between 20 and 40°C. The development of specific activators as well as chemical modification of the enzymes themselves will also influence the speed of these highly desirable economies.

Industrial and institutional laundry operators have already indicated a positive response to new bleaching systems included in the whole detergent formulation, so removing the second chemical addition required for the more traditional bleaching methods. Currently there are no enzymes fully compatible with these additives, so that only where a low temperature stage precedes the release of the bleach agent can an enzymatic benefit be included in the formula. It is a reasonable expectation that attempts will be made to find enzymes that are adequately tolerant of these additives to give good cleaning benefit in single stage washing.

EFFLUENT, BYPRODUCT AND BIOGAS
T. Godfrey

1. Introduction

In response to growing action by industry to produce the maximum economic operating systems, attention has now included the prospect of treating effluent for reduction in external disposal charges. There are a number of techniques available that result in a physical removal of previously soluble or finely suspended natural materials, or use fermentative systems to trap them. Industrial enzymes have found application in various ways to aid these technologies.

Frequently, when an unspecified concentration is exceeded, the effluent is reclassified as a byproduct of the process. It is then assessed as having a potential value, together with conventional byproducts that emerge as 'all the remaining materials' not included in the primary product(s) of the operation. Upgrading processing of byproducts represents a large area of involvement for industrial enzymes when the industry uses natural raw materials. By a combination of factors, materials that could be byproducts, and true effluents are recognized as having a fermentative potential for the production of biomass. This biomass may be a new product in itself or represent a massive concentration of biological oxygen demand, facilitating low cost disposal as sludge or solids to waste or animal feeding. A special case of this type of treatment is the controlled anaerobic fermentation of materials to produce methane fuel gas and simultaneously remove biological oxygen demand from high-volume liquid wastes (*see* below).

2. Treatment of byproduct

Waste materials for processing. A consideration of the industries processing natural 'crop' materials produces an almost unlimited variety of substrates for enzyme modification occurring in their wastes. A representative selection is given in Table 4.8.1; here the materials are broadly classified into the four categories of starch (and sugar); non-starch carbohydrate; proteins and fats and oils.

Table 4.8.2 sets out some available data for major wastes from typical industrial sectors in the UK, giving an idea of the enormous quantities of valuable basic components that are available for improved processing. The figures are for true wastes and do not include already recovered or processed materials.

TABLE 4.8.1

Major waste creating industries producing bioprocess wastes

Waste material type	Industrial activity
Starch	Bread, flour, confectionery
	Brewing
Cereal	Cereal foods
	Distilling
Sugars	Finished compound foods
	Food ingredients
	Paper, adhesives
	Sweeteners
	Textiles
Cellulose	Paper, timber
Lignocellulose	Brewing, distilling
Proteins	Abattoir
	Butchery
	Cereal extraction
	Dairy
	Poultry
	Finished compound foods
	Brewing, distilling
	Fish processing
	Vegetable processing
	Leather
	Gelatine
	Single cell fermentation
	Oil seeds processing
Fats	Abattoir
	Butchery
Oils	Poultry
	Dairy
	Fish processing
	Oil seeds processing
	Finished compound foods
	Cereal foods

Starch and sugar residues represent large amounts of waste from a large part of the food and beverage industries. Large amounts of proteins in a variety of states ranging from edible to contaminated and fermenting suspensions, are generated from the slaughter, oil seed extracting, fish, gelatine, leather and dairy industries. Obtaining some return on the disposal of wastes is a characteristic of many of the larger traditional processing industries. Examples include

TABLE 4.8.2

Annual bulk wastes from key industries (UK 1979)

Industry	Protein	Carbohydrate	Sugar	Fat	Cellulose etc
Abattoir	Solids 30,000 tonnes Blood 25,000 tonnes at 18 per cent protein	Fermentables equal to 60,000 barrels fuel oil		15,000 tonnes	
Brewing	Yeast 2000 tonnes				Grains 70,000 tonnes
Confectionery	3000 tonnes	Fermentables equal to 9000 tonnes absolute alcohol	20,000 tonnes sucrose dry solids equivalent	120,000 tonnes	
Dairy	15,000 tonnes	Whey 600,000 tonnes at 4.5 per cent		200,000 tonnes	
Distilling	Yeast 1500 tonnes	Fermentables 10–15,000 tonnes			Grains 100,000 tonnes

feeding livestock from dairy, brewing and distilling wastes and the rendering of slaughter waste for the fat content for soaps. Many potential extensions of this concept are restricted by the energy and capital costs involved with the result that proposed novel treatments for wastes can produce wide ranging effects. If the waste is processed and so becomes a raw material, it may reduce the supply to the traditional outlets, and create a price rise which points to small benefit for disturbing established industrial equilibria.

The suggested conversion of starch wastes and whey carbohydrate to sugar syrups is both feasible, and, in small operations actually practiced. A simple and short treatment borrowed from primary enzymic processing uses minimum energy and chemicals to produce a highly acceptable syrup to be used directly or as a fermentation feedstock. The capacity for the supply of both these requirements is underutilized throughout the world, and it will require a most favourable price for new materials to secure a market share. This low price will act against the prospects for economic waste utilization unless specialized outlets can be found. High-quality syrups are generally produced centrally and then concentrated and tankered to customers. If waste is processed locally and utilized without concentration, it may be economic enough to generate a product of value. Where waste is plentiful but diverse in its geographical production, the degree of cooperation needed to coordinate its use is not anticipated from the industrialized society for many years yet.

To establish an upgrading bioprocess for waste materials, there are many factors to consider before creating the case for its treatment. These can be divided into three main sectors: (a) the chemical and physical nature of the waste; (b) the source and quantity; (c) the pressures on the viability of a processing scheme. These are listed in Table 4.8.3.

Where added value products are envisaged as a result of waste processing, consideration must be taken to predict how the waste may be handled. If it is highly dilute and heavily contaminated by microorganisms, it can be expected to ferment rapidly in an uncontrolled way. Steps will be required to arrest or preferably prevent

TABLE 4.8.3
Considerations for a waste bioprocess treatment

Chemical and physical nature	Is it subject to rapid biodeterioration?
	Is it simple or complex mixture?
	What are the relative concentrations?
	What is the contamination and its level?
	Is it a genuine waste or underutilized byproduct?
Source and quantity	What are the costs and logistics of collection?
	What existing applications will compete?
	What alternative new uses will compete?
	What factor of concentration is required?
	Can the envisaged market pay back the capital investment?
	Are some sources technically or quantitatively better?
Pressures influencing the viability	Existing treatment costs
	Potential increased treatment costs
	Pollutant grade of waste
	Market for product or reuse value
	Politico/economic direction and finance
	Long term policy regarding source processing
	Legal overview of source and product

this decay if useful products are to be separated. Direct discharge effluents are the least attractive but largest bulk natural wastes. New techniques for water conservation and reuse are increasing the range of potentially useful wastes for processing. A large amount of valuable material is to be found in the spent materials from brewing and distilling, but their recovery will depend on engineering skill and a changed attitude among the producer industries to accept a further wet process step and to seek to cover the redrying costs in added value from the recovered products.

Table 4.8.4 illustrates a number of processes that use enzymes to some extent in the generation of added value as recovered raw materials, products or energy economy.

TABLE 4.8.4

Enzyme aided processing of wastes and byproducts

Economic recovery and reuse	Sugars (direct and by hydrolysis of starch) in confectionery and foods returned to incoming sugar and invert stages
Energy conservation and materials economy	Extraction of high value products from natural sources (e.g. flavours, colours, spices)
Process economy and upgraded byproducts	Reduction of viscosity for evaporation of liquors (e.g. fish products and soluble coffee) Nondestructive oil seed processing Meat and bone scrap processing Reuse of gelatine discards
Alternative higher value derivatives	Fermentation of carbohydrate rich wastes for medical or fine chemical products
Production of new resource materials	Whey carbohydrate to syrups Whey proteins to functional products Blood proteins to food ingredients Straw to nutritionally improved feeds
Net energy gain and effluent reduction	Low grade high biological oxygen demand liquids to biogas Combined and concentrated fermentable wastes to alcohol or other organics

Economic recovery and reuse. Starch and sugars from food processing and confectionery industries are processed to yield a syrup suitable for incorporation into main line products and to minimize raw materials waste. A slurry of waste in water at 60 °C is adjusted to pH 4.5–5.5 and dosed with α-amylase and glucoamylase according to the typical starch industry recommendations given in Chapter 4.15. Doses may be reduced for economy if extended holding times can be accepted. Subsequent heating to 80 °C will destroy the enzymes, complete the release of fats to the surface for physical separation, and then either addition of flocculating chemicals or heating to boiling will coagulate proteins for filtration of centrifugal removal. The resulting syrup is then either used directly, or treated with carbon and ion exchange resins to purify it for more demanding specifications.

Energy conservation and materials economy. A number of examples of enzymic processes for the improved preparation of flavours, colours and spices are discussed in Chapter 4.9 'Flavouring and colouring'. In certain cases, the recovery of more than one valuable substance from the same raw material is facilitated by degradation of interfering substances. Cellulases, pectinases and starch degrading enzymes find application in this area and also in the improved recovery of bulk oils from oil seeds and fruits (*see* Chapter 4.19, 'Edible oils'). The leather industry has increasingly turned to industrial enzymes to reduce waste and to lower the process costs by using less additional chemistry, which in turn results in less pollutant wastes (*see* Chapter 4.11).

Process economy and upgraded byproducts. In many traditional industries the product or byproduct is initially an aqueous solution or suspension that is concentrated by evaporation. In some circumstances, as concentration rises dissolved materials contribute excessive viscosity and prevent the most effective evaporator operation. Use of small amounts of galactomannanase in the aqueous extract of coffee solubles at around 0.05–0.1 per cent on solids prevents this effect and allows economy in the evaporator that is transferred to a lower water removal charge at either spray or freeze-drying.

In the fish meal industry, the use of proteolytic enzymes reduces the viscosity of 'stick water' during evaporation. By rerouting the sequence of passage through the stages of multieffect evaporators, thin stick water enters the third stage where the temperature is low. The addition of neutral or alkaline bacterial proteases such as *Neutrase®* 0.5L or *Alcalase®* 0.6L (Novo) at 0.2–0.5 per cent of dry matter will produce rapid reduction of protein viscosity as the concentration of solids is raised to around 20 per cent. Passing to the first stage for evaporation at around 120 °C up to 30 per cent dry matter, and the second stage at 100°C, the concentration to 50 per cent dry matter for discharge as concentrate is completed. Although information regarding actual savings in total fuel use is specific to each plant, the uptake of this process scheme in many countries indicates that the economy is valuable.

The application of enzyme treatments for the release of oil in oil seed processing has been established. Further extensions into the recovery of both native protein and useful carbohydrate streams are being actively pursued by several researchers with a view to economic and technical improvements in that industry. These are discussed further in Chapter 4.19, 'Edible oils'.

The recovery of valuable upgraded byproducts from blood, meat scrap, bone, collagen and gelatine, together with highly attractive processes for processing photographic film wastes for silver content, have been discussed in Chapter 4.14, 'Proteins'. In each case, the material represents loss of raw material to the primary industry if it is not processed. The processing of these materials produces a reduction in effluent content of environmental and economic value.

Alternative higher value derivatives. Where the raw waste is complex and separation of a single valuable component remains uneconomic, it is possible to introduce enzymatic hydrolysis of the higher polymers to produce a fermentation feedstock. Wastes rich in carbohydrate and with some protein content are readily converted by adopting the concept of the brewing process (*see* Chapter 4.5). Since most carbohydrate wastes will contain gelatinized starches, it will not require a high temperature stage to render them open to amylase attack. Where ungelatinized starches are present, the use of a continuous jet cooker for liquefaction may prove valuable as this method, including the use of thermostable amylases, is fuel efficient.

Material which has undergone jet cooking, or already gelatinized starch substrates, can be treated with α-amylase and glucoamylase at 60°C. Doses will depend on starch concentration and the time course selected. In general, two kilogrammes of bacterial amylase and two litres of glucoamylase preparations per tonne of starch content will provide satisfactory conversion in four hours.

Protein will have been substantially denatured during the heating stages and may be hydrolysed by adding bacterial proteases simultaneously with starch hydrolysing enzymes if the pH remains between 5.0 and 7.0. If lower pH values occur, it may be necessary to raise the pH to neutral. Cooling to fermentation temperature will allow satisfactory protein degradation. Doses around two per cent of proteolytic enzyme on protein weight are typical. While suited to many fermentations, it is often found that such materials are in fact used as the basis of yeast production, with the concentrated or dried yeast being sold to a traditional processor.

Production of new resource materials. The conversion of slaughter blood to new ingredient materials has been mentioned already and an elegant enzymic process for this is described in Chapter 4.14, 'Proteins'.

The application of cellulases and pentosanases is being tested for the upgrading of straw to provide sufficient nutritive improvement to enable it to be considered a new animal feedstuff. The chemical and mechanical pretreatments required to facilitate the optimal action of these enzymes are costly, and to date they have not produced a cost effective system. By addition of cellulases, pectinases and pentosanases to mixed silage systems containing high levels of straw, it has been shown that increased microbial activity, facilitated by enzymic degradation of grass, clover, corn silage etc, enables the straw to be incorporated into good quality silage feed.

A large amount of work has been done to establish the economic routes for the utilization of cheese whey and current opinion suggests that the use of membrane filtration techniques to recover the native whey proteins is important. The recovery of these proteins is of intrinsically high value, and they are also being subjected to proteolytic hydrolysis with plant, animal and microbial proteases in the search for novel functional character that can futher raise their market value. Emulsifying and foaming character has been produced at low degrees of hydrolysis, together with new applications for fully hydrolysed materials (*see* Chapter 4.14, 'Proteins').

The conversion of lactose in the remaining fluid by the action of lactase (β-galactosidase) enzymes is well-established (*see* Chapter 4.6, 'Dairy'). The resulting glucose and galactose sugars represent added value potentials as sweeteners and also fermentables with some markets for the syrups developing in ice cream and yoghurt manufacture. Considerable effort has been directed to the continuous hydrolysis of whole whey and filtration permeates by these enzymes, and both hollow fibre and packed bed immobilized systems are undergoing extensive trials. The soluble enzyme still represents a simple route for waste whey conversion requiring only stirred tanks for the reaction.

3. Direct effluent treatment

The effective breakdown of solids and the clearing and prevention of fat blockage or filming in waste systems is important for many industrial operations. In many cases, neither microbial action nor added enzymes separately provide satisfactory answers when treating pipework or storage tanks and lagoons. In recent years, several specialist companies have produced combination products that improve the separate treatments. Usually offered as powders or pellets, these products contain large numbers of *Bacillus subtilis* spores, inorganic nutrients and declared amounts of industrial enzymes that include amylase, cellulase, lipase and protease. They are formulated on bran to give a stable mix that degrades organic debris and accelerates microbial growth to further attack organic materials with a marked effect on fats. The best results are obtained at efficient oxygenation rates as the developing microbial content will have a high oxygen demand. One such product is *Actizyme®*, from Southern Cross Laboratories, Australia, who have many agents throughout the world. Some guide to use rates can be given, although much will depend on the ambient temperature, oxygenation levels, and the actual load of organic material in the effluent.

For sewerage treatments, doses of 1–6 kilogrammes per 45,000 litres are suggested to start the treatment, followed by maintainance levels of 0.1–1 kilogramme on a daily basis.

For lagoons and holding ranks, very low treatment levels are suggested in the range 150 to 200 grammes per million litres. When treating at institutional establishments, such as hospitals and hotels with connection to sewerage systems, it is suggested to use 15 grammes per toilet per week, while for septic tanks the treatment ranges from a single weekly treatment of 60–100 grammes depending on tank capacity.

Grease trap treatments can be established by an initial dose of around 400 grammes per 1000 litres capacity and maintained with 80 grammes per week.

These treatments can be adapted for industrial processing conditions in abattoirs, the food processing industry, the leather industry and processing of poultry wastes. In some cases, these treatments may assist the regular performance of anaerobic digesters.

4. Biogas production

Anaerobic methanogenic fermentation. In its most simple and practical form, biogas is produced in China and the Indian Subcontinent from animal manure by small communal systems numbering many millions of units. The gas is used with little or no purification as heating, lighting and cooking fuel, and it is not anticipated that higher technology will be introduced to alter its operation or efficiency.

Larger-scale manure fermentation plants have also been established in other parts of the world where sufficiently high mean temperatures allow the fermentation to operate. In a few instances, the utilization of waste low-grade heat from other industrial operations has made the process feasible. With operating temperatures in the range 30–40°C it will be seen that heating is required for successful operation in many industrial regions of the world.

This section will assume that adequate heating is available and that heterogeneous wastes are to be processed in a single or two-tank system with the introduction of cultures of appropriate methanogenic bacteria, at least to initiate reactor operations.

The anaerobic digestive system. The anaerobic digestive system yields methane, carbon dioxide and a number of other gases which may include hydrogen and hydrogen sulphide. A mixed microbial population operates in three consecutive steps which each involve different microbial types.

(*a*) The hydrolytic breakdown and fermentation of organic polymers (protein, starch, cellulose, lipid and fat) to form lower molecular weight materials (glucose, amino acids and glycerol) and finally to organic acids, ethanol, carbon dioxide and hydrogen.

(*b*) The acids and alcohols from the first stage are converted to acetate, carbon dioxide and hydrogen.

(*c*) Specific bacteria convert acetate and hydrogen to methane while using and producing carbon dioxide.

Enzymes aiding anaerobic digestion. Enzymes aiding anaerobic digestion will participate almost entirely in the first stage of the system. Most heterowastes consist of protein, carbohydrates and fats in very widely different proportions and usually in fluctuating amounts and concentrations. Typically, the acid producing fermentation will produce 15 per cent propionic and 20 per cent acetic acid with the remaining 65 per cent consisting of alcohols, aldehydes and long-chain fatty acids. The biological oxygen demand for this stage is almost constant, as molecular rearrangements are required with minimal population growth. Industrial enzymes will be able to accelerate the depolymerization of many of the waste constituents in the tank that first stage if they are selected according to knowledge of the waste components and can function at the pH. Bacterial amylases and proteases are metered into the incoming effluent at rates of 0.05–0.15 kilogrammes per cubic metre. Fungal enzymes would be used where pH values below 5.5 are found, and generally at half the rates for bacterial enzymes.

At present, there is little evidence that microbial lipases or animal extracts are sufficiently inexpensive to permit fat degradation to be aided by their addition, although fungal lipases are becoming more available.

In cases where significant amounts of cellulosic materials are present in the wastes, many projects have been undertaken to provide enzymic pretreatment to accelerate the digestion. Ligno-cellulose is not readily decomposed by industrial cellulases, partly because of the poor accessibility of water to the fibres due to lignin. Enzymic degradation of lignin is known, and the white rot fungi may be considered as sources of complex systems that demethylate, hydroxylate and oxidise the phenolic components of lignin. Addition of industrial cellulases has not yet proved an effective treatment for biogas production.

Pretreatment of complex wastes containing starches, pectins and proteins can be carried out using industrial enzymes under their optimal operating conditions to produce feed to the digester that is rapidly converted. The choice of enzymes and of the treatment needed will be made on the merits of the results and the nature of the waste materials. In all cases, the action of added enzymes will be to increase the availability of waste components to the digester organisms and not in any way to alter the overall yield of fuel gas. The most cost effective treatments with enzymes have been: (a) pretreatment of intractible wastes; (b) compensation for load shocks; (c) compensation for temperature drop; (d) startup stimulation; (e) retention time reduction and consequent smaller vessel design; (f) reduction in sludge volumes and consequent design improvement and sludge disposal needs.

5. Detoxification by digestive processing

Recent interest for the use of microbial fermentation using selected organisms to remove toxic materials from effluents has demonstrated that tolerance of these organisms to toxic loads can be increased by enzymic pretreatment of the organic waste components. It is assumed that readily available nutrients from carbohydrates and proteins encourage rapid growth and multiplication. Dilution of toxic wastes with such materials from nontoxic sources further improves the operation.

References

Birch, G. G., Parker, K. J. & Worgan, J. T. *Food from Waste* (Applied Science, London, 1976).

De Ronzo, D. J. *Energy from Bioconversion of Waste Materials* (Noyes Data Corporation, Park Ridge, New Jersey, 1977).

Webb, B. H. *Byproducts from Milk* (AVI, New York, 1970).

FLAVOURING AND COLOURING
T. Godfrey

1. Introduction

A specialized sector of industry is devoted to the supply of flavourings and colours for food and beverage producers. Whilst many of the compounds used are produced from synthetic chemistry, there is a further range which is produced by the extraction and modification of natural materials. Many of the sources are exotic and costly to harvest and deliver to the processing factories of the world, and so great care is taken to achieve maximum yields and high quality during processing. Undoubtedly, industrial enzymes are used in these extractions, but for the most part the details remain in the confidential records of the processors. In some cases, the enzymes are applied in similar fashion to the methods used for plant tissue modification (*see* Chapter 4.13), whilst the production of flavourings from yeast is described in detail in Chapter 4.24.

This section will endeavour to set down some of the principles and practices for a number of other examples which, together with data from other chapters, should provide information to assist those seeking to modify or design their extraction procedures and flavour development systems. The section concludes with some observations and suggestions for the future direction of enzyme applications in these areas.

2. Flavour systems using industrial enzymes

Flavouring substances represent between 10 and 15 per cent of the weight of world use of food additives, representing up to 25 per cent of the value of the total food additives market. Industrial enzymes find a variety of applications, ranging from the general processes for the production of raw flavour ingredients, such as sugars and protein hydrolysates, through to highly specific enzyme reactions targeted at identified flavour compounds such as 5′ nucleotides. Up to 10 per cent of the total bulk weight of industrial enzymes produced is now used to create flavour.

Sweetness. Although traditionally the province of natural sugars such as fructose (honey) and sucrose (cane and beet), a wider range of sweet substances is now produced. It is possible that the intense sweeteners which are derived from extracts of exotic plants are already being processed with the aid of enzymes degrading the tissues of the source materials, but there are no confirmed reports to hand (*see* Chapter 4.13, 'Plant tissues', for possible enzyme treatments).

Sucrose production utilizes several enzymes which facilitate the recovery of maximum yield by attacking starch residues, and also the degradation of mucilages produced by microbial contamination of juice prior to processing (*see* Chapter 4.18). Fructose is becoming increasingly available by either the processing of sucrose or the isomerization of glucose from starch processing (*see* Chapter 4.15 for this latter enzymic process). Recently, the enzymic hydrolysis of inulin has been proposed as a potentially useful industrial process and enzymes to carry out the conversion are now commercially available.

Inulin is the energy storage carbohydrate of many members of the Compositae family of plants, rich sources being the Jerusalem artichoke and chicory. The inulin, a β-linked homopolymer of fructose, will be extracted by hot water diffusion from crushed, sliced or homogenized plant tissues, and this may be aided by the action of cellulases and macerating enzymes. Microbial enzymes capable of hydrolysing inulin occur in yeasts and some filamentous fungi. The *Aspergillus* spp. enzyme which is commercially available is used at 60° and pH 4.5, and acts on the aqueous juice, which contains 13–20 per cent inulin. With enzyme doses of 2 units per gramme of inulin, 98–99 per cent hydrolysis can be achieved in 48 hours.

Dextrose and maltose/dextrose syrups are key sources of flavouring for many applications covering baking, confectionery, processed fruits, jams, soups, snack and convenience foods, dietary products and breakfast cereals. The syrups prepared from starch are made according to the processes outlined and discussed in Chapter 4.15. For many applications, the standard syrups are selected according to the sweetening and body-giving character they contribute. The more sweetening the effect per unit weight of syrup, the nearer to pure dextrose it will be, whilst the almost non-sweet maltodextrins, with dextrose equivalents of around 26 units, will contribute texture, mouth feel and bulk to the product.

Malt and barley syrups are generated largely according to the basic brewing processes described in Chapter 4.5. However, where a strong malt flavour is required, often together with a good potential to generate browning when heated, the enzyme spectrum used during mashing of the selected cereals will be changed. More nitrogen compounds will be required in the syrups than are needed by the brewer, and also a higher saccharification which will intensify browning and sweetening character.

Typically, neutral proteases unaffected by cereal protease inhibitors will be used at two or three times the doses needed in the brewing industry; for example *Neutrase®* 0.5L (Novo) would be used at levels of two or more kilogrammes per tonne of cereals mashed. Bacterial α-amylase would be used at the upper brewing levels and be supplemented with glucoamylases to accelerate and complete the conversion of dextrins to glucose. Examples of doses would be: *BAN* 240L (Novo) at three to four kilogrammes and *AMG* 200L (Novo) at up to four litres per tonne of cereals mashed.

Where unmalted cereals are used to produce a 'malt' syrup, it is necessary to introduce a maltogenic enzyme, an amylase from fungal *Aspergillus* spp., to the mash system and at the same time reduce the amount of glucoamylase. The most effective method is to delay the addition of the glucoamylase until the mashing temperature is above 65 °C and only allow reaction for a period of 45 to 60 minutes before heating to a temperature high enough to inactivate this and other residual enzymes. This method will produce syrups with a highly satisfactory 'malt' flavour. Maltogenic fungal amylases such as *Fungamyl®* 800L (Novo) need be added in low amounts of around 0.02 to 0.03 per cent of cereal weight to achieve adequate levels of maltose in the final syrups. A typical flavour syrup would have the following sugar ranges:

Glucose	5–10 per cent total dry solids basis
Maltose	35–50 per cent
Maltotriose	10–15 per cent
Higher dextrins	20–30 per cent

The characteristic brown colour, together with the flavour derived from the appropriate spectrum of sugars used, that create the required results when foods are cooked or baked, are also generated by the use of caramels. These are mostly used as colourants, but in many applications are produced *in situ* by the use of syrups. Considerable skill is used by the syrup manufacturer in the selection of type and quantity of both protease and amylase enzymes for these speciality syrups.

Savoury flavours. Meat flavours and similar condiments are produced from many sources and their production often involves a natural microbial fermentation. Amongst these can be included the oriental products based on soya protein, and the use of single cell biomass, the source of which was until recently primarily yeasts but which is now widening to include other organisms. Wheat and maize glutens also find application by their hydrolysis to flavour intense peptides.

The soya bean proteins are hydrolysed either by whole fermentations or by added proteases. In modern processing methods, the duration of the fermentative step is much reduced by prior application of plant or microbial proteases which act on the cooked mash before inoculation. Extension of soya protein by enzymic hydrolysis of other proteins is also practised using enzymes and methods described in Chapter 4.14. The utilization of soya proteins for both flavour and nutrition is the fastest expanding sector of protein use, and the signs are that this will continue for a few years yet.

All the industrial proteases have some effect on the flavour character of soya proteins. Bond specificity is very significant, together with the aggressiveness of the hydrolysis, and this is used to establish the highly individual flavours characteristic of certain producers. Information on the specificity and potency of examples of the protease types are given in Table 5.1 of Chapter 5. When soya, or other protein, is to be hydrolysed, the amount of enzyme used will depend on the degree of hydrolysis required. For conversion to highly soluble, small-peptide fragments, it is common to use up to two per cent enzyme by weight of protein. If more subtle modification is required and emphasis is placed on the control of bitterness and other undesirable flavours, then more narrowly specific enzymes will be used, in doses of the order of 0.1 – 1 per cent on protein weight.

Gentle treatment with highly specific and protease-free, carbohydrases is used to bring about the hydrolytic solubilization of undesirable oligosaccharides commonly present in soya preparations. The need for multiple attack in order to obtain the full effect is well understood, although the reason for it is not, despite the clearly known structures of these compounds, which include stachyose and raffinose. Binding to other substances or perhaps simple entrapment in the matrix of cell structures might explain the need for a broader action.

The production of soya milks includes treatment to eliminate the strong characteristic bean flavour and this is usually achieved by heating the milk. However, the use of bacterial neutral proteases at low temperatures of around 45 – 55 °C will remove this taste without causing yield losses in the milk production.

High glutamic acid levels are characteristic of soya flavour products and this has been enhanced by the application of specific bacterial peptidoglutaminases that do not degrade the peptide bonds but deaminate the α-glutamyl residues of glutamic acid in the chain. Direct addition of these enzymes is at present being evaluated in complete food formulations as a means of flavour enhancement, as well as in the production of high potency soya condiments and other speciality protein hydrolysates. In many cases, the results suggest gains in glutamic levels of 30 per cent over control values obtained without enzyme addition.

Wheat gluten has also been processed with acid proteases from *Aspergillus* spp. at pH 5 and 30–40°C to generate a flavour hydrolysate comparable to traditional acid hydrolysates, but of lower salt content. The inclusion of the specific peptidoglutaminase enzyme improves the strength of the resultant product. The use of fungal acid proteases at 0.03–0.1 per cent on weight of protein, together with 0.015–0.5 per cent of the peptidoglutaminase under the above conditions produces a satisfactory product from the gluten mash after 60–70 hours.

The 5′ nucleotides and monosodium glutamate are highly intense and valuable flavour sources and are obtained from both fermentative and partly chemical conversion processes. Industrial enzymes are little used in the production of monosodium glutamate, but 5′-phosphodiesterases from *Aspergillus* and *Penicillium* species are increasingly being used to degrade isolated pure ribonucleic acid (RNA) to yield flavour enhancing nucleotides based on guanosine, adenosine and inosine. However, details of the activity levels of the enzymes used in this industrial process are not available.

Flavour obtained from the hydrolysis of yeast and more recently from other single cell biomass cultures follows the traditional pattern of protein hydrolysis largely induced by release of the cell's own enzymes. However, this autolytic process can be accelerated by the addition of industrial enzymes, of which papain is a typical example. The flavour-inducing action of papain hydrolysis is well known and may be accounted for by its high preference for bonds containing glutamine and glutamic acid. The accelerated autolysis, however, is more likely to be due to the activity of other enzymes present in industrial grades of papain. Tests with glucanases and cell lysing enzyme complexes, which are essentially without proteolytic action, have been shown to improve the action of papain on this stage, while other industrial proteases can be used to advantage for subsequent protein attack in the preparation of defined flavours, and are particularly effective in eliminating the high salt levels produced after acid hydrolysis.

In yeast processing (discussed in Chapter 4.24), depending upon the source of the yeast or other biomass and the choice of lytic enzymes, the treatment to produce high yields of flavour materials will use from 0.8–2.5 per cent of lytic enzyme to achieve lysis. For example, the use of brewing yeast would be based on a slurry of the yeast at about 20 per cent, with pH around neutral and temperature between ambient and 40°C. Addition of a combination of bacterial glucanases and fungal enzymes to a total of one per cent on yeast dry solids and agitation for two hours will solubilize up to 80 per cent of the total cell proteins. If a high percentage of this protein is required to be acid soluble (i.e. of low molecular weight distribution), then selected proteases such as papain or bacterial neutral protease are added at the end of the first hour of agitation. A sharp rise of pH to 8.5–10.0 may be introduced (if the proteases are alkali tolerant) after the two hour incubation and the reaction continued for a further hour. This stage enhances the flavour value but raises salt levels. The total acid-soluble fraction of the protein hydrolysate is raised to 80 per cent, from traditional values of around 20 per cent, and the process time is very much reduced, thus lowering the risk of microbial contamination. A further benefit noted from the release of ribonucleases during the lytic stage is the lower nucleic acid content of hydrolysate.

There is some evidence that the optimum flavour profiles for protein hydrolysates contain a predominance of peptides which are 4–5 residues long, together with free amino acids at around 25 per cent of the total soluble nitrogen. The final development of specific meat flavours includes combination with selected sugars and then heating under controlled pH and temperature.

Cheese flavours. The traditional and modern processes for commercial cheese production are discussed in Chapter 4.6. Here we will consider the way in which the strength of the flavour is developed as an ingredient. The development of strong flavours in certain specialized traditional cheeses has long been identified as being due to the action of many enzymes from both the starter organisms and other additions made during their production. Thus, characteristics of Italian-type cheeses are the result of the heavy bias towards the use of esterases and lipases from unweaned animals, applied in the form of pregastric esterases. Their action can be mimicked by selected esterases from species of the *Aspergillus, Mucor* and *Rhizopus* moulds. The use of moulds for blue-veined cheeses confirms the role of these enzymes in contributing flavour.

In practice, strongly flavoured cheese is produced by adding protease and lipase mixtures to the scalded curds and then curing at 10–25°C for 1–2 months. The ratio of esterase to lipase activity should be as high as possible and the preferred enzyme products have ratios of two or three to one. The esterases hydrolyse short-chain water-soluble fats, whilst the lipases act on long-chain water-insoluble fats. In all cases, these lipases need to be free of coagulating proteases if they are to be added to milk prior to renneting, as the contaminating proteases are unlikely to have the narrow specificity required of a rennet. Over-treatment results in excessive development of methyl ketones in the cheese, especially 2-

heptanone and 2-nonanone. It has recently been shown that the application of bacterial neutral proteases to the curd used to produce cheddar-type cheeses results in increased rates of ripening. The flavour development appears normal and the concept may be adopted for production of intense flavours.

Concentrated cheese flavourings are produced by the rapid modification of slurries of milk solids, or casein, various fats and emulsifiers, and the use of both proteolytic and lipolytic enzymes. Pancreatic extracts and also fungal enzyme preparations are used to produce 5 to 15 times more flavour, but few data are available regarding doses and reaction conditions. However, the use of fungal proteases with acidic pH optima from *Aspergillus* species is known to facilitate the development of intense flavour when reworking pasteurized and processed cheese wastes. Doses from 0.001–0.01 per cent by weight of the emulsifier to be used have proved effective. Mechanical mixing followed by pasteurization produces a highly flavoured product.

A sour dough flavour for baking is produced by preparing a renneted curd from skim milk and lactic acid, separating the curd, resuspending it and inoculating with microorganisms producing extracellular esterases, lipases and proteases such as *Citrobacter* or *Micrococcus* species. Aerated and agitated culturing for three to five days produces a highly flavoured product that can be spray dried.

Plant oils and resins. The extraction of desired flavours and spices from many plant tissues and seeds generally involves an aqueous or alcohol and water infusion of the finely divided materials over long periods. The application of cell-wall-degrading enzyme systems such as those described in Chapter 4.13 is widespread because it facilitates shorter infusions and higher recovery levels. In many cases the enzymic treatment can be given during the recovery procedures so that aromatic volatiles are not lost. The rehydration of dried source materials is also accelerated by the application of these enzymes, which are typically from the pectinase, cellulase and hemicellulase groups and used at levels of 0.05–1.0 per cent of the dry weight of material. Similar use rates also improve the effectiveness of infusion. Where the aromatic oil is associated with a valuable oleoresin, it may be advantageous to recover both substances from the same batch of raw materials. Where steam distillation is used to recover the oils, however, it is often found that intractable materials remain associated with the resin residues, resulting in poor yields or the need for impractical processing to recover the resin. Where this is due to starch, the steam distillation can include a thermostable bacterial amylase which will degrade the starch and facilitate the second recovery. For example, the use of *Termamyl*® 120L (Novo) at 0.01 per cent of total dry solids in the distillation will be expected to eliminate any starch residues. With other residues, such as glucans, pentosans, pectins and proteins, it is necessary to process the distillation residues in aqueous suspension with the appropriate enzymes at pH and temperatures to suit their action. Subsequent solvent recovery of the resins is then facilitated.

A further route is to treat the crushed, macerated or ground raw materials with the appropriate enzymes in aqueous suspension before the start of the distillation. It should be noted that starch is not degraded under these conditions as it will not have gelatinized at the temperatures that must be used to retain the oils intact.

Increased release of oils from the distillation of spent fermentation yeasts is obtained by pretreatment with lytic enzymes and proteases. The release of higher oil levels from cold pressed materials such as citrus fruits is obtained by treatment with cellulases and pectinases at levels of 0.02–0.15 per cent of dry weight over periods of 4–6 hours prior to pressing Only 30–60 minutes should be used for orange oils, as they are subject to rapid souring.

Debittering. Debittering by the application of naringinase and limonase to prepare concentrated juices of grapefruit and orange takes advantage of the presence of these enzymes as contaminants of pectinases. They are more heat-stable than the pectinases and so a heat treatment to control the action of pectinase can be regulated to leave the debittering enzymes still functional. A further and higher heat treatment will then inactivate them. Pure naringinase is available for industrial use as a debittering enzyme.

3. Enzymic processing for colour

Colour materials represent 0.5–0.8 per cent of the total weight of food additives and some 5 per cent of the value of food additives. The use of enzymes in the extraction of natural colours is not very diverse at present, but it is receiving considerable attention at the research and development level following the rapid increase in the availability of complex enzyme preparations for the degradation of plant tissues under mild conditions.

The sensitivity of natural colouring agents to degradation has always made them unsuitable for large scale industrial use. Many of the carotenoid substances are extremely insoluble in water but solvent extractions have produced low yields. Examples of this include the yellow cheese colour, annatto (bixin), and β-carotene from many sources. The anthocyanin pigments of many fruits are firmly bound to structural elements of the skins and may be released to water-soluble products by the multiple enzyme attack of a fermentation process. Beetroot red is often released by fermentation to remove sugars and starches before extraction. Where fermentation is a standard step in colour recovery, it is generally possible to substitute a short incubation with appropriate enzymes using the basic methods discussed in Chapter 4.15.

For all the desired natural colours of plant origin it is expected that degradation of the cell structure by enzymic action on the divided, crushed or milled materials along the lines described in Chapter 4.13 will be helpful. Work on development to date has given optimistic results with the use of cellulases, hemicellulases, pentosanases and amylases. Pectinases are often useful for recovery of colour from fruit skins, but the presence of anthocyanase activity must be treated for and avoided. Although the degradation of tissues providing approved sources of chlorophyll is easily achieved with carbohydrases and cellulases in particular, the fragile nature of this molecule still poses many problems for the technical improvement of its extraction.

4. Future applications of flavour and colour production

Areas of most immediate development will be those resulting from our increased understanding of the specific substances responsible for colour and flavour in foods. Particularly in the flavour area, the identification of actual flavour substances will provide guidance to the selection of enzymes able to release or produce them from economic sources. Examples are already appearing in the selection of esterases, lipases and proteases with specific action in cheese flavour development. Peptidases will contribute to the subtlety of control of bitterness and flavour components in the hydrolysis of low cost legume and cereal proteins. Similarly, the direction and regulation of colour and flavour resulting from sugar-protein/peptide interaction in Maillard reactions will be clarified and lay emphasis on the enzymes to be used to prepare the reactants.

The detailed biochemistry of flavour components and legislative pressures to reduce the levels of salt and additives in foods will encourage flavour production by *in situ* modification of food substances and the enhancement of flavour in processed waste materials from primary food production. Some evidence for this is seen in the increasing use of enzymes to prepare materials from the blood and viscera of slaughtered animals as well as the hydrolysis of chicken skins and similar byproducts.

For both colour and flavour derived from plant tissues, there is significant activity in seeking to understand the biochemical pathways leading to the final active substance. In many cases, the source plant tissue contains large quantities of these precursors at stages in the growth cycle that do not coincide with the highest levels of the desired substance. The enzymic triggering of the last, or last few steps, to produce the target compound in high yield, from other tissues, or from a wider range of growth or ripeness stages, all represent interesting and commercial potentials.

It is to be expected, however, that the economic cost of the development of the knowledge required will not be easily accepted by the flavour and colour industries until there is more evidence that it will prove to be commercially worthwhile. The most likely route will be through the activities of the pharmaceutical and pharmacological extractives industries, whose target products have far higher intrinsic value.

Further reading

Chase, T. Jr in *Food Related Enzymes* (ed. Whitaker, J. R.) Ch.10 (Advances in Chemistry Series, no.136, American Chemical Society, 1974).

Northcote D. H. (ed.) *Plant Biochemistry* (Biochemistry Series I, Vol. 11, Butterworth Press, London, 1974).

FRUIT JUICE
W. Janda

1. Introduction

The first enzymes to be used by the fruit juice industry were the pectolytic enzymes for the clarification of apple juice. They were introduced in 1930 simultaneously by Z. J. Kertesz (1930) in the USA and by A. Mehlitz in Germany. Since then, fruit juice processing has developed into a major, highly technological industry, covering not only core fruits but also stone fruits, citrus and tropical fruits, berries, grapes and even vegetables. The functions of pectolytic enzymes have become more specialized and other enzymes such as amylases and cellulases also now form an integral part of today's fruit juice technology.

2. The different enzymes

Pectolytic enzymes. The most common enzymes in use in the fruit juice industry are still the pectinases. Their substrate, pectin, is an essential structural component of fruits where, with hemicellulose, it binds single cells to form a tissue. In the immature fruit, pectin is mainly insoluble and becomes partially soluble as the fruit matures and becomes softer. During the juice processing, when the plant tissue is disintegrated, some of the pectins go into solution, some become saturated with juice and some remain on the cell walls. Pectins from some fruits can hamper the processing or lower the quality of the juice, so that in most cases it is desirable at least to modify the pectins or even to break them down completely.

Amylases. In the late sixties and early seventies, in addition to fruits grown especially for their juice, the apple and pear juice industry began to process table fruits from cold storage warehouses. This fruit is normally picked before it is completely mature to ensure its firmness, and is then ripened under a controlled atmosphere in the cold store. Because it is not completely mature, it still contains starch, and this becomes gelatinized during juice processing and may lead to problems with filtration or haze formation. The use of amylases, especially amyloglucosidases, has become a routine way of overcoming such problems.

Cellulases. Cellulases, either as a single preparation or as part of the pectolytic enzyme preparation, are useful tools for speeding up the extraction of colour from the fruit or for the total liquefaction of the plant tissue. For example, the colour in blackcurrants is a valuable component of the fruit, but it is located in the skin cells, which are harder to disintegrate than the juice-containing cells in the fruit flesh. Extraction is also hampered by the poor permeability of the cell walls and membrane. This can be accelerated by improving the cell wall and membrane permeability either by performing plasmolysis at temperatures of 60°C and higher or by the use of cellulases in addition to pectolytic enzymes at temperatures of 50°C or lower.

When used with the total liquefaction method, the effect of the cellulases is much more pronounced, so that the plant tissue becomes completely macerated and the separation of the liquid and solid parts of the fruit can be carried out by filtration, centrifugation or static decantation instead of by pressing.

3. The use of enzymes in the production of different fruit juices

Core fruits, stone fruits and berries. Figure 4.10.1 is a typical flow sheet of a fruit juice processing line. After the fruits have been washed and sorted, and where applicable destoned and destemmed, they are disintegrated in a mill and heated to the temperature required for enzyme-prepress treatment. This is typically performed with *Pectinex® 3XL* (Swiss Ferment Co.) at 3–20 grammes per 100 kilogrammes fruit, and *Celluclast® 2.0L* (Swiss Ferment Co.) at 0.2–2.0 grammes per 100 kilogrammes fruit). The optimal temperature for enzyme-prepress treatment of core fruits is 30°C, while for stone fruits and berries the optimal temperature is 50°C when cellulases are used to improve colour extraction, and 60–65°C when extraction is performed by plasmolysis.

The prepress treatment also helps to break down the insoluble pectin, which occurs as small, slimy jelly-like particles. These

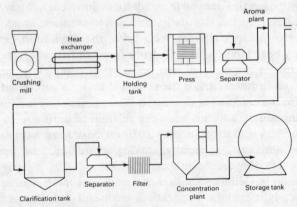

Figure 4.10.1   Fruit juice processing line.

hamper the extraction of the juice in two ways. First, they become saturated with juice, which is then not available for pressing, and second, the particles block the small drainage canals in the pomace through which the juice must run.

Prepressing can be considered to be complete when the juice reaches the desired colour level and when its viscosity has returned to at least its initial value, or even lower (at the beginning of the enzyme reaction, the viscosity increases due to the solubilization of the insoluble pectin). The core fruit crush should not be heated beyond 30 °C because this can destroy its physical structure, which is essential for the pressing operation.

After the mash has been pressed, the aroma of the juices is stripped off and the juice goes into the clarification tank. Juices from core fruits are passed through a centrifuge before they go to the aroma plant in order to separate the main part of the unjellified starch. The remaining part of the starch (some five per cent) is jellified in the aroma plant. Depectinization, breakdown of starch and clarification of the juices can be performed at 20–25 °C or 45–50 °C, typically by the addition of *Pectinex® 3XL* at 1.5–3.0 grammes and *Amylase AG 150L* at 0.5–2.0 grammes per hectolitre (Swiss Ferment Co.). The higher temperature gives an advantage due to the fact that enzymes in general, including the pectolytic and amylolytic enzymes, are more active at higher temperatures. The temperature range 25–45 °C must be avoided, however, because it creates ideal conditions for the growth of microorganisms, especially yeasts.

Depectinization has two effects: it causes coagulation of the cloud, which is stabilized by insoluble pectin, and it breaks down the viscosity-causing soluble pectin. As mentioned above, when juices from core fruits are being processed, the remaining five per cent of the jellified starch must be broken down at this stage by means of an amyloglucosidase.

After the fining of the juice, which must not be performed before the juice is completely depectinized and free of starch and dextrins, the clear juice is separated by means of static decantation, centrifugation and/or filtration from the fining precipitate, and either concentrated or pasteurized and bottled at single strength.

In the USA, where most of the fruit juice is bottled at single strength and where the consumer will accept a slightly hazy juice, the juice from the press is brought directly to the clarification tank, and the pectins are broken down only as far as required for the juice to be filtered. Fining is normally performed with gelatine alone, at the same time as the enzyme reaction. The fining precipitate is then filtered and the juice is pasteurized and bottled.

Several years ago, a new process, originating from the sugar beet industry, was introduced, which replaces the operation unit pressing by countercurrent extraction (Schobinger, 1978). Juices obtained from this process (presently mainly apple juice) normally have lower pectin levels than those obtained by pressing, but are depectinized and clarified in the same way as are pressed juices.

A future trend in this branch of the fruit juice industry will be the total liquefaction of fruits. As mentioned above, the objective of this method is to macerate/disintegrate the fruit tissue with pectolytic, cellulolytic and other enzymes to such an extent, that the liquid and solid parts of the fruit can be separated by methods other than pressing. The final product would then contain almost everything originally present in the whole fruit.

Grape juice. White grape juice and its concentrate are produced by similar methods to those used for core fruits. The grapes are crushed and destemmed, and the crush is prepress-treated with pectolytic enzymes in the dejuicer to increase the yield (typically with *Pectinex® 3XL* at 1.6–5.0 grammes per 100 kilogrammes fruit). After the free-running juice has been drained off, the remaining crush is pressed. The juice obtained is then depectinized in clarification tanks, sometimes fined, stabilized, centrifuged and/or filtered, pasteurized and concentrated or bottled single strength.

Red grapes, especially concord grapes, have a high pectin content, which makes the crush slippery and difficult to press. Furthermore, an objective of the processing is to extract almost all the colour from the skin. Therefore, the method is somewhat different from that used to obtain white grape juice. The crushed and destemmed red grapes are brought to 60–65 °C, where plasmolysis is performed to accelerate colour extraction and where the crush is treated with pectolytic enzymes for approximately half an hour to eliminate the slipperiness and increase the yield. In some cases, the crush is heated beforehand to 80–82 °C to destroy the fruit oxidases and so prevent oxidation and loss of colour.

The free-running juice is then drained off and the remaining crush pressed. Free run and press juice is depectinized with pectolytic enzymes, gently fined if necessary, stabilized, centrifuged and/or filtered and either concentrated or bottled at single strength.

Citrus industry. The four main applications of enzymes in the citrus industry today are: (*i*) in pulp wash; (*ii*) to lower the viscosity of orange juice concentrate; (*iii*) in the preparation of natural cloudifiers; (*iv*) for the clarification of lemon juice.

(*i*) *Pulp wash.* During the processing of citrus fruits, juice, peel and pulp are obtained. The pulp, which forms roughly a quarter of the total fruit, contains considerable amounts of juice which cannot be easily extracted by pressing. Methods have therefore been developed to obtain this juice using a three to five step countercurrent extraction (pulp wash) of the pulp with water. Pectolytic enzymes are used in this process to treat the pulp before the extraction and so increase the washable solids, and also to lower the viscosity of the pulp wash juice, so that a concentrate of 60 °BX can be obtained without risk of jellification.

The treatment of the pulp before extraction is done either in a continuous or a batch process. Pectolytic enzymes are thoroughly mixed with the pulp and allowed to act for about 30 minutes to break down the insoluble part of the pectin and release the trapped juice (typically *Pectinex® 3XL* (Swiss Ferment Co.) at 2.5–5.0 grammes per 100 kilogrammes pulp is used). In the treatment of the pulp washing liquid, a very limited breakdown of the pectins is carried out, just sufficient to reduce the amount of soluble pectins and so lower the viscosity without attacking the insoluble pectin fraction which maintains the stability of the cloud.

(*ii*) Lowering the viscosity of orange juice concentrate. Orange juice, prepared from certain types of fruit, can have a high viscosity and so may undergo jellification if concentrated up to 65 °BX. As with the pulp wash liquid, these problems are overcome by the use of small amounts of pectolytic enzymes (typically *Pectinex® 3XL* (Swiss Ferment Co.) at 1.6–3.5 grammes per hectolitre juice). It should be mentioned, however, that this application of enzymes is illegal in some countries, as, for example, in the USA.

(*iii*) Preparation of natural cloudifier. Because the use of brominated oils and artificial cloudifiers in citrus beverages is prohibited in several countries, the demand for suitable, natural cloudifiers, originating from the citrus fruit itself, has increased considerably. Among the different processes available today for the production of peel extract concentrates, the most effective are those using pectolytic enzymes. A typical flow sheet for such a process is given in Figure 4.10.2.

The raw material, citrus peel, to which pulp and rags may also be added, is ground to an average particle size of 3–5 mm. It is then mixed with water at 1:1 to 1:1.5 on a weight basis, heated to 95 °C to destroy the fruit pectin esterase and cooled to 50 °C. (If the raw material has a low content of pectin esterase, it need only be heated to 50 °C.) Pectolytic enzymes are added and allowed to act either batchwise or continuously for a half hour to an hour (typically *Pectinex® 3XL* at 3.5 grammes per 100 kilogrammes peel is used). During that time, the enzymes bring about a kind of maceration of the peel and release cloudy material such as cellulose, pectins, hemicellulose and cell organelles into the liquid. This

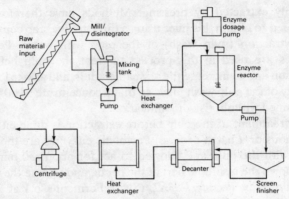

Figure 4.10.2   Plant for citrus peel extraction.

liquid is then separated from the solids, pasteurized and concentrated.

(*iv*) Lemon juice clarification. The traditional way of clarifying lemon juice involves protecting the juice from microbial spoilage by using some 1000–2000 milligrammes of sulphur dioxide per litre to allow self-clarification to occur. Depending on the pH of the juice and the temperature, this can take some 4 to 16 weeks. Such a process therefore requires high storage capacities.

Recently, pectolytic enzymes showing improved activity at very low pH values (2.2–2.8) have been available which modify the pectin in lemon juice within 3 hours at ambient temperature to such an extent that the cloud can be coagulated and precipitated with an agent such as silica sol (typically *Pectinex* ® *3XL* at 11–20 grammes per hectolitre is added). The whole process, from the fresh fruit to the final concentrate, is shortened to some 6 hours.

Future developments in the citrus industry seem to lie with debittering enzymes. Citrus bitter principles like naringin and limonin occur in certain geographical areas, and in certain fruits make it difficult to produce a juice with a pleasant but not too bitter taste.

Naringin, one of the bitter principles of grapefruits, can be converted almost completely enzymatically to its aglycone. This is not possible with limonin, which forms the main bitter substance in grapefruits and oranges. Several groups of scientists, however, are now working on the modification of limonin and its precursor, and may soon come up with an industrially feasible solution to the problem.

References

Kertesz, Z. J. *Bull. N.Y. St. agric. Exp. Stn.* **589** (1930).
Mehlitz, A. *Biochem. Z.* **221,** 217–31 (1930).
Schobinger *Frucht- und Gemüsesafte* (Ulmer, Stuttgart, 1978).

LEATHER
T. Godfrey and J. Reichelt

1. Introduction

The manufacture of leather by treating skins and hides to make them both supple and resistant to decay, or tough and hard-wearing, is one of the oldest craft industries. From the scientific point of view, there is much to learn about the nature of the empirical methods and chemical actions that have grown up over the centuries, and the section presented here will not attempt to make new claims to that understanding. This chapter attempts to set down the points where industrial enzymes are used, the nature of those enzymes and a few of the data on the use rates and choices available. In many cases, the final chemistry used by the tanner will be a closely guarded secret of his own development and knowhow, or perhaps of the chemical supplier who has also devoted much effort in making an effective product. The intention is not to make public any of those details, but rather to set down the basic philosophy from which the individual may develop his own leather processing knowledge.

2. Skin components

Leather processing techniques are all intimately associated with the structure of the animal skins as raw materials. The skin is composed of three distinct layers, the epidermis, the corium and the connective tissues of the under surface. The hairs are embedded in the epidermis within the hair follicles, and there are also glands producing oils and sweat within this layer.

The corium represents the major portion of the skin substance and consists of masses of collagen fibres in bundles, interspersed by the reticular tissue that contains other proteins and proteoglycans. This layer is bound firmly to the inner tissues of the animal by the connective tissue layer, which carries the blood vessels to the corium for nourishment.

The tanner generally considers the outer part of the skin from the epidermis to the beginning of the corium to be the 'grain area', while the surface of the corium on which the epidermis rests is the 'grain surface'. The distinction between 'skin' and 'hide' is not absolute, but for the purposes of this section, we shall adopt the general guidelines that put the smaller and younger animals into the 'skins' definition (e.g. goat, sheep and calves) and the larger animals (e.g. cow, horse and buffalo) into the 'hides'.

3. Processing steps to be considered

The six clearly identified steps in the preparation of leathers will be briefly described before the roles of enzymes in some of them are discussed.

Curing. The long-established and traditional use of the sun to dry the raw skins, the addition of salt to the flesh side as well as drying, and the steeping in a brine bath before drying, all amount to the preservation of the animal skin until it can be economically processed to leather. While there is little advantage in using enzymes in this first stage, the successful control of microbial spoilage right from the flaying of the animal onwards is becoming increasingly important in the enzymatic treatment of leather. The random and often damaging attack of microbial infections on the proteins of the skin increases the risk of producing lower quality leather and thus detracting from the advantages to be gained by the use of the more modern enzymatic methods.

Soaking. On arrival at the tannery, the cured skins or hides are rehydrated and washed to remove the gross soils. Again, the inclusion of antimicrobial compounds in these soaking stages has the same relevance as described for the curing stage. There is a trend towards reducing the use of sulphides and increasing that of alkalis and surfactants. The quaternary surfactants have been found to be excellent for improving the sanitation of the skins, but cause some problems when enzymes are also to be used. The use of non-ionic and to some extent anionic surfactants is, however, compatible with the enzymes, and the sanitizing can be done with sodium chlorite.

Dehairing and dewooling. The traditional methods used for this stage are the development of an extremely alkaline condition in the soaked and swollen epidermis and corium of the skin, then to attack the hair proteins with sulphides to break the bonding of the hair protein fibrils within the hair and to solubilize the proteins of the hair root. In many cases, the presence of these chemicals in the effluent from the tannery has proved to be a very severe pollutant problem as well as causing annoyance to people living nearby because of the odour. Reduced lime and sulphide levels were first supported by deliberate and powerful infections of the skins at this stage, but this proved far too unreliable and wasteful of leather surface by uneven attack. The incorporation of enzymes in these processes has proved very successful both in improving leather quality and in reducing pollution.

Bating. The objectives of this stage are first to de-lime and de-swell the collagen of the skins, and second to degrade the protein fibres partially so that they become soft, supple, and able to accept an even dye and demonstrate the grain in an acceptable manner. The early methods for this process were unpleasant and unreliable, as they were based on the use of animal faeces, the source of the microbial and extracellular enzyme activity required to attack the skin proteins. However, many different industrially produced enzymes are now used in the bating of leathers.

Tanning. Chrome tanning requires a pre-treatment called pickling, in which acid solutions are used to produce a further deliming without re-swelling the collagen fibres. For vegetable tanning, this pickling is not needed and the process moves directly to the introduction of the preservative and staining chemicals that give the quality and long-lasting character to leather. Enzymes have no direct role here, although there are enzymatic stages in the extraction and preparation of the vegetable extracts used in these tanning agents (*see* Chapter 4.13, 'Plant tissues').

4. Industrial enzymes used in stages of leather production

Soaking. In addition to the general washing away of dirt and excess fats from the skins, the rehydration and swelling of the layers of the skin form an important aspect of this first stage at the tannery.

Water uptake can be increased by the addition of small amounts of microbial proteases to the soak liquor bath. The alkaline proteases generally found in bio-detergents are very successful in assisting this stage as they combine well with surfactants and tolerate sodium chlorite well. The lowest level of enzyme activity that gives the water uptake in the time available should be used, and for the simple stagnant soak bath this will be of the order of 70–150 Anson protease units per 100 kilogrammes hides or skins. The activity of bio-detergent proteases can be defined in many different units according to the company which produces them, but can generally be converted into Anson Units (AU) for comparisons. An example of such enzymes is *Alcalase®* 1.5 From Novo Industri, which requires a rate of between 45 and 100 grammes per 100 kilogrammes hides or skins to give the necessary activity. *Milezyme®* 8X from Miles Laboratories would be used at approximately the same dose rate, as it has comparable activity per gramme.

The float volume for the stagnant soak would be in the range 500–1000 per cent. The use of paddle or drum operations reduces the time of contact or enzyme dose some 20–30 fold for the same effect. Figure 4.11.1 illustrates how water uptake increases in

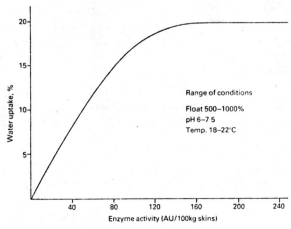

Figure 4.11.1   The increase in water uptake with increasing activity of bacterial alkaline protease in the soaking of sheep skins

relation to the amount of enzyme used.

There is some evidence that water uptake and subsequent proces-sing are improved if fats and gums can also be solubilized and washed out at this stage. Accordingly, the use of pancreatic trypsins at similar levels to those quoted above find regular application. In the leather industry there is a traditional definition of enzyme activity that is specifically used for pancreatic enzymes. This is the Loehlein–Volhard Unit (LVU), which has an approximate rela-tionship to the Anson Units of 116,000 LVU = 1 AU. This rela-tionship varies slightly depending on whether the enzyme is of either bovine or porcine origin, and also on the ratio of trypsin to chymotrypsin in the extract. Bacterial enzymes, which have a strongly alkaline optimum activity, are outside the operational equivalence range and cannot readily be quoted in LVU. If they are, however, it should be borne in mind that the Loehlein–Volhard assay at near neutral pH will give a very low activity for these enzymes, but that they will nevertheless be very potent when used at their intended pH.

The use of pancreatic trypsins for the soaking stage has the additional benefit of the amylase and lipase activities that are present, in undeclared and usually small amounts, and which may be valuable in the preparation of particularly fatty flesh side skins. In their optimal pH range, the pancreatic trypsins would be used at 3500–14,000 KLVU per 100 skins or hides. This would be equival-ent to the use of 10–40 grammes of a pancreatic trypsin having an activity of, say, 350 KLVU per gramme (about 3.0 AU per gramme).

Dehairing. A strongly alkaline medium is generally considered valuable for swelling the hair roots and follicles and so easing the removal of the hairs and soluble debris. These conditions can be obtained by soaking the skins for several days in a bath containing 10–12 per cent hydrated lime, inorganic sulphides and organic amines. Now, however, techniques have been devised which give more specific and controlled degradation that yields bright and clean hides and skins. The hair is removed at the root rather than broken off at the skin surface as is the case with the lime–sulphide methods. The reduction in the amount of lime used has a significant advantage, since the enzymes function at pH's in the range 7–11. Up to 50 per cent of the 'sharpener' chemistry can also be omitted and the replacement with alkaline proteases of suitable specificity also brings economic advantages. For example, in the presence of enzymes at an activity of 1 AU per litre of float volume, the dehairing bath would contain only half the amount of lime and sulphides required for a non-enzyme treatment, have a pH of 8–9 and bring about dehairing after about 6 hours' contact at 35–40°C. Typically, *Milezyme® 8X* and *Novo Unhairing Enzyme® No. 1* would be used at these levels, being added at 0.5–1.0 grammes per litre.

This treatment will result in significant hair-saving, an increased yield of leather area, more even dyeing later on, reduced bating treatment and significantly less polluting waste liquor with lower biological oxygen demand.

Other recent developments include the introduction of proteases selected for their specificity, as particular enzymes are more effective for some animal species than others. These selectivities must always take into account all the chemicals used in the process, for these will modify the enzyme action. Evenness of unhairing is very important and often difficult to achieve, for the way in which the liquor penetrates the skin depends upon many factors in the skin chemistry, particularly the matrix material between the collagen fibres. Some interesting improvements can be obtained by a cold soaking, which allows penetration while having low enzyme activity, followed by warming and a brief hold of a few hours to give even action throughout the skin area.

Proteases optimally active at acid pHs have often been considered, but only recently have they become available in a quantity and at a price suitable for the leather processing industry. Unhairing at an acid pH with the total elimination of lime–sulphide would undoubtedly be a satisfactory answer to the effluent problem, and work is under way to evaluate the best systems and enzymes that are most compatible with the rest of the leather treatments.

Another interesting innovation is the preparation of the unhairing enzyme in an immobilized form. One such preparation, a product called *Rapidepilase® No. 7*, has a protease bound to a clay that allows it to be mixed with the rest of the chemicals necessary for unhairing. The use rates for the immobilized enzyme preparation have not been completely defined, but it is likely that the appropriate amount of activity would be derived by immobilizing 1.25–1.5 times the amount of free enzyme needed to produce the same effect. This activity would be expected to bring about rapid depilation within four to five hours in a tumbler. According to results mentioned in a patent for this preparation assigned to Rhone-Progil (France), hides prepared in this way are characterized by well-defined hair follicles without debris, and an undamaged epidermis (US Patent 3,806,412). It remains difficult to see how an immobilized enzyme can function when the substrate is the protein of the hair follicle, but the claims are very specific nevertheless.

Elucidation of the structure and chemistry of skin to reveal the presence of carbohydrate molecules in combination with proteins in the form of proteoglycans, offers an explanation of how bacterial α-amylases act in improving depilation and bating preparations, as these enzymes may well have the ability to degrade such molecules in the skin. The consequent breakdown of some of the adhesive properties of the matrix materials can be considered to aid the action of the proteases in releasing the hair, especially when enzymes are used at the moderately alkaline pH values of between six and eight.

Dewooling. The accepted method of dewooling coarse-wooled skins is by applying dewooling paints to the flesh side and storing the painted skins at 20–35°C overnight. Higher temperatures or elevated enzyme levels in the paint aimed at reducing the contact time usually result in damaged skins. Table 4.11.1 illustrates a typical dewooling paint involving the use of alkaline bacterial proteases.

TABLE 4.11.1
Typical paint for dewooling

Hydrated lime	300–400 grammes
Sodium chlorite	2.5 grammes
Proteolytic enzyme (*Milezyme® 8X* or *Novo Unhairing® No. 1*)	50–70 AU
Water	1 litre

This would be sufficient paint for four to seven skins.

For fine-wooled skins, lower levels of these enzymes or a less active enzyme, such as the bacterial neutral proteases, are required, and the prepared formulation is often used in powder form, sprinkled evenly over the flesh side of the skin. Table 4.11.2 illustrates a typical powder for this treatment, using the appropriate enzyme. If alkaline proteases are used, they should be present at an activity of only 5–8 AU in the same basic formulation. The use rate of the powder would be of the order of 50 grammes per skin, and the skin should then be hung in the conditioning room at 25–30°C for 24 hours before the wool is pulled.

TABLE 4.11.2
Typical dressing powder for dewooling

Sodium sulphate	50–60 grammes
Sodium sulphate	10–15 grammes
Ammonium chloride	10–15 grammes
Ammonium sulphate	9–12 grammes
Enzyme:	
Alkaline	
(*Milezyme*® *8X* or	
Novo Unhairing® *No. 1*)	8–10 AU
or	
Neutral	
(*Neutrase*®; Novo)	12 AU

This would be used at at a rate of 50 grammes per skin.

A more recent development for dewooling is float soaking, which combines soaking and dewooling, and cleans the wool. It utilizes the conventional alkaline proteases with pH optima in the region of eight to nine. Mildly alkaline conditions brought about by the addition of sodium sulphate and a non-ionic surfactant, and 0.15–0.5 AU enzyme activity (for example, *Alcalase*® *Novo* or *Milezyme*® *8X*) per litre of float volume, would be a suitable mixture for an overnight soak at 25–30°C.

The removal of wool grease and gums to clean the wool and aid in the bleaching can be achieved by washing with a suitable surfactant at a pH around five to seven, with the inclusion of lipolytic enzymes from pancreatic extracts. This is known as scouring the wool. More economic lipases are now being produced from fungal cultures and are used at rates of 0.05–0.15 per cent of the float volume, depending on the activity of the enzyme product selected.

Bating. The first pancreatic trypsins to replace faecal materials were patented by Otto Röhm in 1908. Shortages of trypsins in the 1940s stimulated the development of extraction methods used to obtain insulin from the pancreatic glands which would simultaneously extract trypsins. The action of these enzymes was controlled by the levels to which they were included in the bating mixture, and dosing was aided by adsorption of the enzymes onto wood flour to dilute them. So called 'strong' bates are used for soft, supple and flexible leathers, generally from skins, while 'weak' bates are used for hard, rigid hide leathers.

The action of pancreatic trypsins, in combination with the added chemicals, is to remove any hair residues, allow water penetration but de-swell the collagen fibres, and have a minimal attack on the collagen.

Interfibrillar matrix proteins include elastin and keratin, upon which a mild attack is desirable. Fats are also present, but it has not been established whether lipase action is a positive requirement for a successful bate. The bate should also contain buffering salts such as ammonium chloride and sulphate.

The microbial or pancreatic enzymes used for a bate are selected to some extent for their pattern of specificity towards the various proteins of the skin and especially the matrix materials. Table 4.11.3 indicates the relative specificity of several proteases, including those regularly used in bates.

TABLE 4.11.3
Relative activity of different proteases on selected leather proteins

	Muscle	Collagen	Elastin	Keratin
Bacterial, neutral	++	– – –	– – –	– – –
Bacterial, alkaline	+	– – –	Trace	– – –
Bacterial, highly alkaline	++	Trace	Trace	+
Fungal, acid	+++	Trace	Trace	Trace
Trypsin preparations*	+	Trace	Trace	– – –
Papain preparations*	+	+	+	Trace
Ficin preparations*	+	+	++	+
Bromelain preparations*	Trace	++	+	– – –

*Unpurified.

The way in which the skin or hide is pretreated will affect its responses to the bating chemicals. Acidic treatments produce the most sensitive collagen because the fibres bonding to each other are damaged during swelling. Alkaline treatments afford better control, and only hides exposed to strong alkali for long periods show significantly increased sensitivity to proteases. The different areas of the hide also vary in sensitivity, with the flanks being the most strongly affected. The use of more specific and less aggressive dehairing systems, as previously described, will increase the control of the bating process.

Bating practice. Not only do the characteristics of the skins differ between species, but also different characteristics exist within a species that mean bating remains a definite craft skill, requiring knowledge of all the controls available to the experienced tanner. Temperature, pH, time, chemistry and enzyme concentration can all be varied. It is not possible to make specific recommendations. Usually, the temperature is between 25 and 35°C and the pH between 7 and 9.5 due to the action of the de-liming and buffering salts. For fine leather with strong grain, the pH is usually almost neutral. Good penetration gives controlled bating of the surface grain and also improves the bating of the inner layers.

By using acid conditions of around pH five the use of tryptic enzymes can be retained, albeit below their optimum, in preference to the use of acid optimum enzymes whose action can be too aggressive. Under these conditions the inner layers are strongly attacked, but the surface grain is only mildly bated.

Bating preparations. The early preparations devoloped by Röhm and Haas with pancreatic trypsins included ammonium sulphate and ammonium chloride in ready-to-use formulations. These enzymes were subsequently combined with bacterial or fungal proteases to give a wider bating activity, and probably also to reduce the requirement for trypsin. Many such combined preparations are in current use by bate manufacturers. The tanner's art retains the skill of selection of the finished bate to suit his raw material and the desired leather product, and the bate producer must be aware of the small variations in the preferences of his clients. The finished bates are therefore grouped into categories as those with mild, medium and strong action, with specialities designed for particular skins also available.

The enzyme content, as well as the chemical content, will be varied, and the actual enzyme type or types will not usually be known to the tanner. Typically, the pancreatic trypsins will be used at rates to suit the three main categories:

Weak bate   250– 500 K LVU per kg bate
Medium bate   800–1000 K LVU per kg bate
Strong bate   1200–1500 K LVU per kg bate

Where higher bating of the elastin component of the hide or skin is desired, all or part of the pancreatic trypsin is replaced with alkaline bacterial protease. If Loehlein–Volhard Units are used to determine the activity of these bates, the levels would be approximately half of those given above, since these enzymes are more aggressive than the Loehlein–Volhard assay would indicate.

When moderately neutral bacterial proteases are used at pH 7.5–9, the bate will contain similar Loehlein–Volhard activity to the standard trypsin bates but will require shorter bating contact times.

Use rate for bates. The dose of bate preparation will depend on the conditions chosen and the float volume. Typically, at 25 °C and with a float of 200 per cent, the bate will be used to give enzyme levels of 50–400 KLVU per kilogramme of hide or skin when trypsins are the enzymes used.

With bacterial and fungal enzymes, the use rate is less directly related to LV Units, although a rough halving of the dose in LV Units can be used as a starting point for trials. It may well be necessary to vary the contact time as well as the enzyme content to obtain optimal results with different raw materials, and this is where the microbial enzyme bates very often show advantages in rapid bating systems. Table 4.11.4 gives the range of enzyme content in the bating liquor and the contact times for a typical bate containing microbial enzymes. The treatment is for a soft and supple leather at 28–30 °C.

TABLE 4.11.4
**Range of bating protease content
(microbial) and contact times
for different materials at 28–30°C**

	Kilo LVU/kg skin or hide	Contact time (minutes)
Sheep	3–8	30–120
Calf	3–10	15– 90
Cow	4–20	30–120
Pig	10–50	30–360
Goat	10–65	30–360

5. Enzyme developments in the leather industry

Current activity is concerned with establishing the improved bating performance of enzymatically dehaired and dewooled skins and hides. The economic benefits of the reduced use of pollutant chemicals in liming and de-liming, together with very tight control of the bating process by enzyme bating, will also be of great interest. The influence of regulatory pressure in this sector can be expected to encourage further developments along these lines. These steps will not require the development of new enzymes, but it is assumed that interest in proteoglycan and fat degradation will result in the application of enzymes already used in other industries, or provide the target for new enzyme specifications.

The enzymatic processes used afford significant energy saving. It is likely that enzymes will be produced, or modified from present types, that will give satisfactory and economic action at 12–18 °C.

PAPER
J. H. Mayatt

1. Introduction

The paper industry uses vast amounts of naturally occurring raw materials, the most important of which are cellulose fibre, china clay or chalk and starch. As china clay and chalk are inorganic they have no direct role in biotechnology. In contrast, cellulose fibre and starch in their natural state are not generally suitable for the paper making process and must therefore be modified by mechanical, chemical or biochemical techniques.

In the Western world cellulose fibre is usually derived from trees to give various types of wood pulp. In other parts of the world grasses such as esparto, cereal straw or sugar-cane residues are utilized.

To obtain the cellulose fibre it is necessary to delignify the raw material source, a process which involves the use of chemicals and heat, and which is both costly and produces toxic and corrosive pollutants. For lower-grade pulps thermomechanical methods are used to 'free' the cellulose fibre bundles from the lignin and cellulose.

The use of biotechnology to replace the chemical process is under study. Although the process is still in the very early stages of development, it has already been found that the use of a cellulase-free mutant of *Saccharomyces pulverulentum* improves the burst and tear characteristics of thermomechanical pulp by some 20–30 per cent.

Starch has for many years been known to impart many beneficial properties to paper, including strength, stiffness and erasability. However, when starch is to be used for surface sizing or as a coating binder its viscosity must first be reduced. Its high viscosity is due to the fact that native starch is a polymeric substance; nature condenses the monomeric glucose unit into macromolecular chains, and enzymatic conversion is in effect the reverse of this process.

The first practical method used to reduce the viscosity of the starch was dextrinization – the treatment by acid and heat of the dry ungelatinized starch granule, followed by oxidation with sodium hypochlorite. An alternative method for molecular degradation is straightforward hydrolysis. By adding water molecules, with the catalytic assistance of enzymes, the natural process of starch granule formation is reversed. The enzymes used for this are of the α-amylase type.

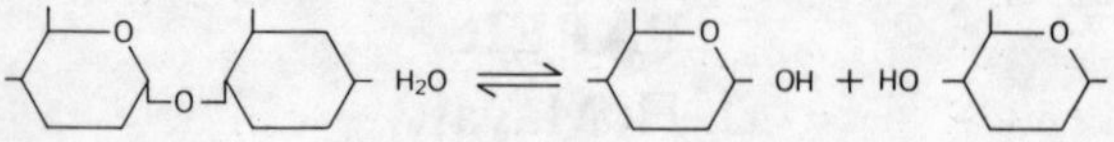

Figure 4.12.1

The ability to carry out starch conversion in the paper mill itself benefits paper makers in two ways: (a) it means that starch costs are reduced to approximately two-thirds those of specifically made starch; and (b) it allows the viscosity of the starch solution to be varied at will from only one raw material stock.

It has been found in practice that the higher the molecular weight of the starch the greater the surface strength improvement imparted to the paper substrate. Molecular weight is measured by intrinsic viscosity, which is itself a physical measure of the number of units in a polymeric chain, known as the degree of polymerization. The degree of polymerization is directly related to the molecular weight of the polymer. Therefore, the greater the degree of conversion, the lower the degree of polymerization and intrinsic viscosity.

The degree of starch conversion achieved is a balance between the type of strength properties required and the practical limitations of the paper machine.

2. Enzyme conversion of starch in the paper mill

The technique of enzymatic conversion of starch was first introduced into the paper industry in the early 1960s, although at that time the process could only be poorly controlled, and was soon overtaken by a technique known as thermochemical conversion. In this process, which has the advantage of being readily automated, starch suspensions in water are jet cooked under super atmospheric conditions in the presence of chemical oxidizing agents such as hydrogen peroxide. Over the past decade, however, enzyme technology has advanced considerably, and the α-amylases now being produced for use under specific conditions are once again being seriously considered by both the technologist and the cost-conscious paper makers of the third world.

The α-amylases available to the paper industry are produced by both fungi and bacteria. In general, the fungus-derived types are less heat stable than those produced by bacterial methods. Enzymes currently available for starch conversion are both reliable and cost effective. The activity rate of each enzyme differs at different temperatures and thus it is possible to carry out conversions at any temperature up to the point where the enzyme is denatured by heat.

The degree to which enzymes are used for starch conversion varies throughout the world. The developing countries hardly use them at all, while in the UK there has been a considerable involvement followed by a steady decline due to the increased use of thermochemical techniques. Recently, however, there has been a resurgence, and today perhaps 30 per cent of all starch used is enzyme converted. In Europe and the USA thermochemical techniques have been less successful than enzyme conversion, probably less for scientific and economic reasons than due to sales marketing policies.

When considering application techniques a number of fundamental points which involve the actual addition and application of the starch must be remembered. The major use in paper making is for size pressing of paper, that is, applying a surface coating during the paper making process. The binding of coating materials such as china clay both to themselves and to the paper substrate is less important although it accounts for considerable starch usage. To develop any adhesive property the starch must be solubilized and, as mentioned previously, its viscosity must be adjusted to a compromise between binding or adhesive power and application practicalities.

Viscosity not only depends upon polymer chain length and solids but also upon temperature. It follows, therefore, that a starch with a high degree of polymerization can be used at high temperatures to give equivalent viscosity to a starch with a low degree of polymerization used at low temperatures. Alternatively, higher starch solution solids can be used at high temperatures. Thus the paper maker can either save energy costs on drying the paper by using solids with a high starch content, or obtain a stronger paper by using starch with a higher degree of polymerization.

In general it has been found that the following conditions are suitable for size pressing

Solids	5–20%
Viscosity	5–120 mpas measured with a Brookfield viscometer at 50 °C
Running temperature	45–60 °C

The size press is a large press section consisting of two rollers through which passes the formed and almost dry (7–12 per cent water) sheet of paper. The paper, on passing through the nip of the press, picks up starch solution and is subsequently redried. The preparation of the starch solution therefore involves solubilizing the starch granules by cooking in water and reducing the viscosity.

In conjunction with Figure 4.12.2 we can consider suitable cooking/conversion plants and decide on the type of enzyme and raw starch which should be used. Enzymes stable at high temperatures cannot be used in the open tank conversion system.

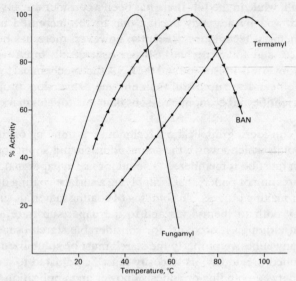

Figure 4.12.2   Activity versus temperature.

3. Comparison of systems

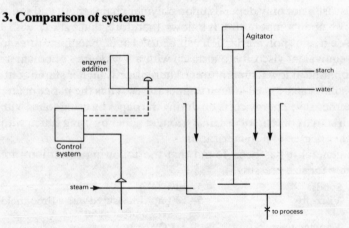

Figure 4.12.3   Batch conversion system.

Batch

Advantages Low capital costs

Simple control system for enzyme addition and reaction line

Medium to high solids conversion possible 25–40%

Low maintenance costs

Low enzyme demand

Disadvantages   Batch process lengthy
                   Not easy to change viscosity/solids rela-
                       tionship once process has commenced
                   Powerful agitator required to withstand
                       high peak viscosity
                   With many enzymes care required to
                       denature enzymes completely; often
                       involves addition of chemicals

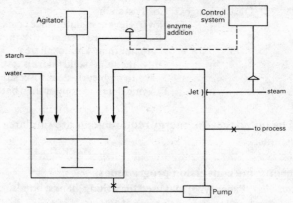

Figure 4.12.4   Recirculatory batch system.

Recirculatory batch

Advantages   Moderate capital costs
                   Simple control system for enzyme addi-
                       tion and reaction time
                   High solids possible – 50% since no
                       severe peak viscosities
                   Low agitator power
                   Enzyme kill easy by jet cooking
                   Low enzyme demand

Disadvantages   Batch process lengthy
                   Not easy to change viscosity/solids rela-
                       tionship once process has commenced

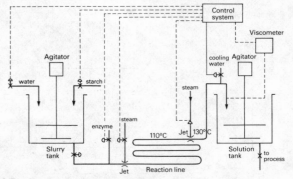

Figure 4.12.5   Continuous conversion system.

Continuous system

Advantages	Continuous process, easy to automate
	Short reaction times
	Easy control of enzyme addition
	Possibility of changing viscosity/solids relationship once the process has commenced
	More uniform steam requirements
Disadvantages	Costly system
	Not irrelevant maintenance costs
	High pressure steam requirement, greater than 4 atmospheres
	Only medium solids levels attainable due to high early peak viscosities
	Enzyme demand greater than batch processing

It should be noted that the energy requirements are similar in all the above systems.

4. Typical enzyme conversion programmes

Example 1. Paper type: that used for writing, banks, bonds.

Final starch solids required	12%
Final starch viscosity required	25 mpas (equivalent to a dextrose equivalent of 6–8)
Final starch temperature required	50°C
Use system detailed in Figure 4.12.3	
Water	4000 litres
Starch (potato)	1000 kg
Enzyme – fungal – *Fungamyl*® (Novo)	50 ml
	Adjust pH to 5.5–6.5

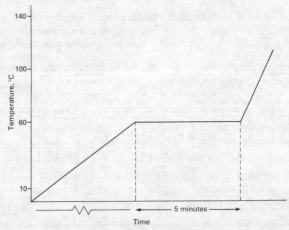

Figure 4.12.6 Cooking cycle: enzyme conversion of potato starch with *Fungamyl*®

The procedure is as follows (*see* Figures 4.12.6 and 4.12.7). (*a*) The slurry containing the enzyme is heated by steam to 60–62 °C at which stage the starch swells and begins to gel. The steam is automatically turned off. (*b*) Hydrolysis takes place with resulting thinning of the starch paste. (*c*) Steam is then added at a faster rate to raise the temperature rapidly, thus denaturing enzyme and solubilizing starch fully. (*d*) The starch solution is then diluted to obtain the desired solids and temperature. The required final viscosity is obtained by adjusting enzyme quantity or hold time.

If maize starch is being used a bacterial enzyme would be needed

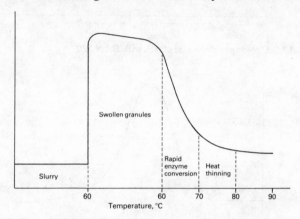

Figure 4.12.7 Viscosity change: enzyme conversion of potato starch with *Fungamyl®*

> *Example 2.* Paper type: fluting medium.
>
> | | |
> |---|---|
> | Final starch solids required | 8–10% |
> | Final starch viscosity required | 70 mpas |
> | Final starch temperature required | 50 °C |
> | Use system detailed in Figure 4.12.3 | |
>
> | | |
> |---|---|
> | Water | 4000 litres |
> | Starch | 750 kg |
> | Enzyme – *BAN 120* | 1 litre |
> | Adjust pH to 6–7.5 | |

as the peak activity period for fungal enzymes is generally well below the gel point of the starch.

The slurry is heated as for example 1 except that the hold period is at 75 °C, and it is necessary to retain the temperature at 95 °C for at least 20 minutes to denature the enzyme. Alternatively, the pH may be lowered to 3.5 or an oxidant such as sodium perborate added.

For the batch/recirculatory process similar quantities of enzymes are used as for the open tank system; however, for the continuous system, enzyme quantity will depend upon the production rate.

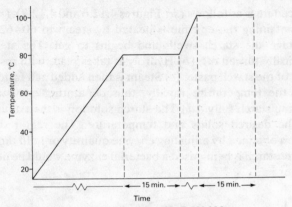

Figure 4.12.8. Cooking cycle: maize starch with *BAN 120*

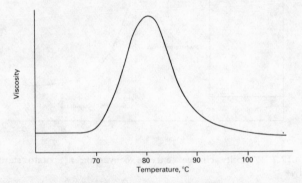

Figure 4.12.9 Cooking cycle: maize starch with *BAN 120*

5. Choice of system and costs

It is not only economic factors that decide the choice of enzyme and conversion systems used but also the type of starch available, the equipment on site, the degree of technical competence available to maintain the equipment at the paper mill and the steam pressures available for the process.

In Europe one can assume that all types of enzyme are available in adequate quantity and are of commercial quality. The starch to be converted – be it potato or maize – is similarly available. Arguments in favour of each type exist but generally it can be said that both find their place in the market.

The high cost of the continuous conversion system would preclude its installation in small mills as would probably its requirement for high pressure steam. Similar arguments exist regarding the batch recycling system.

The batch system shown in Figures 4.12.3, 4.12.6 and 4.12.7, being inexpensive and simple to maintain, would be ideal for a production unit with a starch usage of up to 1 tonne per hour with a time cycle of less than 50 minutes, using potato starch and *Fungamyl®* and an enzyme cost of less than £1 per tonne of starch converted. Such a system would probably offer a repayment on investment within 3 months for any user changing from oxidized starch and using 20 tonnes per week. The payback period for the batch recycling system would be longer.

The continuous system, although still showing savings over oxidized starch, would probably only be a sound investment if 30 to 40 tonnes per week were used.

PLANT TISSUES

T. Godfrey

1. Introduction

Throughout this book there are many examples of the enzymic modification of plant tissues relating to well-defined applications. The production of fermentable substrates for the brewing and distilling industries are major examples. Much more diverse interest is growing for the modification of many different plant materials and with far wider horizons in terms of target product. These new targets are frequently the subject of extreme confidentiality, as they relate to the production of new, commercially valuable substances, the dramatic improvement in a process efficiency, the utilization of a very inexpensive or waste raw material or the development of substitute products. Accurate and complete description of enzyme applications is thus restricted in this section.

The majority of the enzymes used are standard industrial preparations and are slotted into well-defined applications, so that degradation of structural cell wall materials, or carbohydrates or proteins, can be achieved in the appropriate sequence, or individually and exclusively, according to the overall process objective. For some applications the definitive enzymic purity of the product used will be vital if losses or modification of the desired end product are to be avoided. For example, the carbohydrases to modify soya must not interfere with the valuable protein during its isolation. In many cases, as long as there will be no damage to the target product, the more heterogeneous a collection of other activities that can be used the better and the more rapid will be the degradation process. There is clear evidence that both pure and 'cocktail' preparations of enzymes have their markets, but very different considerations are needed with regard to specification and potency guarantees. In many cases, the degree of uniqueness of the enzyme package will require that the user combines standard commercial enzymes to his own recipe.

2. Fermentation feedstocks

The general preparation of suitable feedstocks for brewing and distilling is covered in Chapters 4.5 and 4.1, respectively, and they deal effectively with the processing of starch and proteins. It remains to consider the effective modification of structural plant materials for improved fermentation potentials.

Silage processes. It has long been recognized that silage processes are the product of enzymic modification and parallel microbial fermentations, under sequential and specific conditions created by the succession of invading microorganisms. Workers in Europe and North America have markedly improved the rate of ensilage processing by the use of added cellulases when treating grass, lucerne and red clover. By using up to 4 kilogrammes of conventional industrial cellulase from *Trichoderma reesiei (viride)* per tonne of harvested material, and spraying it with the dilute formic acid traditional treatment, they showed that fermentable sugar and nitrogen products were released more rapidly and in up to 15 per cent greater total amount.

The application of hemicellulases and pectinases in association with cellulases has been the subject of much investigation for agricultural silage production. The presence of the cellulase appears to be the most reliable indicator of improvement in fermentation, but in the case of silages containing large amounts of fruit waste and pulp, the inclusion of pectinases is beneficial. Similarly, the processing of large amounts of straw or cereal husk will be aided by the action of hemicellulases and pentosanases.

In all considerations of the silage process the cost of treatment is very critical and must be related to the potential use of the finished silage. In many cases the digestibility is so improved by cellulase action that non-ruminant animals can be persuaded to make good growth on high dietary levels. The tendency for a higher water-soluble loss to occur with the increased degradation of the cellular structure of plant materials has encouraged the application of cellulases in the studies to recover leaf proteins of high value.

Leaf protein isolation. Although not yet of major commercial importance, the improved release of solubles from leaf materials by incubation of the minced tissues with 2–5 grammes of cellulase preparation per kilogramme dry weight at 40 °C for 30 minutes is marked. It is also noted that isolation of the protein is facilitated by this treatment both by flocculation and ion exchange.

Alcoholic fermentation of cellulose. Despite intense study and many ambitious pilot plant developments, the prospects for net energy gain by the conversion of cellulose to alcoholic fuels is still remote. Firstly, lignocellulose is the common source, with pure cellulose being a rare natural product. It is resistant to microbial breakdown by virtue of the hydrophobic and toxic phenolic lignin polymers as well as the predominance of β-glucose links in the cellulose itself. The tendency is for nature to produce mobile storage polymers with α-bonds and to construct the structural polymers with β-bonds. Attempts to produce fermentable materials have covered many methods of pretreatment of the raw lignocellulose and it is apparent that some of them must be retained to facilitate the subsequent enzymic hydrolysis of the cellulose.

The cost and practical limitations of these pretreatments are the restrictions to fermentative development. Once in suspension and finely divided, cellulose may be hydrolysed with good efficiency by combinations of cellulase and cellobiase enzymes. The product of cellulase action is the disaccharide cellobiose, which is able to inhibit the further hydrolysis of cellulose. The inclusion of cellobiase ensures that glucose is generated and the thermodynamics remain favourable for cellulose hydrolysis.

The immediate and continuous fermentation of the produced glucose is also considered to be the most likely way of pulling the conversion thermodynamics sufficiently in favour of cellulose hydrolysis to be economic. If the pretreatment of the feedstock is discounted, the conversion to glucose is rapid and efficient under these circumstances with better than 80 per cent yield at 50°C and pH 4.8 over 36–48 hours without fermentation, and 95 per cent yield if alcoholic fermentation is linked to the process, which then has to operate to some extent in a manner suited to the yeast fermentation conditions.

Industrial cellulases and cellobiases are characterized by their good stability, tolerance of low pH values typical of alcohol fermentations and insensitivity to many components of lignocellulase mashes. A number of β-glucosidases are available, but only some show adequate activity towards cellobiose to be suitable for this application.

3. Isolation of primary products
The modification of plant tissues to improve, or render possible, the isolation and extraction of materials of primary commercial value utilizes several enzyme treatments. Table 4.13.1 indicates the range of enzymes that influence these processes in the following ways.

Starch production. The isolation of starches from cereals and tubers requires the release of the undegraded particles as cleanly as possible, and the opportunity to wash and then recover the starch at the lowest energy cost. Protein components of starch sources usually do not interfere with starch isolation and remain undissolved. Where solution does occur, there are often problems of colour and flavour development which cannot be reduced by proteolytic action during processing. For these reasons it is important that enzymes used to assist in starch production should have minimal proteolytic contaminants.

TABLE 4.13.1

Enzymes with known applications in plant tissue modification (and common side activities)

Industrial enzyme Main activity	Common side activities
α-amylase (bacterial)	Proteases (neutral and alkaline) β-glucanase
α-amylase (fungal)	Acid protease, glucoamylase
β-glucanase (bacterial) β-glucanase (fungal)	Protease and amylase β-glucosidases, amylase
Cellulases	Hemicellulase, pentosanase, amylase, glucoamylase, protease, lipase
Cellobiases	Amylase, glucoamylase, β-glucosidases
Galactomannanase	Amylase, xylanase
Hemicellulase	Amylase, cellulase, pectinase, β-glucosidase, protease, xylanase, lipase
Ligninase	Cellulase, β-glucosidase
Pectinases	Cellulase, hemicellulase, galacturonase, arabanase, xylanase, protease, lipase
Proteases (bacterial) Proteases (fungal)	Amylase and β-glucanase Amylase, glucanase, cellulase, peptidases

Cereal starches are released in improved yield and with better processing characteristics when the slurried starch is treated with cellulase and pentosanase before centrifugation and washing. Typical doses are in the range 200–500 grammes of each enzyme per tonne of starch added to the stirring slurry at 20–40°C and without pH adjustment as long as it lies in the range 4–5.8. Contact time varies, according to the starch origin, within the range of 2–10 hours.

Gums of both glucan and pentosan origin can be hydrolysed during starch processing and so lower the viscosity of the slurry. The improved energy economy in separation and washing stages is considered valuable. Care should be used in the selection of glucanases and pentosanases to ensure that starch degradation will not be introduced by contaminant amylases. This is a very low risk if the starch has not been subjected to gelatinizing heat, which is also rare if starch recovery is intended. Use rates for the enzymes depend on the level of gums to be treated, but usually range from 0.01–0.05 per cent of the starch weight when the gum content is not known.

Protein production. In addition to the comments in the chapter on proteins, some further enzymic treatments are used when proteins are isolated from a number of cereals and legumes. In many cases the protein quality, as determined by the absence of denaturation, the retained functional properties and the characteristic digestibility factors must be retained during its isolation. It is necessary to seek very low protease activity in any of the enzymes considered for use in the improvement of protein isolation.

There are four sectors of concern in the non-protein components of plant proteins that can sometimes be treated by enzymes.

(*a*) Starch, cellulose and gums that remain with the protein and reduce its purity. Starch is the most difficult to process, since it is largely undamaged by enzymes until raised through its gelatinization temperature. Damage to protein structure at 58–70°C is common and therefore starch isolation is usually mechanical. Amylases act rapidly if gelatinization has occurred.

Cellulose and gums may be treated by the action of cellulase, with or without cellobiase, and β-glucanases and pentosanases. A stirred tank reactor containing the slurried materials at pH values from 4–6 and temperatures from 20–40°C would be dosed with these enzymes at from 0.01–0.1 per cent of the total material weight and allowed to react for 2–10 hours.

(*b*) Protease inhibitors are often present in cereal and legume products and are commonly removed, or rendered ineffective, by heat treatments during protein isolation. Where it is not suitable to apply heat to the protein, it is sometimes possible to deactivate the inhibitor by the action of carbohydrases. Some inhibitors are composed of glycoproteins and can be attacked from the glucose or other sugar end of their structure.

β-glucosidases, pentosanases and hemicellulases have been effective when used at levels of 0.01–0.03 per cent of the protein to be treated. The higher dose levels are needed when the pH optimum for the enzyme lies close to the isoelectric point of the protein and so has to be used at a pH well away from optimum. This factor is important in all treatments of proteins and influences enzyme dose and nature of action quite markedly.

(c) Off flavours in many plant proteins can be removed by the action of carbohydrases, in particular by β-glucosidases, although testing to establish doses and efficacy should be carried out on each material.

(d) Many plant proteins are isolated in forms that do not eliminate the small amounts of oligosaccharides responsible for flatulence and other gastric disorders. These include stachyose and raffinose, which can be degraded by the action of α-galactosidases. The use of more general carbohydrases, including the pectinases and β-glucosidases, has proved effective in eliminating these materials during the isolation of proteins from soya and faba beans. Enzyme treatments in the range of 0.05–1.0 per cent on protein weight at 20–40°C and pH to suit the protein extraction will usually be effective over periods of 30 minutes to 2 hours.

Isolation of oils and fats is discussed in more detail (in Chapter 4.19, 'Edible oils'). Essentially, if a wet process, or adequately water active stage, can be introduced to oils and fat recovery, then enzymic hydrolysis of proteins and carbohydrates can be used to increase yield and quality.

Production of pectins. The removal of contaminating proteins and starches from extractions for pectin production is performed by the use of standard industrial enzymes at doses that relate to the amount of contaminant to be removed. These proteases and amylases are not usually contaminated with pectinolytic activity but should be checked if pectin strength falls.

More recently, the cellulases and β-glucanases have found application in the treatment of fruit skins and extracted pomaces for the release of pectin for processing. Pulped skins or resuspended pomaces are stirred with cellulase and glucanase enzymes at 20–45°C for periods of 2–6 hours at doses of typically 0.01–0.03 per cent of solids weight.

Sucrose industry. The application of enzymes to aid the recovery of cane and beet juices and control of spoilage problems is discussed in Chapter 4.18, 'Dextranase and sugar'.

4. Specialized products and processes

There is a rapidly-expanding range of uses of various enzymes for processing plant materials to obtain special and unique effects, or to extract particular substances. The following information is an attempt to identify the enzymes used for the items listed in Table

4.13.2. The process conditions and enzyme dose can be given in only a few cases, but many examples can be obtained from recent patent literature.

Alginates. The use of complexes containing cellulase, pentosanase and pectinase is indicated to improve the yield of alginate base materials from harvested seaweeds. Treatment requires a shift from alkaline to acidic conditions which does not entirely agree with the industrial process optima. Treatment of dried seaweed during hydration seems that most advantageous application, as it increases hydration rates as well as subsequent extraction yield.

Analytical sample preparation (see Chapter 4.3, 'Analytical applications'.). The breakdown and solubilization of many substances for analytical work is currently the topic for many enzyme research workers. Many international agencies and national organizations now operate procedures that use enzymes to prepare the samples. In some cases the ability to establish regulations covering the use of a food colour or flavour depend upon a reliable recovery assay for that substance. The use of the main range of proteases and carbohydrases has been developed for sample preparation and more recently supplemented by the introduction of highly potent proteases such as Subtilisin A, and the microbial lipases to improve the release of fat-bound substances. The detection of drugs in small samples of blood, urine or for forensic work from tissues or soils, for example, has been greatly enhanced by the action of the Subtilisin protease in releasing protein bound materials to give high recovery levels and greater sensitivity.

Animal feeds. In addition to silage production, the application of commercial enzymes directly in animal feeds is now developing. The improved digestability of cereals for feeding to poultry has been achieved by the inclusion of β-glucanase, cellulase and pentosanase in pelletized barley feeds. Treatment of cereals to heighten the nutritional conversion of proteins by young animals has been an application for tryptic proteases. The use of plant and fungal proteases has not been so successful to date. The inclusion of forage materials in the diets of nonruminants is a well-established target for which the cellulases are considered vital. The improved conversion achieved to date is not yet regarded as adequate. Dose levels and enzyme types are not yet available for commercial applications, and may be awaiting the development of systems to deliver the enzymes in fully active form without inactivation by the conditions used to produce pelleted feeds.

Baking additives. Recent studies have shown a wide range of carbohydrases which have a modifying action on flour constituents and influence baking character. The traditional use of fungal amyl-

TABLE 4.13.2

Some proven applications of enzymes for industrial processing of plant tissues

Application	Action (enzymes used where known)
Alginates	Extraction and/or rehydration (cellulase, pentosanase, pectinase)
Analytical samples	*See* Chapter 4.3, 'Analytical applications' Solubilization of complex materials, unbinding of protein or carbohydrate complexes (subtilisin, proteases, pectinases lipase, amylases)
Animal feeds	Silage production (cellulase, pectinase) Direct feeds (amylase, glucanase, protease, cellulase, xylanase)
Baking	Dough and gluten modification (amylases, proteases, cellulases)
Citrus cloud	Partial degradation of pectin (modified pectinases)
Citrus juice	Reduction of viscosity of gums for concentration (cellulase, pectinase)
Coffee processing	Removal of mucilages at fermentation (pectinases) Reduction of extract viscosity for concentration (galactomannanase)
Colour extraction	*See* Chapter 4.9, 'Flavouring and colouring' Release of coloured compounds (cellulase, amylase, protease, β-glucosidase)
Cooking	Cereals, beans and pulses pretreated to give rapid cooking (amylases, pectinases, cellulases)
Enzymes	Isolation and extraction from plant tissues (cellulase, pectinase, macerating systems)
Fibre	Deactivation of protease inhibitors, processing and analyses (cellulase, pectinase, hemicellulase)
Flavour	*See* Chapter 4.9, 'Flavouring and colouring' Release and improved extraction (mainly amylase and protease, but most are used in some examples)

Gums	Isolation and purification (amylase, protease)
Maceration	Production of purées, release of pharmacological products, genetic engineering (macerating and cell wall lysing preparations – largely cellulolytic complexes)
Pharmacologicals	As for maceration
Plant crops	Breaking dormancy, spray seeding, seed cleansing (pectinases, cellulases, hemicellulases, amylases)
Soya products	Improved milks and fermentations (cellulase, protease, amylase, pectinase)
Tanning agents Tea production	Extraction, fermentation extract viscosity control (pectinase, cellulase, protease, hemicellulase)

ases to regulate the gassing power and texture development of fast-processed doughs and the altering of gluten strength for biscuits by bacterial proteases is well-established (*see* Chapter 4.4, 'Baking').

Citrus cloud products. With the increased legislative pressure against brominated vegetable oils, the use of citrus peels for cloud production for beverages has increased (*see* Chapter 4.10, 'Fruit juice', Section 3). The main concern is to have pectinolytic action with the lowest pectin esterase activity consistent with adequate yield. Pectinolytic enzymes will be selected against these criteria and dosed at between 20 and 100 parts per million on peel weight. Mashing with finely-milled peel will be about 45–60 minutes at 50°C. Pasteurization of the cloudy filtrate must be adequate to ensure complete inactivation of pectin esterase activity and to ensure cloud stability.

Citrus juice concentration is assisted also by the action of cellulases and pectinases that reduce the molecular weight of gums that are isolated with the juice and which cause excessive viscosity when the juices are evaporated.

Coffee extraction. Fermentation to remove the mucilage is aided by the application of pectinases, cellulases, β-glucanases and hemicellulases. Typically a pectinolytic enzyme with the remainder of the above activities present as side activity will be used at around 20–50 grammes per tonne of coffee, 2–10 grammes per tonne of cherries in water at 15–20°C and a contact time of 10–20 hours. Mucilage breakdown is at about double the normal rate and the coffee is left very clean.

Coffee extract concentration is aided by the use of galactomannanase to reduce the molecular weight of a number of soluble gums that are coextracted with the coffee. The lower viscosity of the extract facilitates evaporation to higher solids levels as feed to spray and freeze-drier equipment.

By holding the extract for 15–20 minutes at 65–75 °C and dosing with a suitable enzyme at 0.05–0.1 per cent on weight of solids to be treated, the optimum improvement in evaporation can be achieved so that substantially lower water removal loads reach the expensive final drying stages.

Other treatments.

In colour extraction carbohydrases can facilitate the release of colours from, for example, grape, blackcurrant, rose hip, and stability may be improved by removal of parts of molecules not responsible for colour but subject to degradation by, for example, β-glucosidases or proteases (*see* Chapter 4.9, 'Flavouring and colouring').

Cooking treatments of plant materials, especially cereals, can be improved by a pretreatment with enzymes such as pectinases or amylases. Certain enzymic procedures can reduce the cooking times and temperatures needed for cereals, beans and pulses which find application in convenience food products.

Enzyme isolation is often assisted by the application of tissue-degrading enzymes and the use of the cellulases, pectinases and macerating enzymes all find application here.

Fibre for dietary purposes sometimes contains protease inhibitors and these can be deactivated by the use of pectinolytic enzymes. A similar treatment with mixed pectinases and hemicellulases has been adopted for accelerated retting of flax.

Flavour substances are extracted from many plant tissues with the use of starch hydrolysing and proteolytic enzymes to facilitate the recovery from tissue macerates (*see* Chapter 4.9, 'Flavouring and colouring').

Gums for use in food and specialized oil drilling applications are purified by treatment with microbial proteases and in some cases amylases are used to eliminate traces of starch.

Maceration of vegetables and fruits for the preparation of specialized diets, sauces and infant and junior foods utilize the cell tissue degrading complexes found in some cellulase, pectinase and macerating enzyme products.

Thus some enzymes are now being widely used to facilitate cell wall lysis of bacteria, fungi and plants in specific treatments for genetic engineering and breeding techniques.

Pharmacological products of plant origin can be extracted with improved yield, and sometimes multiple products can be isolated using the macerating enzyme systems.

Plant crop treatments now include enzymes to aid the breaking of dormancy by increasing water permeability of the seed coat. Some seeds are produced in mucilage, for example tomato, and these can be cleansed for dry packaging by treatment with pectinolytic enzymes.

Soya products such as milks and condiments are now produced by accelerated and controlled processes that include proteolytic hydrolysis to reduce unwanted flavours and to increase the yield of solubles. Improved fermentation of condiments is observed when carbohydrases and proteases are used during the preparation of the base medium.

Tanning agents can be extracted with improved efficiency from bark and other tissues by pretreatments with lignases. Subsequent processing and drying can be aided by the action of pectinases to reduce the viscosity of aqueous extracts. Due to the tendency of tannins to inhibit enzymes, the enzyme dosages are often substantially higher than expected from other plant tissue processing levels.

Tea fermentation and the concentration of extractives are both aided by the application of hemicellulases and pectinases.

5. General considerations
The degradation of plant tissues for more defined targets is of increasing importance, and many highly valuable products are being produced. As the application of the available enzymes increases, it is becoming apparent that many enzymes have remarkably different activities. Most of the pectinase preparations, together with the cellulases, hemicellulases and macerating complexes, contain a number of undefined and unspecified side activities and it is these that are providing the different actions. When considering the application of enzymes for plant tissue processing, it is therefore still necessary to make trials with a selection of these enzyme preparations to establish which is the most effective. It should also be remembered that in most cases the chosen product will be standardized on one defined character and not have a reliable side activity content.

When the influence of a particular activity can be correlated to a specific and analysable enzyme action, it is possible to collaborate with enzyme producers to obtain reliable supplies, even though it is a side activity from a primary production.

The separation and identification of these activities in the complex carbohydrases is being undertaken by industrial enzyme companies and a wider range of standardized products can be expected to emerge. Nevertheless, in practical application many activities will be required, acting sequentially or perhaps synergistically to hydrolyse the extremely complex and highly varied plant structural materials. To date the applications that have been described are mostly the result of trial and error investigations using the complex nature of current enzymes of these types to good effect.

References

Pintauro, N. D. *Food Processing Enzymes – Recent Developments* (Noyes Data Corporation, Ridge Park, New Jersey, 1979).

Slesser, M. & Lewis, C. *Biological Energy Resources* (E. & F. N. Spon, London, 1979).

Stumpf, P. K. & Conn, E. E. *The Biochemistry of Plants* (Academic, New York, 1980).

Whitaker, J. R. *Food Related Enzymes* (American Chemical Society, 1974).

PROTEINS
D. Cowan

1. Introduction
The theory that large sections of the Earth's population were suffering from a shortage of protein has been modified by the discovery that the shortage was primarily an energy shortage and that protein malnutrition played a secondary role. Animal protein is, however, costly and its production requires much greater input in terms of energy and agricultural land than does that of vegetable proteins.

The concept of enzymatic modification of protein is not new, as illustrated by the dairy and leather industries. However, the high cost of animal protein has led to development of the enzymatic treatment of proteins to increase their value and availability and for their recovery from scrap or waste materials. A wide range of enzymes and substrates are currently available for examination and the number of modified proteins resulting from these processes grows daily.

2. Enzymes
Industrial proteolytic enzymes are derived from a wide range of sources and may have widely differing pH and temperature optima and specificities. They are also one of the few groups of industrial enzymes which are still obtained from animal and plant as well as microbial sources. When examining this group, it is important to differentiate between those products with a genuine application in industrial enzymology and those which, whilst possessing interesting properties, may not be used for technical, legal or commercial reasons. For example, there is a constant demand for a true collagenase, but to date the only organisms found to produce this enzyme (*Clostridium perfringens* and *Clostridium histolyticum*) cannot be used due to the prevalence of this genus to produce pathogens and toxins.

The choice of protease for a particular use depends primarily on its specificity, but other criteria such as pH and temperature conditions and the necessity for activators must also be taken into account. A summary of the properties of the various industrial proteases is given in Table 4.14.1 The range of pH shown is that in which the enzyme shows greater than 80 per cent of the maximum activity. The maximum operating temperature is the maximum

TABLE 4.14.1
Properties of various industrial proteases

Enzyme	Source	Protease type	pH optimum	Maximum operating temperature	Number of bonds attacked
Plant proteases					
Papain	*Cariaca papaya*	Sulphydryl	5–7	65–70	9
Ficin	*Ficus glabrata*	Sulphydryl	5–7	50	4
Bromelain (stem)	Pineapple	Sulphydryl	5–7	50	–
Animal proteases					
Pepsin	Bovine, porcine stomach	Acid	2.5–3.0	60	5
Chymosin (rennin)	Calf stomach	Acid	3.5–6.5	45	2
Trypsin	Bovine, porcine pancreas	Serine	6–9	45	2
Chymotrypsin	As trypsin	Serine	6–9	38	3
Microbial proteases					
Fungal acid	*Aspergillus saitoi*	Acid	2.5–4.0	45	9
Fungal neutral	*Aspergillus oryzae*	Unclassified	4.5–7.0	45	9
Fungal alkaline	*Aspergillus oryzae*	Unclassified	8–9	45	5
Fungal milk coagulant	*Mucor miehei*	Unclassified		55	2
Bacterial neutral	*Bacillus subtilis*	Matalloproteinase	5–7.5	50	6
Bacterial alkaline	*Bacillus licheniformis*	Serine	8–9	55	7

practical temperature at which the enzyme may be used for reaction periods not in excess of four hours. For longer reaction periods a lower temperature should be used, and this is also the case if the enzyme is operated outside the optimum pH conditions.

In an attempt to allow comparison of different proteases, their mode of attack on a well characterized substrate (β-chain of oxidized insulin) has been studied (Svendsen, 1976). Both the number and the types of bonds attacked provides information as to the specificity of each enzyme and how this might best be fitted to the choice of protease. The coagulation of milk in cheese making is the most widely quoted example of this principle. A subtle modification of the casein is required and whilst almost any protease can destroy the physico-chemical equilibrium in milk, causing it to clot, only rennin and the specially selected microbial coagulants (e.g. Novo's *Rennilase®*) are able to give clotting without subsequent extensive curd hydrolysis. Indeed, some proteases dissolve the milk clot within seconds of its formation.

At the other extreme, protein hydrolysates having maximum solubility require broad spectrum proteases which give extensive amounts of short chain peptides and amino acids. As no single industrial enzyme is capable of effecting complete hydrolysis, proteases are often combined to increase the range of bonds hydrolysed and to allow for synergism between the differing products.

However, enzyme specificity is not the only criterion of interest; as noted above, pH and temperature effects are also of great importance. Further more, the normal pH of the material to be treated will also determine the choice of enzyme, as it is not always desirable, or feasible, to alter the pH of the system to suit an enzyme. In many hydrolysis procedures, alteration in pH results in increased salt content and/or changes in flavour. There is a cost penalty and often a flavour change to consider.

Operating temperatures are also important, as a high system temperature reduces the problems of microbial contamination. However, this may be a disadvantage when enzyme inactivation is required, as extra energy input is needed, which itself may have a deleterious effect on flavour or functional characteristics.

The end result of these considerations is that the potential user makes an initial choice of enzyme based on the specificity required and the conditions necessary for operation. He can thus draw up a short-list of possible enzymes and the suitability or otherwise of these in practice may only be determined by practical trials.

Plant proteases. The main biochemical properties of the plant proteases are shown in Table 4.14.1. Papain and ficin are similar enzymes, although papain is a broader spectrum enzyme than ficin.

This is due in part to the presence of chymopapain in commercial papain, which extends the range of bonds hydrolysed by the enzyme. Bromelain, whilst similar to the other two plant proteases in its action on the β-chain of insulin, is a glycoprotein and shows a differing mode of attack on glucagon. Table 4.14.2 summarizes the main areas of application for these three enzymes.

TABLE 4.14.2

Plant and animal proteases

Enzyme	Distribution within industry	Special factors
Papain	Beer chillproofing Meat tenderization Protein hydrolysates Baking	Wide range of purities Broad spectrum protease
Ficin	Meat tenderization Protein hydrolysates	Similar to papain but higher specificity
Bromelain	Meat tenderization Digestive aid Clinical applications	A more recently developed product than papain or ficin but with a long history of practical use
Trypsin	Leather manufacture Protein hydrolysates Clinical application Membrane cleaning	Range of purities Narrow spectrum enzyme
Pepsin	Functional protein hydrolysates Cheese production	Acid protease used primarily as rennin extender
Rennin	Cheese production Casein production	Historical milk coagulant

Supplies of all three products have fluctuated in past years due to environmental and political factors. It is to be hoped that the recent trend towards consistency of supply within this area will continue. Further data on these products are given by Leiner (1974) and Caygill (1979).

Animal proteases. The range of activity types within this group is considerably larger than that of the plant proteases (*See* Table 4.14.1.) Trypsin is available in a number of forms ranging from pancreatin, which is a crude pancreatic extract containing trypsin, chymotrypsin, lipase and α-amylase, through to pancreatic trypsins containing trypsin and chymotrypsin. In addition, there are highly purified trypsins and chymotrypsins for the specialized production of dietetic protein hydrolysates.

Pepsin, the most acid-tolerant protease, is not used as a single enzyme to the same extent as trypsin, but the enzyme has found application for the production of specifically modified foaming proteins. In addition, it is often found as an extender of rennin in the dairy industry. Rennin is also an acid protease but is used exclusively for casein modification in the dairy industry. In addition to clotting activity caused by disruption of the casein colloid in the milk, it also produces a general proteolytic hydrolysis of the casein. The exact functioning of this latter hydrolysis is open to some doubt and it has been suggested that the observed hydrolysis is not enzymatic in action. The picture is further complicated by the inclusion within commercial rennets of small amounts of pepsin; levels of up to 20 per cent have been encountered in commercial calf rennet.

The availability of these enzymes is dependent upon a supply of organs from abattoirs. Trypsin is now produced following extraction of pancreatic glands for insulin and is therefore readily available. Local conditions may restrict rennin availability due to the high cost of the slaughter of young calves (see Chapter 4.6). In addition, certain religious practices forbid its use and microbial coagulants are used instead. Detailed reviews of the action of the animal proteases may be found by reference to Boyer (1971) and Hoffman (1974).

Microbial proteases. Although a large number of organisms producing proteases have been described, relatively few are of industrial importance (see Table 4.14.3). It is in this group, however, that the largest single production of industrial enzymes may be found: the detergent alkaline proteases from *Bacillus licheniformis*. Both fungal and bacterial extracellular enzymes are available and the majority of them are produced by submerged fermentation processes.

Fungal proteases acting optimally at acid and neutral pH are used in the modification of wheat protein for baking. The acid optimum protease may be used in silver recovery in the photographic industry and in the production of protein hydrolysates in the fermentation industries. They also have a long history of use in the production of Oriental fermented foods, as they are secreted by the microrganisms carrying out the fermentation.

Fungal milk coagulants are now well established for use in the production of cheese. Early criticism of the higher thermal stability of these products compared with calf rennet has been overcome by the production of destabilized second generation microbial coagulants such as Novo's *Rennilase*® XL (see Chapter 4.6).

The use of bacterial proteases in the food industry has grown from their initial application in the brewing industry, where they

TABLE 4.14.3
Microbial proteases

Enzyme	Distribution within industry	Comments
Fungal acid	Baking industry Photographic industry Brewing and fermentation industries Protein hydrolysates	Low purity enzyme with other activities. May be produced by surface fermentation
Fungal acid (milk coagulant)	Dairy industry for cheese and casein only	Second generation enzymes now identical to calf rennin
Fungal neutral	As for fungal acid	As for fungal acid
Fungal alkaline	Little application within industry	
Bacterial neutral	Brewing industry Gelatine industry Photographic industry Protein hydrolysates Cheese ripening Meat industry	High purity, medium specificity, well characterized enzyme. Well supported by applications data
Bacterial alkaline	Gelatine industry Leather industry Meat industry Protein hydrolysates Enzyme detergents	As for bacterial neutral protease

provided a source of assimilable nitrogen for yeast nutrition. Protein hydrolysates are now being prepared from a wide range of substrates using both alkaline (e.g. *Alcalase®*) and neutral (e.g. *Neutrase®*) optimum proteases (*see* Table 4.14.3). Outside the food industry, bacterial proteases are used in silver recovery from photographic waste, leather preparation, sugar recovery from confectionery and the cleaning of plant and equipment with protein-bound soiling (*see* Chapter 4.7 'Detergents').

3. Applications of proteases

Enzymatic treatment of protein is carried out for a number of different reasons. The protein may be modified so as to increase its value; alternatively, the aim may be extraction of the protein to turn it into a more usable form.

Estimates of protein wastage vary from country to country, but in the UK it has been put as high as 15 per cent. Recovery of these proteins would provide a useful source of revenue and contribute to the overall efficiency of protein utilization.

This section on the applications of proteases will examine practical examples of how these enzymes may be used to fulfill the above aims. The reader should bear in mind that, whilst a representative cross-section of such applications will be presented, the purpose here is to provide guidelines as to what is achievable and to suggest lines of investigation, but not to cover every possible permutation of enzyme, substrate and process.

(*i*) Bitterness. Whenever protein hydrolysis is considered, one of the first questions raised is that of bitterness. It is an accepted fact that proteolysis can give rise to a bitter taste; this is associated with hydrophobic peptides. Methods of bitterness control are of three types: masking, removal and prevention. Combinations are also used.

Tokita (1969) was able to mask bitterness in casein hydrolysates by the addition of polyphosphate. Presumably, it reacted with the bitter peptides and prevented them from having a taste effect. Schwille *et al.* (1976) have shown that the addition of gelatine to the hydrolysis reaction will also mask bitterness. Stanley (1981) has suggested that this masking effect is due to the high concentration of glycine (an amino acid known to be sweet) in gelatine hydrolysates.

Various schemes for the removal of bitter peptides have been proposed, including chromatography and butanol extraction. Such methods are both costly and unsuited to food processing. However activated carbon and glass fibre filters are also good removers of bitter peptides (Helbig *et al.*, 1980). Treatment of hydrolysates with activated carbon for colour removal is a normal practice and i would now appear that it serves more than one purpose. Adler-Nissen and Sejr-Olsen (1982) have demonstrated that bitter peptides may be reduced by acid precipitation at pH 4.5. It is significant that acid inactivation of the protease is a preferred process step in the production of a highly soluble soy hydrolysate. The acid may also help to mask the bitter taste (Sejr-Olsen and Adler-Nissen, 1979).

The third method of control is avoidance or prevention. This may be achieved by control of the hydrolysis reaction; for milk protein, for example, a shorter reaction period gives less bitterness. In addition, the specificity of the protease should be such that the production of hydrophobic peptides is minimized. For example, *Alcalase*® (Novo) is known to cleave proteins at the carboxylic side of hydrophobic amino acids, which should result in less bitterness than if the cleavage was at hydrophilic amino acids (Svendsen, 1976).

(*ii*) Methodology. The successful modification of proteins depends upon tight control of the process. There are two main control parameters relevant to this situation: protein yield and degree of hydrolysis. Protein yield is simply measured by determining the amount of nitrogen that the enzyme has brought into suspension (e.g. by Kjeldahl digestion) and then multiplying by the appropriate conversion factor to obtain protein concentration. A further refinement is to determine what percentage of this is soluble, by centrifuging the suspension and then measuring the nitrogen content of the supernatant.

The second, and perhaps more important, parameter is degree of hydrolysis (DH). Measurement of protein in solution gives no information as to the state of the protein other than that it is soluble. As functional properties and bitterness are associated with peptide chain length, it is important to be able to measure this parameter. The degree of hydrolysis of a protein is that percentage of the total number of amino bonds in the protein that have been hydrolysed by the enzyme. If the pH stat concept of Sejr-Olsen and Adler-Nissen (1979) is used, the carboxyl groups liberated may be titrated with base to keep the pH constant. It follows that a defined degree of hydrolysis will be obtained when a calculated amount of base (equivalent to a certain number of carboxyl groups) has been consumed. This may then be used to indicate that the reaction has reached a desired end-point and the proteolysis may then be terminated. However, it must not be forgotten that the degree of hydrolysics is an average of all the peptides generated and that proteolysis does not produce peptides all of the same chain length.

Direct treatment of protein substrate. (*i*) Soy protein. Although widely used in the food industry, the functional properties of soy protein are not always those desired. Enzymatic modification can be used to produce products of improved functionality compared with the untreated protein. There are two classes of soy protein hydrolysate, the highly functional hydrolysate and the highly soluble hydrolysate. For the first class, it is suggested that the low rate of inclusion of this product reduces the impact of any off or bitter flavours. The highly soluble hydrolysates, on the other hand, are used at relatively high concentration and therefore strict control of bitterness is necessary.

Highly functional hydrolysates may be prepared from soy isolate or concentrate and the method of production of the protein affects the final result. Traditional extraction methods produce greater protein degradation and hence a greater susceptibility to hydrolysis.

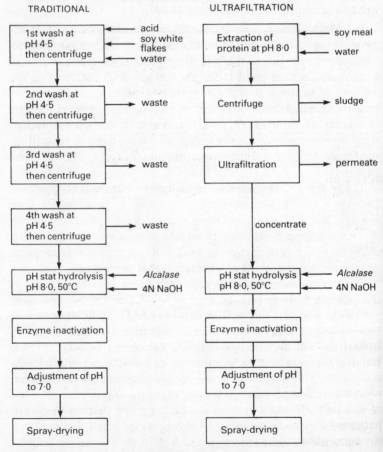

Figure 4.14.1 Soy protein treatment by traditional and ultrafiltration processes.

However, recent studies have shown that soy protein produced by membrane isolation methods yields, after hydrolysis, material of improved functional properties. Flow diagrams for the two extraction and hydrolysis procedures are shown in Figure 4.14.1. In both cases, the hydrolysis step was performed at 50°C, pH 8.0 in a pH stat and with an enzyme dose of 20 grammes *Alcalase®* 0.6L (Novo) per kilogramme of protein. The functional properties of the hydrolysates, and the isolated proteins processed in the same manner but not hydrolysed, are shown in Table 4.14.4. Hydrolysis to a degree of hydrolysis of three resulted in little difference between the emulsification and iso-electric solubility values of the traditionally isolated protein and the membrane isolated protein. However, the whipping expansion was far higher in the membrane isolated protein. Improved whippability is also claimed for soy protein modified by papain, when the soy protein is also produced in a less denatured form (Gunther, 1979).

TABLE 4.14.4

Functional properties of soy hydrolysate

| | Soy isolate | | | |
| | Traditional | | Membrane | |
	Before hydrolysis	After hydrolysis[*]	Before hydrolysis	After hydrolysis[*]
Iso-electric solubility	5%	42%	20%	44%
Emulsification capacity	100 ml/g	280 ml/g	130 ml/g	220 ml/g
Whipping expansion	20%	200%	380%	1100%

[*] Degree of hydrolysis 3.

Protein hydrolysates with a much higher degree of solubility require more extensive hydrolysis. A degree of hydrolysis of 10 has been found to give high solubility, good protein yield and low bitterness when soy protein is treated with *Alcalase®*. The process is summarized in Figure 4.14.2. The resulting hydrolysate has been used to produce protein-fortified soft drinks and in the construction of special dietetic feeds for the hospital sector, in particular for patients suffering from cancers which affect their eating patterns. If soy concentrate is the starting material, then the resulting hydrolysate may be incorporated into calf milk replacers.

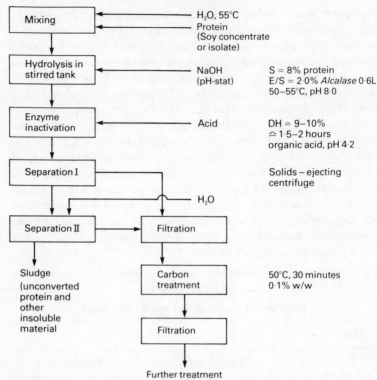

Figure 4.14.2   Iso-electric soluble protein hydrolysate. The flow sheet shows the production of a non-bitter soluble soy protein hydrolysate suitable for incorporation into soft drinks and other low pH foods.

(*ii*) Milk and whey proteins. Hydrolysis of milk proteins presents problems because of the nonspecific clotting action of the protease and the tendency for casein hydrolysates to be very bitter. Hydrolysis of casein is, however, possible and casein hydrolysates can be readily prepared by the use of proteolytic enzymes such as *Alcalase*® and *Neutrase*® (*see* Figure 4.14.3.).

Casein hydrolysates are largely prepared for incorporation into dietetic products. In cystic fibrosis, enzyme secretion by the pancreas is impaired and tryptic and peptic hydrolysates of casein are used in the treatment of this illness. However, the difficulty in controlling bitterness with this protein may make other milk proteins more desirable for this application.

Caseinates may also be used to prepare hydrolysates, the sodium type having less tendency to clot. Van Kremenburg (1974) describes a process by which proteinases, including *Alcalase*® and *Neutrase*®, are used to stabilize calf milk replacers containing fat (20 per cent), soy protein concentrate (15 per cent), whey powder

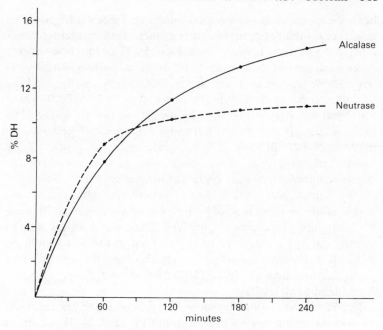

Figure 4.14.3 Hydrolysis of casein with *Alcalase*® and *Neutrase*®.

(55 per cent), sodium caseinate (9 per cent) and minerals (1 per cent). An enzyme dose of 0.05 per cent by weight *Neutrase*® 0.5L or *Alcalase*® 0.6L at pH 6.6 or 7.5, respectively, was used to treat a 10 per cent suspension of the components. The hydrolysis resulted in the production of a stable homogeneous suspension with no tendency to separate.

TABLE 4.14.5
Solubility improvement of various milk proteins following hydrolysis with *Alcalase*®

	Whey protein		Milk protein		Skim milk		Casein	
	a^*	b^*	a	b	a	b	a	b
Degree of hydrolysis (DH)	0	12	0	14	0	14	0	15
Protein solubility % at pH	7	90	0	90	10	90	0	90

* a, Before hydrolysis; b, after hydrolysis.

Enzymatic modification of whey protein is now being investigated as the membrane isolated, and other whey proteins, have been found to have less than the required functionality. Hydrolysis of whey protein with *Alcalase*®, *Trypsin* and *Esperase*® have all been shown to result in increased solubility. Table 4.14.5 gives the results of solubility improvement of a range of milk proteins following hydrolysis with *Alcalase*® at pH 8.0, 50 °C for four hours and at an enzyme dose of 20 kg enzyme per tonne of protein. An alternative enzyme (*Esperase*®) was used by O'Keeffe and Kelly (1981) to solubilize denatured whey protein by hydrolysis at pH 8.0–9.0 at 50 °C and one per cent enzyme by weight. Ammonia was used to maintain the pH in excess of 8.0 and it was eventually removed by the evaporation process used to concentrate the product. The hydrolysed whey protein was then used as a replacement for skimmed milk in a calf milk replacer formulation.

(*iii*) Wheat gluten has for many years been subjected to acid hydrolysis to yield a range of hydrolysates for flavouring. Because of the nature of the process involved, these materials are high in sodium chloride, a mineral whose content in the diet should be reduced. Enzymatic hydrolysis of gluten offers the possibility of producing low salt hydrolysates with a spectrum of flavours ranging from bland to highly flavoured.

(*iv*) Yeast and single cell protein. The production of yeast extracts by proteolytic enzymes is examined in Chapter 4.24. However, in addition to the extraction possibilities of proteolytic enzymes, these

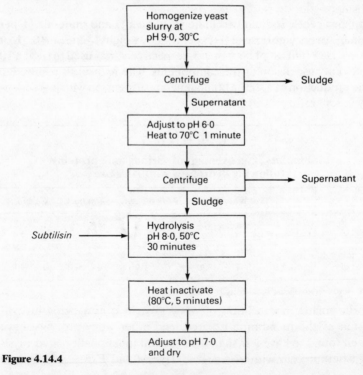

Figure 4.14.4

materials can be used to change the functional properties of the extracted protein. Proteolytic enzymes have been used to modify heat precipitated protein from various *Saccharomyces* and *Candida* yeasts (Anon, 1978). Figure 4.14.4 outlines the process used to prepare the functional protein. The resultant protein is suggested as a replacement for other proteins such as sodium caseinate, non-fat dry milk, gelatin and egg albumin.

(*v*) Collagen and gelatine. Gelatine is produced from collagen contained in bones and animal hides. It is found in two forms: type A which is prepared by acid treatment of collagen, and type B, derived from alkaline treatment. The alkaline process using hides as the collagen source requires that the hides be soaked for 5–12 weeks in a solution of 5–15 per cent lime. Petersen and Yates (1979) described an enzymatic conditioning process in which the lime is replaced by an alkaline protease (*Alcalase®*) and the process time reduced to 24 hours (*see* Figure 4.14.5). Following conditioning, the gelatine is extracted into hot water and then dried normally. Gelatine of high enough quality for use in food may be produced as long as the enzyme used is fully inactivated to prevent it from subsequently degrading the gelatine.

Gelatine hydrolysates have also found application in the food and cosmetic industries. The low level of some amino acids (e.g. tryptophan) in gelatine prevents its use as a sole protein source, but in combination with other proteins it has dietetic applications. The

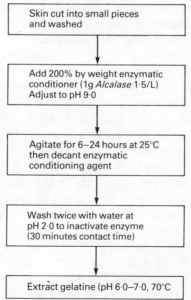

Figure 4.14.5   Collagen conditioning.

cosmetic industry uses hydrolysed gelatine as a component of shampoos and ointments and recently manufacturers of wine and juice have found it to be a useful wine fining agent.

For maximum viscosity reduction, a combination of enzymes is used. A mixture of gelatine and cold water (30 per cent gelatine, 70 per cent water) is slowly heated to 50–55 °C to dissolve the gelatine. The pH is adjusted to 7.0 and *Alcalase®* and *Neutrase®* added, both at one per cent of the weight of gelatine. The reaction is allowed to continue for up to four hours to obtain the maximum decrease in viscosity, although the gel forming ability is very rapidly destroyed. By shortening or extending the reaction period, hydrolysates of differing properties may be prepared.

(*vi*) Brewing. The brewing industry is one of the major users of proteolytic enzymes. In the production of the brewery wort, neutral optimum bacterial proteases (e.g. *Neutrase®*) are used to solubilize protein from barley adjuncts and to activate β-amylase (*see* Chapter 4.5). Papain is used extensively for removing proteins that precipitate during the cold storage of beer, by hydrolysing them to such an extent that they remain in solution.

Recovery of scrap protein. (*i*) Blood proteins. The slaughtering of animals for meat results in the production of the so-called 'fifth quarter' of offals, bones and blood. This blood may be collected hygienically and the plasma separated and used in food processing. Although accounting for only 40 per cent of the volume, the red cell fraction contains 75 per cent of the total blood protein. However, the intense colouration that develops on heating has limited the application of this material in food products.

Enzymatic decolourization of the red blood cells yields a colourless or light coloured highly soluble protein product (Hald-Christensen *et al.*, 1979), suitable for incorporation in a variety of food materials. The process is outlined in Figure 4.14.6 and is applicable to all kinds of animal blood. Optimum results are obtained by carrying out the hydrolysis using 40 kg *Alcalase®* 0.6L per tonne blood cell protein, with a reaction period of four to five hours at 50–55 °C and pH 8.5.

Drepper and Drepper (1981) describe a similar process based on the use of fungal proteases of acid pH optima, and they claim improved colour removal and reduction in bitterness compared with the use of alkaline protease (Hald-Christensen). Using a pH of four to five, a temperature of 45 °C and an enzyme dose of 12 microtyrosine units per gramme of red blood cells, the reaction was complete in six or seven hours. The original fungal enzyme was said to contain 2092 mTu/g when the substrate was haemoglobin.

Not every abattoir, however, has the facility for hygienic collec-

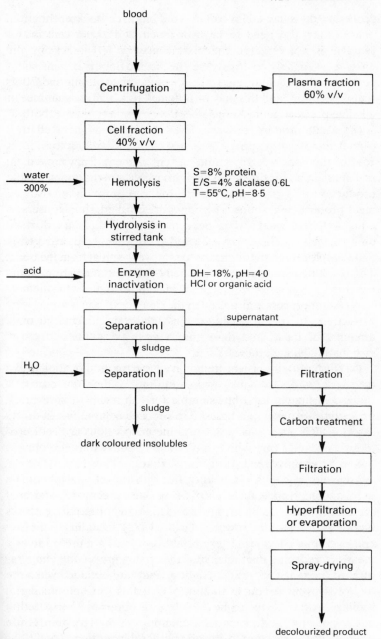

Figure 4.14.6   Enzymatic decolorization of blood.

tion and separation of slaughterhouse blood into cells and plasma.
For non-human food applications (e.g. animal foods), enzymatically decolourized whole blood may also be produced. The basic

process is the same as for isolated red blood cells, except that the plasma may also need to be hydrolysed. If 100 per cent colour removal is not required the plasma may be left intact and the enzyme used only to hydrolyse the haemoglobin. If high solids contents or high colour removal is the target, then plasma hydrolysis is also necessary and this may be obtained by a high enzyme dose or alkaline pretreatment (Anon, 1979).

(*ii*) Meat protein recovery. Even the most efficient abattoir cannot remove all the meat attached to bones, and up to five per cent of the bone weight can be present as meat. Enzymatic meat recovery allows for the removal of this meat from the bones and the production of a gravy-like product for incorporation into canned meat products and soups. The bones are crushed, slurried with an equal weight of water and 0.3 per cent *Alcalase®* 0.6L added, based on bone weight. The slurry is heated to 60 °C and held, with gentle agitation, for three to four hours to remove the meat from the bone. The resulting meat slurry is separated by settling and then pasteurized for 15 seconds at 98 °C to inactivate the enzyme.

A similar process is also used in the cleaning of bones for gelatine extraction. In this application, it is important to remove trace amounts of meat, and these may readily be solubilized by an enzymatic soaking process.

(*iii*) Fish protein. Fish protein is available for modification in two forms, as waste from fish processing and as fish not currently utilized in human food but suitable for fish meal. In both cases, proteolytic enzymes can be used to produce soluble fish hydrolysates of high nutritional value. As the endogenous levels of proteolytic enzymes vary with fish type and season, external proteases are used to control the hydrolysis reaction.

To avoid problems of rancidity, fish with low oil levels should be used, and in practice, oil levels of below one per cent are required if the use of antioxidants is to be avoided. Many different proteases have been used for hydrolysis (Mackie, 1982), including both plant and microbial enzymes. Thermostable enzymes are preferred as a means of reducing microbial spoilage, giving processing temperatures in the range 55–65 °C. Highest yields are obtained when the hydrolysis is carried out at alkaline pH and this also limits microbial spoilage. Although the highest solubility is obtained when reaction periods of three to four hours are employed, the risk of infection often limits the reaction to 30–60 minutes. Under these conditions significant solubilization of protein occurs (*see* Table 4.14.6).

The resulting protein hydrolysate has been evaluated as a component of calf milk replacer where it has been found to replace successfully 100 per cent of the milk protein for lambs, and 50 per

TABLE 4.14.6

Fish protein hydrolysate

Enzyme	% Dose	+ °C	pH	Time (minutes)	% Protein yield
Papain	0.05 (on fish weight)	65	6.5	30	85
Alcalase®	2 (on protein)	55	8.5	30	93
Neutrase®	2	50	7.0	30	92

cent of the milk protein in calf feeding. Some modification in feeding pattern was required as the hydrolysate does not coagulate, which delays absorption. It is also expected that the fish protein hydrolysate will be able to compete economically with skimmed milk and thus enter the calf milk replacer arena. The development of fish protein hydrolysate as a human food, however, awaits research into the functional properties which may be obtained on proteolysis of fish proteins.

Miscellaneous applications. (*i*) Cleaning applications. Proteolytic enzymes of medium to high purity are used in the removal of proteins from manufacturing equipment through to delicate instruments. In the meat processing industry, protease-containing detergent products are used to remove meat and blood residues and enzyme-containing foams are now available. Certain ultrafiltration and reverse osmosis membranes which cannot be cleaned with alkaline detergents require proteases to remove protein-based fouling. For delicate laboratory equipment, such as auto analysers and pH meters, protein soiling (e.g. blood) is also removed by the use of this enzyme group. For this application, enzymes in liquid form that do not contain suspended solids are preferred, and in many cases because of the material to be cleaned, they will need to be food grade. Their major advantage is that they can remove protein under mild conditions of pH and temperature, thus causing less damage to sensitive surfaces.

(*ii*) Viscosity reduction. Gelatine is a component of a number of confectionery products and is used for its gelling properties. Recovery of sucrose and glucose from out of specification materials requires the destruction of this gelling ability. Proteases such as *Alcalase®* or *Neutrase®* rapidly destroy this gelling ability, allowing the manufacturer to recover valuable confectionery ingredients. A similar application is that of the reduction in viscosity of fish stick water (*see* Chapter 4.8).

(*iii*) Photographic silver recovery. The photographic industry uses large quantities of silver in the light-sensitive emulsions that it produces. When such film is processed to recover the expensive silver, the procedure involves separating the silver-containing gelatine from the film base. The aqueous solution that results contains both gelatine and silver, but the presence of the protein hinders the separation of the silver. The addition of proteolytic enzymes such as *Alcalase*® or *Esperase*® at a temperature of 50 °C and pH 8.0 rapidly degrades the gelatine and allows the silver particles to separate out. A suitable dose of enzyme is 0.5–1.0 kg of *Alcalase*® 1.5M or *Esperase*® 4.0M per cubic metre of suspension. The enzymatic degradation of the gelatine is normally complete within 30–60 minutes. Following hydrolysis, reduction of the pH to 4.0 inactivates the enzyme and encourages settling of the silver particles. The treated material is normally transferred from the enzyme reactor to a settlement tank, where the separation of the silver occurs over a four to six hour period. The silver slurry is then further processed, either by incineration or electrolysis, to yield the metal for further refinement. The acid effluent is neutralized before discharge either to sewer or further on-site effluent treatment.

(*iv*) Preparation and treatment of culture media. Many microbiological media contain hydrolysed proteins as sources of growth factors, peptides and amino acids. These protein hydrolysates or peptones are prepared by the action of pepsin, trypsin or papain on both animal and vegetable proteins. The protease is reacted with the protein so as to produce maximum solubility and a range of amino acids and peptides. As with many such processes, a compromise is reached between low dose with extended reaction periods and microbial spoilage. The properties of a number of different peptones are shown in Table 4.14.7.

Proteolytic enzymes of high purity have found an application in

TABLE 4.14.7
Typical microbiological peptones

Protein source	Protease used	% Total nitrogen	% α-amino nitrogen
Meat	Papain	14.5	1.7
Casein	Trypsin	13.0	1.6
Soy	Papain	10.1	1.0
Meat	*Alcalase*® + *Neutrase*®	14.9	1.2–1.8

the removal of cell monolayers from culture bottles. When it is desired to leave the cell intact, the less aggressive enzymes, such as purified trypsins, are used. When cell degradation is needed to free essential metabolites or to extract drugs or other pharmacologically active materials, the more broad spectrum protease, *Subtilisin A*, is used (Osselton *et al.*, 1977).

4. Future developments in proteases

Extraction of oil seeds. Current technology for the extraction of oil from oil-rich seeds such as soy, rape and sunflower, involves solvent extraction. These processes require a high capital investment in solvent recovery equipment and the use of highly inflammable solvents increases the overall risk of the operation. In the near future, enzymatic routes for oil extraction and the separation of the other desired components should be developed.

It has also been found possible to treat fat-containing soy flour with a proteolytic enzyme and obtain separation between the oil and aqueous phase (Sejr-Olsen, 1979). In this process, fat-containing ground soya is washed at pH 4.5, which removes some oil and soluble carbohydrate. The remaining, partially defatted soy protein is then hydrolysed with *Alcalase®* to yield a soluble protein fraction, an oil fraction and a sludge which may be recycled. Although the process still requires improvement, possibly by the use of acid proteases and carbohydrases, the basic principle has been proven and thus points the way to future processes without organic solvents and the risks associated with their use.

Keratinase and collagenase. Keratin is a nutritionally useful waste protein which to date has not succumbed to enzymatic hydrolysis. Present utilization of keratin-containing waste products (e.g. hair and feathers) depends on the use of high temperatures and pressures to disrupt the structure and allow the release of the amino acids. These procedures, by their very nature, require large energy inputs and result in the destruction of part of the available protein. The development of microbial keratinases has been hampered by the fact that the most promising sources are pathogenic dermatophyte fungi, whose cultivation on a large scale would neither be permitted nor feasible. Recently, however, new sources of this activity have been developed from more acceptable sources, and we may therefore expect their introduction in the future.

The collagenase enzymes are in a similar position to the keratinases. Whilst many proteases (e.g. *Esperase®*) will degrade collagen, this is a nonspecific attack and if other proteins are present, as in meat, then these are also degraded by the enzyme. Enzymes with specific collagenase activity have been isolated, but to date it has not been possible to obtain this activity from an acceptable source. Whilst it would be extremely unwise to predict that this goal could never be achieved, it would appear that the only route likely to succeed would be that of transferring the desired activity to an acceptable host by genetic modification.

Acid proteases. In the field of proteolytic enzymes, there are very few truly acid optimum proteases. Of the fungal enzymes, a large percentage of the so-called acid proteases will not operate at much below pH 4.0, and should therefore be reclassified. Operation of processes at pH below 4 offers a number of advantages in terms of microbial stability, and hence processing times. As the enzyme dose required to reach a given end-point is inversely proportional to the time available for the reaction, the ability to extend process time can result in lowered enzyme costs.

Also in the field of effluent treatment, food wastes tend to become acid during fermentation processes and we may therefore expect to see extension of acid proteases into this area. The third area in which these enzymes can be predicted to function is in the production of flavour hydrolysates. It is well known that glutamic acid has a flavour-potentiating property and that flavours are generated on the reaction of amino acids and reducing sugars. The combined action of acid optimum carbohydrases and proteases is a route by which the base material for flavour products may be produced without the yield loss and side reactions inherent in acid hydrolysis.

Exopeptidases and aminopeptidases. Exopeptidases have not so far been available in commercial quantities, although they have generated much interest within the food industry. It has been suggested that if exopeptidases were used in conjunction with endopeptidases, the exo-enzyme could degrade bitter peptides formed by the endo-enzyme. This has been demonstrated by Clegg (1978) working on casein hydrolysates, but work on soy bean protein has also been reported (Ishida and Yamamoto, 1976). However, by far the greatest potential for the use of exopeptidase enzymes would appear to be in the field of accelerated cheese ripening. Recent developments have shown that the neutral protease, *Neutrase®*, will accelerate ripening of cheddar cheese by several months without drastically altering the structure of the cheese. However, some slight modification of structure has been observed, and it is suggested by Law and Wigmore (1982) that the combined effect of the endo- and exopeptidases will give the desired decrease in ripening time without causing this modification of cheese structure.

Aminopeptidases are a second area of development that may

be expected to become more significant in the near future. A process has now been developed where an aminopeptidase (*Novozym®* 180), obtained from *Pseudomonas putida*, is used in the production of D-phenyl glycine. The enzyme is used to hydrolyse the L-form of phenyl glycinamide in a racemic mixture of the two isomers, leaving the D-form to be isolated by chemical methods. The D-phenyl glycine is then produced from the amide by acid hydrolysis and the unwanted L-phenyl glycine is recycled via racemization to the start of the process. In addition, the same enzyme can be used to prepare D-*p*-hydroxyphenyl glycine. Both of these products are important to the pharmaceutical industry, as they are side-chain components of different semi-synthetic pencillins.

Aminopeptidases are thus beginning to find application in the synthesis of optically active molecules of specific interest to the pharmaceutical industry. In this connection, aminoacylases have been used to produce optically active amino acids for a number of years, and this area can be expected to grow as more and more semi-synthetic antibiotics are produced.

Protein synthesis. Enzymes as catalysts will reach a reaction equilibrium which may be affected by concentration of product and substrate. The problem of product inhibition can be of major importance in enzymatic processes and has practical significance in starch hydrolysis, to name but one example. If the amount of available water is reduced, and amino acids are supplied, then proteases can be forced into a synthesis reaction and proteins produced. The plastein reaction is a well known example of this, although it has not yet found much application within industrial enzymology.

Two recent processes, however, Novo's method for the conversion of porcine to human insulin and the production of the dipeptide *Aspartame* by Toyo Soda in Japan, utilize this technique and may well herald the introduction of more protein produced by synthesis.

5. Conclusions

The protein industry is perhaps unique among those industrial processes that utilize enzymes, in the diversity of substrates that are available for modification. Applications vary from the direct modification of proteins to the recovery of protein wastes, from subtle attack to massive proteolysis. It is the very diversity of these substrates and the variety of enzymes available that offer the industrial enzyme user a range of possibilities not found in any other area. The purpose of this chapter has been to highlight some of the recent developments and applications in this area and thus to offer to the reader some indications of the possibilities of their own protein materials.

References

Adler-Nissen, J. & Sejr-Olsen, H. in *The Quality of Food and Beverages* (ed. Charalambous, G.) (Academic, New York, 1982).

Anon US Patent 1,536,990 (1978).

Anon UK Patent 1,556,439 (1979).

Boyer, P.D. in *The Enzymes* 3rd edn., Vol. 3 (ed. Boyer, P.) (Academic, New York, 1971).

Caygill, J. C. *Enzyme Microb. Technol.* **1**, 233–242 (1979).

Clegg, K. M. in *Biochemical Aspects of New Protein Foods*, 109 (ed. Adler-Nissen, J. *et al.*) (Pergamon, Oxford, 1978).

Drepper, G. & Drepper, K. *Fleischwirtschaft* **61**, 30–33, 104 (1981).

Gunthe R. *Candy and Snack Industry*, Feb. 28–31 (1979).

Helbig, N. D., Ho, L., Christy, G. E. & Nakai *J. Fd. Sci.* **45**, 331–335 (1980).

Hoffman T. in *Food Related Enzymes* (ed. Whitaker, J. R.) 146–185 (Advances in Chemistry Series, no. 136, American Chemical Society, 1974).

Ishida, K. & Yamamoto, A. *Nippon Shokuhin Kogyo Gakkaishi* **23**, 524 (1976).

Law, B. A. & Wigmore, A. (personal communication).

Leiner, I. E. in *Food Related Enzymes* (ed. Whitaker, J. R.) 202–219 (Advances in Chemistry Series, no. 136, American Chemical Society, 1974).

Mackie, I. M. *Process Biochem.* Jan/Feb. 26–28, 31 (1982).

O'Keeffe, A. M. & Kelly, J. (1981)

Osselton M. D. *J. forens. Sci. Soc.* **17**, 189–194 (1977).

Petersen, B. R. & Yates, J. R. UK Patent 1,557,005 (1979).

Schwille, D., Seiz, H., Sorg, E., Sommer, U. & Agfa-Gevaert. US Patent 3,974,294 (1976).

Sejr-Olsen, H. UK Patent 2,053,228A (1979).

Sejr-Olsen, H. & Adler-Nissen, J. *Novo Publ.* A5370 (1979).

Stanley, D. W. *Can. Inst. Fd Sci. Technol. J.* **14**, 49–52 (1981).

Svendsen I. *Carlsberg Res. Commun.* **41**, 237 (1976).

Van Krannenburg, S. UK Patent 2,021,921A (1979).

STARCH
J. R. Reichelt

1. Introduction

During the last decade the liquefaction and saccharification of starch-containing raw materials by enzymes have become increasingly more important than traditional acid and acid-enzyme hydrolysis techniques. Enzyme technology applied to the processing of starch provides higher yields, significant improvements in product quality as well as energy savings.

This chapter describes the basic structure and composition of starch and goes on to give a comparison of various native starches, including their gelatinization temperatures. This is followed by a description of the processes of gelatinization, liquefaction, saccharification and isomerization of starches by enzymes. Process parameters are discussed, including indications of substrate and enzyme use levels for the production of maltodextrins, 42–63 dextrose equivalent syrups, high maltose syrups, glucose syrups and isoglucose (high fructose syrups). The formation of amylose–lipid complexes in starches and their effects on gelatinization of liquefaction processes are considered. These complexes are the cause of many problems encountered at the saccharification and final product preparation stages of starch hydrolysate production. Future trends in enzyme starch processing are considered, including the latest developments in the manufacture of isoglucose, together with the recent introduction of a high productivity glucose isomerase. (It should be noted that the objective has been to present the most recent information on the practice of starch conversions).

Starch is the reserve carbohydrate source of plants and is found, for example, in cereal, roots, tubers and palm stem pith. These varied sources yield starches with significantly different chemical and physical properties. Consequently, a number of different techniques are employed in the industrial conversion of these starches to sweeteners.

Starch is found in plant cells as large granules which can be seen under the microscope. These granules are either arranged in concentric layers, as in cereal, or as eccentric layers, as in potatoes. These layers are clearly visible where starch granules have been exposed to heat treatment. Starch is made up of two types of glucose-linked polymers: amylose consists of an unbranched chain of α-1,4 glucoside glucose linked residues, 250–300 units long in the form of a helix; amylopectin is a branched chain of α-1,4 glucoside and α-1,6 glucoside linked glucose residues up to 1000 units in length. These two polymers are linked together to form a crystalline structure. (*See* Chapter 4.2, Figure 4.2.4.)

In nature amylopectin and amylose are combined in complexes with other cellular components. For example, amylose forms a complex with fatty acids, phospholipids and other substances found in cereal starches. In potato starch amylopectin is joined in a complex with phosphoric acid esters (*See* Figure 4.15.1).

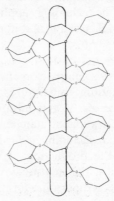

Figure 4.15.1   Schematic representation of an amylose–amylopectin complex.

2. Enzyme processing

The introduction of acid enzyme conversion processes for corn (maize) in the late 1930s provided the technology for the production of noncrystallizing syrups of high sweetness and fermentability for starch processors. In the 1960s amyloglucosidase (glucoamylase) was commercially available and the production of dextrose (D-glucose) followed. During the last decade the liquefaction and saccharification of starch-containing raw materials by enzymes have steadily increased in importance. Because of their greater efficiency and better quality of product, enzyme techniques have largely replaced those of acid hydrolysis.

Enzymes have contributed greatly to the growth of the starch industry by improving the existing processes and also by providing a wide variety of starch hydrolysates with well-defined physical properties and carbohydrate profiles. Glucose isomerase is an excellent example of the application of enzyme technology; its use in the production of isoglucose (high fructose syrups) gives a range of syrups with sweeteners equal to or exceeding that of sucrose. Figure 4.15.2 indicates the basic outline for enzyme starch processing through process sequence and product formation.

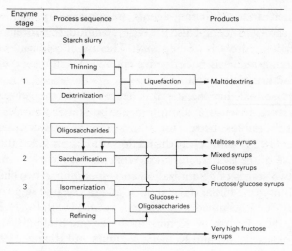

Figure 4.15.2   Main steps in enzyme starch processing.

Table 4.15.1 gives a brief summary of the major starch hydrolysis products and their typical applications.

TABLE 4.15.1
Starch hydrolysis products and applications

Starch product	Typical applications
Maltodextrins	Fillers, stabilizers, glues, pastes, thickeners
Mixed syrups (42–63 dextrose equivalent)	Confectionery, soft drinks, brewing and fermentation, jams, conserves and sauces, ice cream, baby foods
High maltose syrups	Hard confectionery
Glucose syrups	Soft drinks, caramel, wine and juice fermentations
Isoglucose (high fructose syrups)	Soft drinks, conserves, sauces, yoghurt, canned fruits

The most important stages in a successful starch conversion are gelatinization and liquefaction. However, prior to processing starch it must first be mixed with water, to form slurries or paste; in industry these should be 25–40 per cent dry substance basis (DSB).

Gelatinization. The starch slurry is heated to above 60°C so that the starch granules swell and burst (Williams, 1968). This process also releases any adhering protein material which then coagulates. The temperature that is required to produce total gelatinization depends mainly on the source of the starch. For example, for waxy corn starch a temperature of 105–110°C will produce complete gelatinization. This gelatinization process produces extremely high viscosities, and so thinning agents are a necessary addition; they reduce the viscosity and also prevent retrogradation (precipitation) and partial hydrolysis of the starch. The use of cold water to swell the starch granules is not effective enough in distorting the crystalline structure of the granules.

It is essential that the starch is heated above the gelatinization temperature to ensure disruption of the starch granules and to present a suitable substrate for enzyme action. The work carried out by Katz (1928) with X-ray diffraction studies provided the key to the cause of these characteristics. Working with wheat starch, Katz found two stages of gelatinization corresponding to two changes in crystallinity; the initial stage at 60 °C and the second stage between 100–105 °C. Recent work by Kugimiya *et al.* (1980, 1981), Höpcke *et al.* (1980), Eberstein, Konieczny-Janda and Stute (1981), Stute and Woelk (c1983) and Konieczny-Janda and Richter (1982) has supported these facts as expressed in their differential scanning calorimetry thermograms and X-ray diffraction measurements on wheat and other native starches. These results are summarized in Table 4.15.2. Differential scanning calorimetry thermograms are measurements of the heat uptake of starch mixtures or slurries during heating.

TABLE 4.15.2
Gelatinization temperatures of different native starches

Starch	Gelatinization temperatures (°C)[1] Method: DSC[2]			Method: microscope[3]		
	Onset	Peak	Conclusion	Onset	Peak	Conclusion
Corn starch	65	71	77	65	69	76
Waxy corn starch	65	72	80	64	70	78
Wheat starch	52	59	65	55	61	66
Rye starch	49	54	61	51	54	58
Oat starch	52	58	64	54	58	61
Rice starch	70	76	82	72	75	79
Potato starch	61	65	71	58	64	68
Tapioca starch	63	68	79	64	69	80
Arrow root (Maranta starch)	67	75	85	69	76	84

[1]Höpcke *et al.* (1980).
[2]Differential Scanning Calorimetry DSC–111, Setaram/Lyon (heat uptake during swelling of the starch granules).
[3]Hot stage microscope, Leitz/Wetzlar (loss of birefringence of the starch granules).

Amylose–lipid còmplexes. During the initial heating of starch slurries there is an endothermic effect between 55–85 °C due to the breakdown of the partial crystalline structure. With non-waxy creal starches, with normal amylose and lipid content an additional effect between 85–107 °C is apparent. This has been shown to be caused by the dissociation of amylose–lipid complexes, which are naturally present in starch (Konieczny-Janda & Richter, 1982; Kugimiya *et al.* 1980; Konieczny-Janda & Stute, 1981; Stute & Woelk, 1983). The thermostability of these complexes has been shown to increase with increasing chain length and fatty acid saturation number (Konieczny-Janda & Stute, 1981). Recent studies have shown that approximately five to ten per cent lipid material is sufficient to complex amylose almost completely. From this work the complexed part of amylose in native starch granules has been appraised and estimated values of 24 per cent in maize and 33 per cent in wheat starch have been quoted.

For starch processors, the most important factor arising from this work is that the second transition peak measured between 85–107 °C is 'reversible'. This means that after heating the starch paste or slurry to a very high temperature (up to 150 °C) and then subsequently cooling it, the insoluble amylose–lipid complex precipitates again. These complexes are the major cause of hazes and flocculants in saccharification procedures, which are carried out at 60 °C for periods of 48–72 hours. These precipitates are very difficult to remove and usually involve filtration procedures; in some cases they have proved impossible to remove. This is termed 'retrogradation of starch'.

3. Liquefaction

Traditionally the processes of thinning and dextrinization of gelatinized starch were carried out by acid (Palmer, 1981). The starch slurry was acidified to pH 1.5–2 and heated to 140–155 °C for 5–10 minutes. This resulted in complete gelatinization of all starches, and produced hydrolysates which could easily be filtered, however; many reversion products, colour and salts were also produced. This led to the use of acid/enzyme and then enzyme systems for liquefaction, with the use of bacterial α-amylase enzymes from *Bacillus subtilis*. These enzymes are able to operate at temperatures of 85–87 °C and for short periods of time at 90–95 °C.

Initial use of these enzymes presented problems as not all native starches can be gelatinized at temperatures of 90–95 °C, potato and waxy starches being the exceptions. The two stage addition process was introduced to overcome these problems with a pressure or jet cooking stage at temperatures of 140–150 °C for 5 minutes.

Two stage addition batch process. This process involves the addition of bacterial α-amylase (e.g. *Optiamyl®* or *Tenase®* at 0.2–0.4

litres per 1000 kilogrammes starch dry substance basis) to the starch slurry 25–40 per cent dry substance basis to which sufficient calcium ion, 200–400 parts per million (dependent on water hardness), and 300–450 parts per million sodium ion have been added after adjustment to pH 6.8–7.0. The slurry is then heated to 85–90°C and held for approximately 20 minutes. Pressure cooking at 140°C for 5 minutes and cooling to 85°C is followed by the second addition of enzyme (e.g. 1.0–1.8 litres per 1000 kilogrammes dry substance basis *Optiamyl®* or *Tenase®*). This is sometimes called the second liquefaction or dextrinization step. This temperature is held until the desired dextrose equivalent (DE) is achieved. Dextrose equivalent is the term used to characterize the degree of degradation of the starch; it is the reducing power of the starch material as compared with pure dextrose which represents 100 per cent. When the desired dextrose equivalent is reached the starch hydrolysate is heated to 100°C, and held for 10–15 minutes to ensure that all the enzyme is inactivated.

Similar enzyme processes of this type had widely replaced acid liquefaction methods until the development of heat stable, high temperature α-amylases from *Bacillus licheniformis.*

Continuous enzyme starch liquefaction. The development of thermostable high temperature bacterial amylases from *Bacillus licheniformis* which could operate at sustained temperatures above 95°C, and withstand temperatures of 105–110°C for short periods, led to the continuous starch liquefaction process with a single enzyme addition step as shown in Figure 4.15.3.

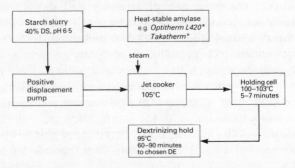

Figure 4.15.3   Continuous enzyme starch liquefaction.

These enzymes have as their cofactor a calcium metalloprotein, in which the calcium is tightly bound to the enzyme. Only low levels of calcium ion are required in processing, 75–100 parts per million, for example. However, allowance should be made for any degree of hardness of the process water. The enzyme is also very low in protease activity, therefore the risk during processing of protein

hydrolysis and subsequent formation of coloured products is reduced. These products increase the number of purification stages
and thus the processing costs. Starch substrates with higher protein
content can therefore be used for syrup production as a result of the
introduction of these thermostable enzymes.

The systems most frequently used have been jet cooker or live
steam injection processes operating at 105–110°C with a single
enzyme addition step prior to the jet (typically 1.8 litres per 1000
kilogrammes starch dry substance basis *Optitherm®* or
Takatherm®, or 0.9 litres per 1000 kilogrammes starch dry substance basis *Optitherm® L 420*). This is followed by a short holding
time of 5–10 minutes at these temperatures and then flash cooling
at 95–100°C for 1–2 hours until the required final dextrose equivalent has been reached (usually 12–15 dextrose equivalent). Heat
treatment at temperatures of 120°C and above with low pH are
required to inactivate the enzyme before saccharification. The jet
cooker is usually described as a venturi tube in which the live stean is
introduced into the starch slurry with a mixing action that allows
instantaneous heating to the required temperature. The length of
pipe at the end of the cooker can be varied to achieve short time
delays, rather than using holding cells where high temperatures are
used (e.g. 140–150°C).

Although this system has been used successfully for potato and
cereal starches, it would not achieve total gelatinization of all other
native starches, as previously described above in the section on
amylose–lipid complexes; amylose–lipid complexes and their associated problems have been experienced with maize and wheat
starches. Miles Kali Chemie, who have overcome these problems,
recommend a dual addition process which ensures complete
gelatinization and minimizes the retrogradation of starch.

Dual enzyme addition jet cooking process. (*See* Figure 4.15.4.)
This process is ideal for the preparation of maltodextrin, 42–63
dextrose equivalent, maltose, high maltose, glucose and isoglucose
syrups. Although most of the world's glucose syrups are produced
from maize starch, usually by wet milling processes, within the
European Community there is a growing use of wheat starch for
sweetener production. These starches require much higher temperatures for complete gelatinization than those used in the single
addition process.

A starch slurry of 35–40 per cent dry substance basis starch is
prepared, and calcium chloride added to achieve 100 parts per
million calcium ion in the slurry, with some allowance for the degree
of hardness of the process water. The pH is adjusted to 6.0–6.5 and
for the first liquefaction step, 0.15–0.3 litres *Optitherm® L210* or

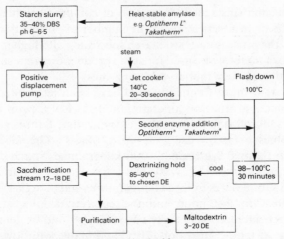

Figure 4.15.4   Dual enzyme addition jet cooking process.

0.075–0.15 litres *Optitherm® L420/Takatherm®* added per 1000 kilogrammes dry substance basis starch. The slurry is then passed through a jet cooker at 140°C and held for a minimum of 20–30 seconds. This time delay can be extended by an additional length of exit tube from the cooker, the time delay being dependent on the process conditions (i.e. rate of throughput). The treated slurry is then flashed down to 100°C and the addition enzyme put in (0.7–1.0 litres *Optitherm® L210* or 0.35–0.5 litres *Optitherm® L420* per 1000 kilogrammes dry substance basis starch). This is held at 98–100°C for 30 minutes. The slurry is then cooled to 85–90°C and held at this temperature until the desired hydrolysate or liquefaction is achieved. This reduction in temperature avoids colour formation and chemical isomerization of maltose to maltulose; also the yield of dextrose following saccharification is higher. Recent work has indicated that the dextrose equivalent level is achieved at a faster rate when the temperature is lowered to 85–90°C and this is currently being investigated.

The slurry is then heated to 120°C and held for 10–15 minutes to inactivate the enzyme; an acid pH 3.8–4.5 will aid this process where applicable. The resulting hydrolysate can be filtered, clarified with carbon, concentrated and dried or spray dried. Where maltodextrin is being prepared of dextrose equivalents of 3–20, or where intermediate dextrose equivalents are required, dextrose equivalents of up to 40 may be achieved using enzyme liquefaction.

The jet process as a continuous process has many advantages for enzyme liquefaction procedures. The process is simple, dependable, offers flexibility, is highly efficient and economical in use. The jet process can increase capacity at minimal capital investment and occupies minimal space. The process improves the control over the liquefaction stage so resulting in improved product uniformity and quality. The resulting low dextrose equivalent hydrolysates contain a minimal consistent level of saccharides which is particularly important where the liquefied starch is to be used for dextrose, high maltose syrup production or other enzymatic conversions.

4. Saccharification

Although the dextrin complex produced from the liquefaction system is commercially valuable for its rheological properties and as a carrier for other food ingredients (*see* Table 4.15.1), it does in fact form the substrate for enzymatic saccharification.

Further hydrolysis of the oligosaccharides is achieved by the use of two saccharifying enzymes: amyloglucosidase (glucoamylase) from *Aspergillus niger* and fungal α-amylase from *Aspergillus oryzae*. These two enzymes used separately or in combination are capable of producing a variety of sweeteners with widely differing sugar profiles (*see* Table 4.15.3).

Amyloglucosidase is an exo-α-amylase and produces glucose from oligosaccharides. It is used for the saccharification of liquefied starch to dextrose (glucose) syrup, and where active on an enzyme liquefied substrate it can produce 96–98 dextrose equivalent syrups.

Fungal α-amylase is an endo-α-amylase and will hydrolyse α-1,4 oligosaccharides to maltose and maltodextriose. It is used where a maltose syrup is required with little dextrose (glucose) production, and is capable of working on low dextrose equivalent substrates. Fungal amylase has a much broader substrate specifity than bacterial α-amylase and is capable of both dextrinizing (liquefying) and saccharifying actions on starch.

Amyloglucosidase and fungal amylase can be used in combination to give high conversion syrups, for example 62–63 dextrose equivalent high conversion syrups with profiles of 30–35 per cent dextrose and 40–45 per cent maltose. The saccharification process should be carried out as soon after liquefaction as practically feasible, and cooled rapidly to the saccharification temperature to avoid retrogradation.

Dextrose (glucose) syrups. (*See* Figure 4.15.5.) For the production of glucose syrups amyloglucosidase enzyme is used for the saccharification of the liquefied starch to dextrose syrup. Refined dextrose syrup with a dextrose equivalent of 97–98 usually has a D-glucose of 95–97 per cent dry substance, with 3–5 per cent higher saccharides, usually maltose or isomaltose.

The final syrup can be spray dried or dried and sold without

TABLE 4.15.3
Summary table for enzyme produced syrups from starch

Process	Glucose	Syrup Maltose	Syrup High maltose	Syrup High conversion	Syrup Isoglucose
Liquefaction	Thermostable bacterial α-amylase	Thermostable or conventional bacterial α-amylase	Thermostable bacterial α-amylase	Acid/conventional/thermostable bacterial α-amylase	Thermostable bacterial α-amylase
Saccharification	Amyloglucosidase (AG)	Fungal α-amylase	Fungal α-amylase	Amyloglucosidase (AG) Fungal α-amylase	Amyloglucosidase (AG)
Isomerization	—	—	—	—	Glucose isomerase
Profile					
Dextrose equivalent	96–98	40–45	48–55	56–68	98
Glucose	95–97	16–20	2–9	22–35	52
Maltose	1–2	41–44	48–55	40–48	—
Fructose	—	—	—	—	42
Isomaltose	0.5–2	—	—	—	—
Maltotriose	—	—	15–16	—	—

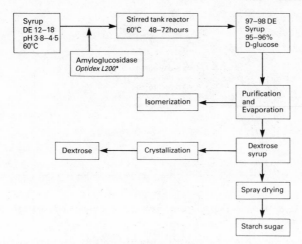

Figure 4.15.5   Production of dextrose (glucose) syrups.

further purification. However, the syrup may also be used for the production of pure dextrose by a two or three stage crystallization processes. This syrup is also used as the starting point for the production of isomerized syrups, for example isoglucose (high fructose syrups).

The saccharification process is usually carried out in tanks equipped with agitators because of the long reaction times used in the process (48–96 hours). These tanks are usually used as batch reactors, although they can also be used in series to form tank reactors, but here it is difficult both to control and to obtain very high dextrose equivalent products.

After starch liquefaction the solution should be between 27–40 per cent dry substance. Following a rapid cooling to 60°C the pH is adjusted to pH 3.8–4.5, usually with hydrochloric acid. The amyloglucosidase enzyme is then added while the tank is filling (e.g. 1.0–1.2 litres *Optidex® L150* or 0.75–0.9 litres *Optidex® L200* per 1000 kilogrammes dry substance basis). The temperature must be carefully maintained at 60°C to optimize the reaction rate. Temperatures above 60°C reduce the stability of amyloglucosidase, while reduced temperatures lead to a drop in the reaction rate and so increase the risk of microbial infection. Gentle agitation should be used and after 48–72 hours a final dextrose equivalent of 97–98 should be obtained using this enzyme process.

The reaction should be stopped when the maximum dextrose level is obtained, for if the reaction is continued the glucose level will fall. This is due to the reverse reaction whereby a condensation reaction produces maltose and isomaltose. If the syrup is not ion exchanged following saccharification then for further processing a

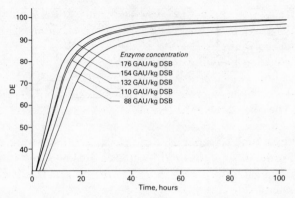

Figure 4.15.6   Relationship between enzyme concentration and dextrose equivalent with time.

heat treatment must be used to inactivate any remaining amyloglucosidase activity. This is achieved by heating to 80°C for approximately 20 minutes, but pH adjustments will reduce this time. The saccharification time depends on the dosage of enzyme used and Figure 4.15.6 shows the relationship between enzyme concentration and dextrose equivalent with time. Low substrate levels are also converted much more efficiently than high concentration. However, *Optidex®* will convert liquefied starch of up to 50 per cent dry substance basis into glucose. For industrial processes 27–40 per cent dry substance basis is used and normal saccharifications are carried out at 30–35 per cent dry substance basis to achieve high dextrose equivalent syrups at economic costs. Low substrate concentration products would involve higher evaporation costs to achieve the same yields. The recommended pH range of 3.8–4.5 based on experience provides maximum conversion with minimal colour formation, and it also reduces the amount of carbon required in any clarification procedures.

During initial purification the syrup is usually filtered or passed through a separator system to remove insoluble materials, such as fat and denatured protein. The syrup can then be further refined by activated carbon and ion exchange treatments.

Lower dextrose equivalent syrups can be obtained by using amyloglucosidase with the liquefied starch as previously described, and stopping the reaction when the dextrose equivalent reaches approximately 36–42. Such syrups contain glucose, maltose, maltotriose and higher sugars and their typical profile is shown in Table 4.15.3. In the past the majority of these syrups were made by acid/enzyme processes.

Maltose syrups. Prior to liquefaction using heat stable bacterial α-amylase, maltose syrups were produced by saccharification using

malt extract on acid liquefied starch. At present the use of fungal
α-amylase provides a process which is more economic than malt
extract (cereal β-amylase).

Low glucose-containing maltose syrups. A high maltose, low
dextrose syrup can be produced using fungal α-amylase on enzyme
liquefied starch suspension and yielding dextrose equivalents of
10–20. The oligosaccharide mixture is concentrated to 38–52 per
cent dry substance basis, the pH adjusted to 5.0–5.3 and cooled to
55 °C. Preparation is recommended in a stirred tank system, similar
to that for dextrose production. Agitation at slow speed is recommended. Fungal α-amylase is added to the tank as it is filled (e.g.
MKC *Fungal Alpha Amylase* at a level of 0.016–0.024 per cent dry
substance basis). This level of enzyme addition will give an approximate conversion time of 40–48 hours. Shorter conversion times can
be achieved by increasing the quantity of enzyme.

After the desired dextrose equivalent has been obtained, it is
essential to inactivate the enzyme to minimize dextrose production.
This can be achieved by raising the temperature to 80–85 °C for
approximately 20–30 minutes. The final syrup should be processed
by conventional plant methods. Finished syrups from this process
should have profiles giving dextrose equivalents between 48–52,
with 48–52 per cent maltose, and 5–9 per cent dextrose. In some
cases maltose levels of 60 per cent have been achieved when using
enzyme liquefied starch with fungal amylases.

High conversion syrups. (*See* Figure 4.15.7.) These syrups are
produced by the use of fungal α-amylase and amyloglucosidase
enzymes. Both these enzymes have similar pH optima and they are
able to act simultaneously on starch hydrolysates to produce high

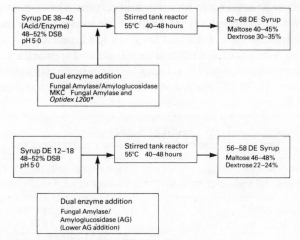

Figure 4.15.7 Production of high conversion syrups.

conversion syrups of 62–63 dextrose equivalents. The syrups are stable enough to resist crystallization at low temperatures and high concentration (80–83 per cent dry substance). With their excellent glucose maltose ratio, colour, flavour, sweetening properties and high fermentability they are widely used in food processing, brewing and the fermentation industries (*see* Tables 4.15.1 and 4.15.3).

In the past these syrups were produced from acid liquefied starch beginning with dextrose equivalents of 38–42. After preparation of a 38–42 dextrose equivalent syrup by acid or enzyme liquefaction the syrup is concentrated to 48–52 per cent (dry substance basis). A lower dextrose equivalent of 14–18 can also be used and these syrups give a higher maltose to dextrose ratio.

The concentrated syrup is adjusted to pH 5.0–5.2 and the temperature lowered to 55 °C. Preparation is recommended in stirred tanks, using low speed agitation. The dual enzyme addition (e.g. 6 grammes per 100 kilogrammes dry substance basis MKC *Fungal Amylase-P 40.000* and 700 Glucoamylase Units per 100 kilogrammes dry substance basis *Optidex®-L*) is made as the tank is filled to give an approximate conversion time of 48 hours.

After the desired dextrose equivalent has been obtained the enzymes should be inactivated to minimize dextrose production. This can be achieved by heating to a temperature of 95 °C for approximately 10 minutes. Heat exchangers can also be used for this inactivation as they minimize syrup discolouration and increase flexibility during conversion. Contact times of two to three minutes at 100 °C result in instantaneous inactivation. Carbon treatment is also very effective at removing fungal α-amylase and amyloglucosidase enzymes for the converted syrup.

However, it is possible by very careful control of enzyme usage, time and plant schedules to produce processed syrups without enzyme inactivation. Table 4.15.4 gives use levels and conversion times required to reach a dextrose equivalent of approximately 62–63 under optimal conditions, using 50 per cent dry substance basis substrate with MKC *Fungal Amylase-P 40.000* and *Optidex-L®*

TABLE 4.15.4
Effect of enzyme concentration on conversion time

Fungal amylase (MKC Fungal Amylase-P 40.000)	*Amyloglucosidase* (Optidex-L®)	*Time* (hours)
9.0 g/100 kg DSB	1050 GAU/100 kg DSB	36–48
6.0 g/100 kg DSB	700 GAU/100 kg DSB	48–60
4.5 g/100 kg DSB	515 GAU/100 kg DSB	60–72

Use levels may vary from plant to plant due to differences in starting substrates and processing conditions.

The syrups are processed by conventional methods and should yield syrups with the following approximate profiles: dextrose equivalent 62–63; dextrose 30–35 per cent; maltose 40–45 per cent; maltotriose 8–10 per cent and the remainder consisting of 20–22 per cent higher saccharides.

5. Isomerization

Isoglucose (high fructose syrups). During the late 1960s in the USA fructose syrups containing 15–42 per cent fructose were produced from starch. These syrups were initially called high fructose corn syrups (HFCS) and more recently high fructose syrups (HFS). Within the European Community these syrups are now called isoglucose. Since 1970 these syrups have been manufactured by enzymatic isomerization of glucose. They have gained an increasing share of the industrial sweetener market, especially in the USA, assisted by the rise in sugar prices in 1974 and 1975. These syrups have approximately the same composition as invert sugar and allow the same sweetness as sugar to be produced from starch sources.

The growth of fructose syrups in the USA has not been parallelled within the European Community. Production of isoglucose was severely limited by the European Commission's quota system and then levies on starch sources, while the Common Agricultural Policy will continue to restrict product development. However, second generation fructose syrups are now being produced with 55–60 per cent fructose in the USA and Japan, and they will

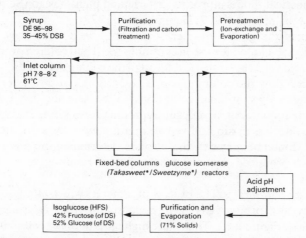

Figure 4.15.8 Typical process layout for isoglucose production.

probably soon be exported as production increases.

Glucose isomerase. Glucose isomerase enzymes catalyse the isomerization of D-glucose to D-fructose, a reaction that is reversible. Fructose formation is favoured by using alkaline pH conditions and at equilibrium, ratios or 52 per cent fructose, 48 per cent glucose are achieved. The enzyme is a thermophillic metalloenzyme and requires traces of magnesium as cofactor. The isomerization process is summarized as the transfer of two electrons from one carbon to the next, and the conversion of aldose to ketose. (For further information in this area *see* Chapter 2, 'Kinetics', pages 35–37).

In nature there is no reason why there should be an enzyme to convert glucose into fructose. To achieve this goal one would have to develop an enzyme specifically for the purpose, which is not yet possible, or use what is available – xylose isomerase. Glucose isomerases should really be categorized as D-xylose isomerases which have D-glucose-isomerase as a side or secondary activity.

This means that with batch reactors enzyme dosages are required using soluble enzymes for the conversion, resulting in a costly process. Glucose isomerases are intracellular enzymes and therefore do not yield the same quantities of product as extracellularly produced enzymes, and therefore have higher production costs. In order to overcome the inefficiency of the enzyme and still run a commercially viable process it was essential to keep reusing the enzyme. This repeated usage was achieved by enzyme immobilization. Now, several of the large enzyme manufacturers produce immobilized glucose isomerases. There are excellent account of these earlier developments in this process technology including mathematical models, process criteria and plant design (Seidman, 1977; Antrim *et al.*, 1979; Hemmingsen, 1979).

Most of the glucose isomerase enzymes used today have been specially developed for use in continous fixed bed column processes with downward substrate flow through the columns. In addition to a continous flow the column process gives short syrup-enxyme contact and allows optimal conditions to be selected for enzyme productivity without significant byproduct formation. This gives a clear colourless syrup and reduces purification costs for the final syrup. Contact time with the enzyme determines the amount of fructose produced and the whole system requires very careful control.

Process parameters. The process parameters which affect activity, stability and productivity have been summarized in Chapter 2, 'Kinetics', Figure 2.61. This figure highlights both the complexity of the system and the need for sophisticated process controls and analytical techniques to ensure adequate substrate quality, and to maintain pH, temperature and product quality.

The activity of the enzyme is usually expressed as Immobilized Glucose Isomerase Column Units (IGICU). For design purposes the reactor columns are assumed to be plugged flow reactors and calculations based on mathematical models have been drawn up to evaluate enzyme performance, using bed height, particle size, pressure drop, pH and temperature. The activity determined experimentally under defined conditions for the substrate is the initial rate of reaction (i.e. the quantity of fructose formed per unit time per weight of enzyme) starting with a fructose-free substrate. Stability represents the amount of activity retained over time. Productivity (total product produced per quantity of enzyme in a given time period) is a result of the combined effects of activity and stability. In industry productivity is usually defined as the kilogrammes of fructose produced per kilogramme enzyme during its lifetime. The productivity of most first generation glucose isomerases is quoted as between 2000–4000 kilogrammes per kilogramme enzyme.

The effects of temperature on activity, stability and product formation are shown in Table 4.15.5. A temperature of approximately 60°C usually ensures adequate activity and stability while reducing the risk of microbial infection and at an economic level. Figure 4.15.11 shows how a change in temperature or ± 1°C from the optimum will affect column productivity significantly.

TABLE 4.15.5
Effects of temperature on activity and product formation

Temperature	$t_{\frac{1}{2}}$ hours	Design enzyme lifetime	Productivity kg DS/ kg enzyme (200 IGICU/g)	Total enzyme Bed volume for 100 tpd-plant
65°C	350	$2 \times t_{\frac{1}{2}}$	1130	9.2 m³
		$3 \times t_{\frac{1}{2}}$	1300	11,7 m³
61°C	800	$2 \times t_{\frac{1}{2}}$	1820	12.6 m³
		$3 \times t_{\frac{1}{2}}$	2100	16.1 m³
60°C	1000	$2 \times t_{\frac{1}{2}}$	2090	13.6 m³
		$3 \times t_{\frac{1}{2}}$	2430	17.4 m³
57.5°C*	1800	$2 \times t_{\frac{1}{2}}$	3100	16.2 m³
		$3 \times t_{\frac{1}{2}}$	3600	20.8 m³

*This temperature is below the 60°C, recommended and is included here only for illustration purposes. (Data from Novo *Sweetzyme Q*®).

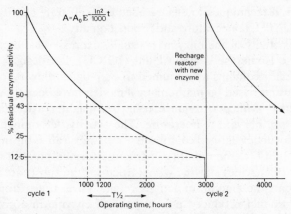

Figure 4.15.9   Activity decay of immobilized glucose isomerase.

pH control is very important and it is recommended that all pH readings are measured at 25 °C. Unfortunately the optimum activity of glucose isomerase and the catalytic stability do not share the same pH value. With soluble enzymes this is not a problem, but in the case of immobilized enzymes in a fixed bed column, the pH of the substrate must be maintained at a constant value to ensure adequate column productivity. In industry this is normally at approximately three half lives (*see* Figure 4.15.9).

The concentration of the substrate syrup must also be controlled. Due to the high solids content there is a tendency of the enzyme particles to show an increased diffusion resistance and this must be reduced. Conversely, low substrate levels also lower activity and increase the risk of microbial fouling in the columns. Dry substance contents of the substrate syrup should be in the order of 38–45 per cent with an optimum level around 38–40 per cent.

The feed syrup should be processed and purified by filtration, separation, carbon treatment and an ion/cation exchange. It is then concentrated by evaporation, which also reduces both the oxygen content and byproduct formation causes of loss of activity of the column. The highest possible dextrose equivalent should be used in the syrup, normally 96–98, for maximum conversion. Reactors run in series or in parallel provide a smooth productivity flow and facilitate process control during column regeneration. The columns are staggered so that they do not all require regeneration at the same time.

Plant design is based on the proposed throughput and this will determine the size of plant the number of enzyme reactors which will give this level of productivity. To achieve an economic operation a number of factors must be taken into account and these include enzyme activity losses, pressure drop over the enzyme bed, syrup residence time, flow distribution conversion rates and column regeneration cycles with process control. Plant design is a very skilled process and it is advisable to contact the respective enzyme manufacturers for specific advice on plant layout and reactor design recommendations.

Production process. The saccharified feed syrup should have a dextrose equivalent of 96–98, and a dextrose content of 94–96 per cent, 93 per cent being considered the absolute minimum. The syrup is refined by filtration and activated carbon treatment, then passed through ion exchange resins. The refined syrup should be free from heavy metal ions, with calcium less than one part per million. It is then concentrated to 40 per cent dry substance bases. The pH adjusted to that recommended (e.g. pH 7.8 for *TakaSweet®* or 8.2 for *Sweetzyme Q®*) and the temperature brought up to 61 °C just before the syrup enters the isomerization columns. The temperature will drop slightly as the syrup passes through the columns leaving the column at a temperature of around 59 °C. An example of a typical process layout is shown in Figure 4.15.8.

During isomerization byproduct formation is a function of temperature, pH and time. Under the reaction conditions described, residence times of 0.8–4 hours provide the recommended minimum and maximum. Any colour generated during the process can be removed by activated carbon treatment during final product purification.

The activity of the immobilized enzyme decreases with time and it is necessary to control the flow rate during the run in order to achieve the correct degree of conversion which is usually 42–45 per

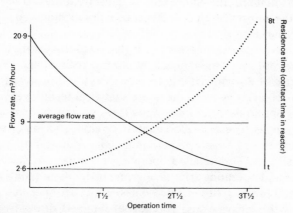

Figure 4.15.10   Flow variation in single column operation.

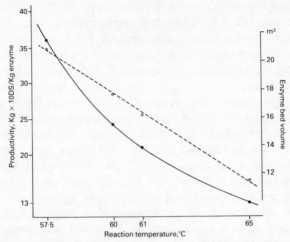

Figure 4.15.11   Effect of temperature on half life productivity.

cent. For production purposes it is essential to know the activity decay curve. This is shown in Figure 4.15.9.

Variation of flow rate (and column residence time) with operation time for a single column is shown in Figure 4.15.10. Figure 4.15.11 shows the effect of temperature on half life productivity and reactor size.

After isomerization, the syrup is blended and the pH adjusted to pH 4.0–5.0. The syrup is then passed through carbon and ion exchange treatments before evaporation to approximately 70–71 per cent dry substance. The syrup is usually sold in bulk and held at 30°C during storage. A typical fructose syrup will have the following approximate profile before ion exchange treatment: fructose 42 per cent dry substance, glucose 52 per cent dry substance, 6 per cent other saccharides.

This process of the isomerization of glucose to fructose by glucose isomerase represents the pinnacle of enzyme technology. The precise control of enzyme activity (productivity) and process control has been achieved by using the latest analytical techniques such as high pressure liquid chromotography (HPLC) and microprocessor control systems. These developments have enabled a low efficiency process to become an acceptable commercial operation.

Editors' note: The detailed operating formulae for calculating analytical and operating criteria, enzyme bed volume, the average activity and productivity of glucose isomerase, are specific to each source of this enzyme and should be obtained direct from the manufacturing supplier. Please refer to Data Index 2.

6. Recent developments

Maltodextrins have recently been defined in the UK by the Ministry of Agriculture, Fisheries and Food (MAFF) through their Food Additives and Contaminants Committee (FACC). The Committee has issued the 'Report on Modified Starches' (FAC/REP/31) in which maltodextrins are defined as starch hydrolysates with dextrose equivalents of 3–20. Maltodextrins are to be excluded from these regulations, but may be only obtained from starch by enzymatic and/or acid hydrolysis, to yield purified aqueous solutions to nutritive saccharides or the subsequent dried product. The enzymes used for the production of maltodextrins, bacterial α-amylasas, are also under review by the Committee, and a summary table of the Committee's recommendations is given in Chapter 3.1.

Miles Kalie-Chemi AG have recently introduced a second generation immobilized glucose isomerase enzyme, *Optisweet® 22*, with a standard productivity of 22,000 kilogrammes dry substance per kilogramme of enzyme. The enzyme has excellent flow characteristics, with smaller particle size and higher space velocities than conventional glucose isomerase enzymes which result in lower fixed bed heights. Current reactors could be used without expensive modifications to increase plant capacity, beyond that of conventional glucose isomerase. New plant designs, however, could use much smaller reactors with reduced capital costs and lower spatial requirements. (A 0.7 metre bed height of *Optisweet® 22* is equivalent to using conventional glucose isomerase with a 3.5–3.7 metre bed height on start-up, with conventional glucose isomerase at pH 7.5 and a substrate inlet temperature of 60°C in a 2-bar test pressure reactor.)

7. Future trends

With the development of a second generation of high productivity immobilized glucose isomerase and the continuous liquefaction processes developed using high temperature heat stable bacterial α-amylases, an immobilized high productivity amyloglucosidase enzyme for saccharification is required for a total continuous system. This enzyme would need to match the current conversion levels attained in batch saccharification using soluble enzyme for high dextrose equivalent syrups to be produced at an economic level.

Several major enzyme manufacturers are working in this field and these developments cannot be too far away. A combined continuous starch process would give starch processors flexibility. With the advances in microprocessor and analytical control systems it would also offer much greater process control.

The development of debranching enzymes such as pullulanase from bacterial sources which have a higher temperature optimum than those currently available could lead to increased dextrose–glucose yields. This occurs during saccharification breaking down any remaining 1,6 linkages, when used with amyloglucosidase.

References

Antrim, R. L., Colilla, W. & Schnyder, B. J. in *Applied Biochemistry and Bioengineering*, Vol. 2 *Enzyme Technology* (ed. Lemuel, B., Wyngard, J., Katchalski Katzik, E. & Goldstein, L.) (Academic, New York, 1979).

Fullbrook, P. D. in *Nutritive Sweeteners* (ed. Birch, G. G. & Parker, K. J.) (Applied Science, London, 1981).

Hemmingsen, S. H. in *Applied Biochemistry and Bioengineering*, Vol. 2 *Enzyme Technology* (ed. Lemuel, B., Wyngard, J., Katchalski Katzik, E. & Goldstein, L.) (Academic, New York, 1979).

Hollo, J. & Szejtli, J. in *Starch and its Derivatives* (ed. Radley, J. A.) (Chapman & Hall, London, 1968).

Höpcke, R., Eberstein, K., Konieczny-Janda, G. & Stute, R. *Stärke* **32,** 397ff (1980).

Howling, D. in *Sugar, Science and Technology* (ed. Birch, G. G. & Parker, K. J.) (Applied Science, London, 1979).

Katz, J. R. in *A Comprehensive Survey of Starch Chemistry* (ed. Walton, R. P.) (Chemical Catalogue Co., New York, 1928).

Konieczny-Janda, G. & Stute, R. *Differential Scanning Calorimetry of Starches, Part 2 Investigation on Starch-Lipid-Complexes* (Paper presented at the 32nd Starch Convention, Detmold, 1981).

Konieczny-Janda, G. & Richter, G. *Enzymatic Degradation of Starch-Lipid-Complexes during Liquefaction of Starch and Synthetic Amylase-Lipid-Complexes*, (Symp. Use of Enzymes in the Food Industry, Paris, 1982).

Kugimiya, M. & Donovan, J. W. *J. of Fd Sci.* **46,** 765 (1981).

Madsin, G. B. & Norman, F. E. in *Molecular Structure and Function of Food Carbohydrates*, 50 (ed. Birch, G. G. & Green, L. F.) (Applied Science, London, 1973).

Palmer, T. P. in *Nutritive Sweeteners* (ed. Birch, G. G. & Parker, K. J.) (Applied Science, London, 1981).

Seidman, M. in *Developments in Food Carbohydrate 1*, 19–42 (ed. Birch, G. G. & Schallenberger, R. S.) (Applied Science, London, 1977).

Stute, R. & Woelk, H. U. German Pat. DE 2231 (c1983).

Takasaki, Y. & Yamanobe, T. in *Enzymes and Food Processing*, 73 (eds Birch, G. G. Blackebrough, N. & Parker, K. J.) (Applied Science, London, 1981).

Williams, J. M. in *Starch and its Derivatives* 4th edn (ed. Radley, J. A.) (Chapman & Hall, London, 1968).

TEXTILES
T. Godfrey and J. Reichelt

1. Introduction

Weaving of fabrics places considerable strain on the warp, and to prevent breaking of these threads they are usually strengthened by the application of an adhesive size. Throughout the world the predominant size is still based upon starch, despite the introduction of other substances, for example gelatine, plant gums, water-soluble celluloses such as methyl and carboxymethyl, cellulose and polyvinyl alcohol.

The economic value of very inexpensive starch has maintained its dominance, and the kind of starch used is really only a matter of geographical availability. Whilst Europe uses mainly potato starch (farina), the North American textile industry uses maize, and the Middle and Far East mainly rice. With the wider use of imported fabrics the nature of the size is often uncertain, and the weaver or processor of woven cloth needs to know what size has been used.

Sized cloth is less absorbent so that uptake of dyes, bleaches and texturizing chemicals can be impaired unless the size is adequately removed. Many garments, especially the ubiquitous 'jeans', are desized after machining. Truly water-soluble sizes may be removed with hot water and detergent washing, but the starch sizes need degrading to make them soluble. Although most size starches are prepared by brief treatment of the raw starch with enzyme or acid to gelatinize and thicken it, the resulting adhesive is still not fully water-soluble. Chemical methods for desizing that use oxidizing agents or sodium hydroxide often include pressure cooking and can frequently damage the fabric. Most textile factories now use amylase enzymes to hydrolyse and solubilize the starch as these do not harm the fibres.

2. Desizing amylases

Plant, animal and fungal amylases. The earliest users of amylases took the knowledge of the brewer and used concentrated extracts of malt, prepared in such a way that the natural enzymes were preserved at their fullest practical activity. During this time there was a parallel increase in the use of the amylases present in extracts of animal pancreatic glands, until the cost, like that of the malt enzymes, became too high to be economical for the textile industry.

In the Far East, the starch hydrolysing enzymes prepared for fermentation products such as 'Saki' were adapted for textile applications. These enzyme concentrates were obtained from the 'Koji' process of cultivating a mixed fungal population on a moist cereal bran and then extracting with a minimum amount of water.

The most noteworthy practical limitations to the use of these enzymes are their relatively slow action for the cost, coupled with their sensitivity to the chemistry of the environment in which they are to function. Most of the problems have been overcome with the introduction of bacterial amylases, and most recently with the advent of thermostable amylases.

Bacterial desizing amylases. With these enzymes it is possible to accelerate the desizing process considerably, largely due to the higher working temperatures of the enzymes and their lower sensitivity to other chemicals present in the desizing liquor. The superior stability of these enzymes in working solutions gives a further processing advantage.

Table 4.16.1 lists some of the typical characteristics of the various amylases available for desizing. The influence of calcium ions is

TABLE 4.16.1

Typical characteristics of various amylases available for desizing

	Operating temperature (°C)	Operating pH	Inhibitors	Activators
Malt α-amylase	55–65	4.5–5.5	Metal ions Alkalis Starch contaminants	Calcium ions
Pancreatic amylases	40–55	6.5–7.0	Metal ions Acids	—
Fungal amylases	50–55	4.5–5.5	Metal ions Alkalis Sequestrants	Calcium ions
Conventional bacterial amylases	60–75	5.5–7.0	Sequestrants Anionic surfactants	Calcium ions
Thermostable bacterial amylases	85–110	5.0–7.5	Anionic surfactants	Calcium ions in very soft waters only

most significant when the conventional bacterial types of enzyme are used, and it is typical to add 0.5 grammes of calcium chloride per litre of desizing bath liquor to provide maximum stability.

Both types of bacterial amylase are described as endoamylases, and they hydrolyse the α-1,4 glucose links in the starch polymers in a random manner to produce water-soluble dextrins and sugars. There are clearly some special characteristics that vary with the microbial source of the enzyme and unfortunately the methods of analysis and application technology do not yet allow accurate predictions of these differences. Practical testing 'on-site', however, will show that some are better at inducing liquefaction of starch than others. Liquefaction is the main action responsible for inducing solubility and the most important economic targets for textile application are that liquefaction should occur in the shortest possible time and with the minimum enzyme dosage rates. The prospective user should evaluate the enzymes by a simple test with a prepared starch paste in such a manner that the reduction in viscosity for a given dose and temperature can be observed.

With the variety of traditional desizing methods still widely used, together with the specificity of some of the modern equipment which is designed for use with bacterial enzymes acting at their optimum, it is important to select the enzyme which will act most efficiently under the conditions of the method being used.

Temperature and pH activity of bacterial amylases. Reference to Figures 4.16.1, 4.16.2 and 4.16.3 will show that conventional amylases exhibit a relatively sharp decline in performance when desizing temperatures exceed 75 °C. This is in contrast to their action in, for example, brewing and distilling applications, where much higher concentrations of the starch substrates provide greater protection from deactivation by denaturation at elevated temperatures. A similar sensitivity to extremes of pH also occurs here

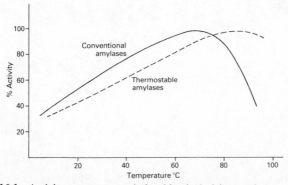

Figure 4.16.1   Activity temperature relationshipa fo desizing amylases.

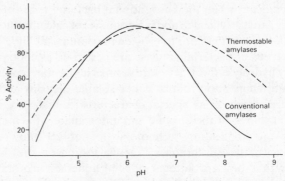

Figure 4.16.2   Activity pH relationships for desizing amylases.

because of lack of substrate protection. With the thermostable amylases the effects are less marked, but still below their maximum performance under stable substrate protection conditions (*see* Sections 1, 2 and 5 in this chapter). Figure 4.16.3 demonstrates that for an equivalent assayed activity the performance of the thermostable amylases is three times greater at 95 °C than at 60 °C, the temperature at which the conventional enzymes show roughly equal performance.

These data form the basis of the use of thermostable enzymes for modern continuous desizing operations (*see* p. 408). In all other methods, the size breakdown may require contact times of up to 16

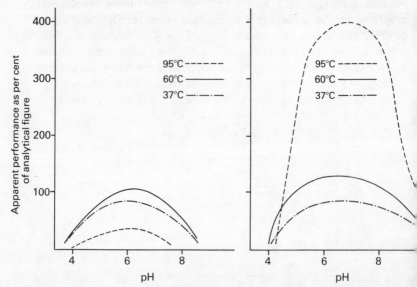

Figure 4.16.3   Apparent performance of equal activity doses of conventional and thermostable amylases at different temperatures (desizing applications).

hours, although 2–4 hours is more generally used. Consequently, the stability of the enzyme in the desizing bath is important.

Enzyme solution stability. A number of factors influence the stability of an enzyme solution. Of these, temperature, pH and the presence of substrate and calcium ions are particularly relevant to desizing operations. The first two factors have been discussed in the previous section, and the levels of starch available will vary with different fabrics, and especially with their origin, if imported ready-woven grey-cloth is being processed. Calcium ions have very strong protecting and activating effects on the performance of the conventional enzymes, and most suppliers of these enzymes include calcium salts in the formulation of their products. In addition, it is recommended that the desizing liquor should contain an additional 0.5 grammes of calcium chloride per litre. Figures 4.16.4 and 4.16.5 illustrate the influence of starch concentration and calcium ions on conventional amylases.

The thermostable enzymes, when used at near their maximum performance temperatures, also show some response to calcium, and this is more marked if the pH is below the optimum. These circumstances are most often encountered when working with very soft water (i.e. water having less than 50 parts per million hardness). It is common to use some sodium hydroxide to bring the pH up to the optimum and then calcium addition will rarely have any significant effect. Figure 4.16.6 illustrates the stability curves for both enzyme types at different temperatures but at optimal pH. The influence of calcium on the conventional amylases is clearly of importance.

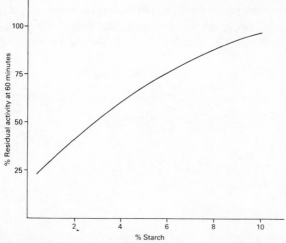

Figure 4.16.4 The influence of starch concentration on the stability of conventional amylases at desizing operational temperatures.

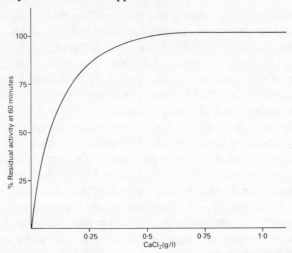

Figure 4.16.5 The influence of calcium salt concentration on the stability of conventional amylases at desizing operational temperatures.

Desizing surfactants and enzyme performance. To ensure even action of the enzymes, it is usual to add powerful surfactants to the desizing liquors. As will be described below (*see* Section 6), the starch becomes swollen by the hot water and must be thoroughly wetted to ensure that water for the hydrolytic reaction is readily available. This wetting stage may last for only a very few seconds and the wetting agents are thus very important. Although the most effective surfactants are generally of the anionic type, these may damage the enzymes. Optimal performance of the enzymes can be obtained by using non-ionic types, although a small amount of cationic surfactant in combination with these is usually acceptable,

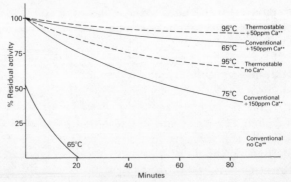

Figure 4.16.6 Stability of α-amylases at operating temperatures and different calcium levels (desizing applications).

and can reduce surfactant costs. Furthermore, the addition of cationic surfactants is sometimes helpful because a totally non-ionic system can tend to precipitate when exposed to the higher desizing bath temperatures.

3. Preparation of desizing liquors

For hygienic considerations in practice, it is usual to steam or otherwise sterilize the vessels and pipework involved in the operation at regular intervals. This should include the liquor preparation equipment. When diluting stock enzyme products, it is advisable to use sodium chloride, or an admixture of sodium chloride and sodium sulphate, as a diluting brine of around 16–18 per cent weight to volume. Care should be taken to ensure that the chemicals used will not contribute heavy metal ions or oxidizing agents to the liquor, and also that the pH of the brine is in the range 6.0–8.0, using dilute sodium carbonate or acetic acid for any necessary adjustment. With continuous mixing, the stock enzyme solution is diluted with the brine to reduce activity to a level appropriate to the operating method. Most reputable enzyme suppliers will provide simple analytical protocols for establishing the activity of prepared liquors.

Any surfactants to be used are usually included in the brine solution, together with a sanitizer, for example 0.05 per cent *Sanitrol®*.

Special additional preparations. It is at this stage that the calcium chloride required to stabilize the conventional amylases should be dissolved and thoroughly mixed into the liquor. The rate of 0.5 grammes hydrated calcium chloride per litre of liquor noted above is sufficient for all normal process conditions.

If the stock enzyme solution is to be diluted to less than 20 per cent of its original activity, it is recommended that a prepared soluble starch is slowly added to the well-stirred liquor to a level of 10 per cent weight to volume to stabilize the enzymes further. This applies to both the conventional and thermostable types of amylase.

4. General stages of the desizing process

The four main stages of desizing will be described briefly here, while Section 6 gives descriptive accounts of specific examples.

Prewashing. For many processes, a rapid wetting bath stage brings about a valuable acceleration to the overall process. This is particularly beneficial to heavyweight fabrics. The prewash is also useful for removing any waxes or non-starch water-soluble additives used when the size was applied, and improves access of the enzyme to the starch particles.

Impregnation. Similar to prewashing but using the enzyme liquor, this stage enables the cloth to take up the active solution, swells the starch if no prewash has been given, and starts the desizing hydrolysis of starch. For maximum performance of the amylases, this stage should be run at a temperature that is as high as possible in relation to the total process time and stability of the enzyme selected. Typically, conventional amylases can be impregnated at 65–70°C and the thermostable amylases at 75–80°C. The uptake of liquor should be between 90 and 110 per cent of the fabric weight.

Starch hydrolysis. Although started during impregnation, this process will have a time course of 2–16 hours, depending on the temperature, pH, enzyme concentration and stability, and the fabric itself. Long reaction times at low enzyme levels can be economical if the stability is ensured. Conversely, very short times can be achieved at high enzyme levels and maximum temperatures.

Care is required to ensure the fabric does not dry out, and in some processes, such as the batch method (*see* p. 406), the roll is wrapped in a plastic sheet and rotated slowly to keep the liquid load even. Where continuous desizing is practised using a J-BOX or steam box, there is no problem of drying out and the reaction times are much shorter (*see* p. 407).

When using conventional amylases, the temperature of this stage should approach the optimum for the enzyme, which will be in the range 70–75°C for operations of one–four hours, but down to 40–50°C when overnight holds are intended. The thermostable enzymes can complete this stage in as little as 20 seconds, with 30–60 seconds being typical. This is achieved using a steam charged vessel or chamber at 90–110°C.

After-wash. Complete removal of the hydrolysis products is vital to the desizing process and washing equipment must be used which gives good agitation of either the fabric or the wash water. The highest temperatures available should be used, ideally 95–100°C. A synthetic detergent and sodium hydroxide added to encourage release of the soluble starch products are usually incorporated in the after-wash waters.

Thorough rinsing follows and often includes the addition of acid to neutralize the alkalis before bleaching begins.

5. Monitoring starch desizing

The simplest method of monitoring starch desizing is based on the reaction of starch with dilute iodine solution. With raw starches, the colour produced is a deep blue-black which pales to blue and then violet as the starch is broken down. A pale yellow-brown colour indicates that all starch has been hydrolysed. The test solution is made to 0.005 N iodine by dilution of stock iodine solution with water at a ratio of 1 part stock to 19 parts water. Stock iodine solution is 0.1 N and is prepared by dissolving 18 grammes of potassium iodide and 12.69 grammes of iodine in about 500 millilitres of water and diluting accurately to 1000 millilitres. Stored in a brown glass bottle away from light, this stock solution can be kept for many months.

To test the progress of the desizing process, a portion of the cloth is treated with a few drops of the diluted iodine solution and the colour noted after an interval of about 30 seconds. As this test is very sensitive, and will be positive to only traces of starch residues, it should be used with reference to the subsequent behaviour of the desized fabric as determined by experience, for which there is, as always, no true substitute.

6. Typical desizing processes

To provide working examples, this series of processes will be illustrated with actual recommended enzyme use rates for typical commercial enzymes: conventional bacterial amylases; for example *Aquazym®* 120L (Novo) and *Optisize®* L (Miles), and thermostable bacterial amylases, such as *Termamyl®* 60L (Novo) and *Optisize®* LT 210 (Miles). The use of other products will follow in similar fashion with the doses recommended by their suppliers.

Desizing on a jig. (*See* Figure 4.16.7.) This is a very simple method that relies upon changing the liquor in the tank for each stage and processes one roll of fabric in a batch.

Prewash: run two ends (pass fabric twice through bath) in boiling water and surfactant at 0.5 grammes per litre.

Impregnation and breakdown: enzyme concentration between 50 and 150 grammes *Aquazym®* 120L per 100 litres (0.05 per cent weight to volume), or *Optisize®* L.

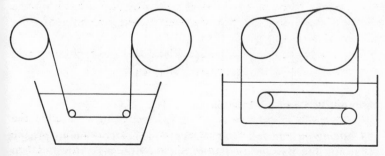

Figure 4.16.7 Desizing on a jig.   **Figure 4.16.8** Desizing on a winch.

After-wash: use water at 95–100°C containing detergent and 5–10 grammes sodium hydroxide per litre.

Desizing on a winch. This is the simplest system for continuously passing the fabric through the liquor but still requires the changing of the liquors (*see* Figure 4.16.8).

Prewash: run twice the length through boiling water with surfactant at 0.5 grammes per litre.

Impregnation and breakdown: enzyme concentration between 25 and 100 grammes *Aquazym®* 120L per 100 litres (0.025–0.1 per cent weight to volume), or *Optisize®* L. Run continuously for 30–60 minutes at 65–75°C.

After-wash: use water at 95–100°C containing detergent and 5–10 grammes sodium hydroxide per litre.

Desizing on pad roll – roll storage. This allows retention and some re-use of prepared liquors, as several rolls can be processed and stored for the breakdown stage (*see* Figure 4.16.9).

Prewash: use boiling water containing surfactant at 0.5 grammes per litre.

Impregnation: enzyme concentration 100–250 grammes *Aquazym®* 120L per 100 litres, or *Optisize®* L (0.1–0.25 per cent weight to volume).

Breakdown: wrap roll pad in plastic sheet and store, slowly rotating for 2– hours at 70–75°C, or overnight at low temperature (below 50°C).

After-wash: use water at 95–100°C containing detergent and 5–10 grammes sodium hydroxide per litre if necessary.

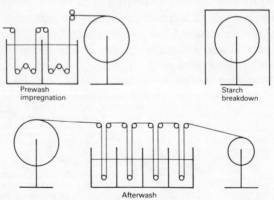

Prewash
impregnation

Starch
breakdown

Afterwash

Figure 4.16.9 Desizing on a pad roll (roll storage).

Desizing on pad roll – pit storage. This is a continuous process regarding the passage of fabric, but requires a hold time in the storage pit which interrupts progress of the fabric (*see* Figure

4.16.10).

Pre-wash: use boiling water containing surfactant at 0.5 grammes per litre.

Impregnation: enzyme concentration 100–250 grammes *Aquazym*® 120L per 100 litres (0.1–0.25 per cent weight to volume), or *Optisize*® L.

Breakdown: store in pit for 2–4 hours at 70–75 °C or overnight at lower temperatures (below 50 °C).

After-wash: use water at 95–100 °C containing detergent and, if necessary, 5–10 grammes sodium hydroxide per litre.

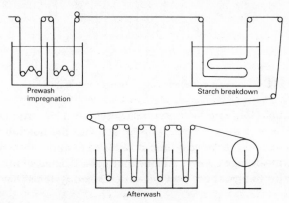

Figure 4.16.10 Desizing on pad roll (pit storage).

Continuous desizing – conventional amylases. This is a fully continuous operation with a total operating time of only around 30 minutes. The equipment includes the use of a J-BOX to provide the holding time, and a steam chamber to complete the operation of the enzyme action (*see* Figure 4.19.11).

Preparation of desizing bath: heat the water to 65–70 °C and add sodium chloride at 300 grammes per 100 litres, calcium chloride at 50 grammes per 100 litres and surfactant at 50 grammes per 100 litres. Check pH and adjust to between 6.0 and 8.0. Add enzyme to the appropriate concentration: *Aquazym*® 120L 250–500 grammes per 100 litres (0.25–0.5 per cent weight to volume), or *Optisize*® L.

Pre-wash: use boiling water containing surfactant at 0.5 grammes per litre.

Impregnation: use desizing liquor as prepared, top up the tank with fresh liquor to maintain level and activity.

Breakdown: treat in a J-BOX (or steam chamber) for 1–4 minutes at 95–100 °C. Steam in the chamber at 70–80 °C for 20 minutes.

After-wash: use boiling alkaline water with 20–35 grammes sodium hydroxide per litre. Rinse in hot and then cold water.

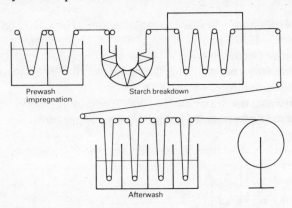

Figure 4.16.11   Continuous desizing (conventional amylases).

Continuous desizing – thermostable amylases. The extremely high performance of these enzymes dispenses with the holding time in the J-BOX and the process time is reduced to about five minutes. The enzyme dosage is lower and the thermal efficiency of the system far superior to any of the previously described systems (*See* Figure 4.16.12).

Preparation of desizing bath: using water at 70–80°C, add 0.5 grammes per litre surfactant, check and adjust the pH to between 6.0 and 8.0. Add thermostable amylase to the appropriate dose: *Termamyl®* 60L at 50–300 grammes per 100 litres (0.05–0.3 per cent weight to volume), or *Optisize®* LT 210. Recheck the pH and adjust if necessary.

Prewash: use boiling water containing surfactant at 0.5 grammes

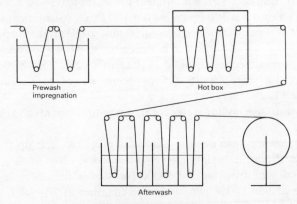

Figure 4.16.12   Continuous desizing (thermostable amylases).

per litre.

Impregnation: the desizing bath is prepared as described previously and kept at 70–80°C.

Breakdown: in a steam chamber for 20–120 seconds at 100–110°C.

After-wash: use boiling alkaline water with 10–31 grammes sodium hydroxide per litre. Rinse with running cold water.

7. Future enzyme developments in desizing

It is not clear which direction enzyme developments in desizing will take, except that thermostable amylases will be more widely used as they provide the most rapid and efficient continuous processes. The return of starch sizing as the most economical size seems likely in the face of rising costs for the polyvinyl alcohol alternatives, although it may be expected that synthetic sizes with other beneficial characteristics will be developed.

The detergent industry may well provide new developments through the application of fabric softeners and the consequent reduction in harshness. These areas are currently giving those involved with enzymes food for thought and some of the newer enzymes could become incorporated in desizing methods, in particular those applied to finished clothing.

If the development of protease enzymes continues, and reduces the unit cost, it can be envisaged that gelatine and other protein sizes could become more generally used. The methods of desizing proteins will probably be with enzymes working at relatively low temperatures, around 50–65°C, and so fitting well to the older and more traditional textile operations.

WINE
R. Felix and J.-C. Villettaz

1. Introduction
The production of juice and wine from grapes uses different tech-nologies with a common aim: the extraction of the relevant com-pounds. In some cases juice extraction is more difficult because of the presence of high molecular weight substances such as pectin and hemicelluloses. Although grapes already contain various enzymes, they occur in such small quantities that they are of little practical value because of the short time for which they are allowed to act in the wine and juice making process. The objective of an enzyme addition is therefore to reinforce and complement the natural enzymatic activities present in the substrate.

A first possibility is to add the enzymes (pectinases, hemicellul-ases, cellulases) during the maceration or pressing process in order to obtain a better initial extraction of the various constituents (colour, juice, flavours etc). Another possibility is the addition of the enzymes to the juice or wine so as to hydrolyse the soluble high molecular substances (e.g. colloids) which could stabilize the cloud-iness in the juice and therefore hamper clarification and cause difficulties in the filtration of the finished product. Such enzymes, mainly the pectolytic enzymes, have been widely used in the grape juice and wine industry for more than ten years.

The first part of this section will describe the application of pectolytic enzymes in winemaking, and the second part will con-sider the use of a specific enzyme, a β-glucanase, and the way in which it helps to solve certain problems in filtration and clarifica-tion. Finally, future developments of enzyme applications in the grape juice and wine industry will be briefly discussed.

2. Pectolytic enzymes in winemaking
Under current legislation in force wine is referred to as a natural product which results from a number of biochemical – mainly enzymatic – reactions that begin during the ripening of the grapes, and continue to their harvesting, throughout their alcoholic and malolactic fermentation, clarification and even after bottling.

Many of these reactions have always been and are still left to nature and the microorganisms present in the grapes, but science has enabled the winemaker to influence a few of the reactions in a useful direction, with the help of pectolytic enzymes. The progressive and economy-conscious winemaker will utilize these biological aids in order to speed up the natural processes of winemaking and to improve the quality of the wines, wherever possible, as well as to make the fullest use of his facilities and equipment. It is generally known that wines, particularly those of high quality, cannot simply be produced and stabilized according to a fixed production scheme. A subjective assessment of the character of the vintage and of the wine to be produced is necessary, and this requires continual monitoring, control, timely decisions and much experience.

The use of pectolytic enzymes makes this particularly feasible. In principle, there are two different applications of such enzymes in winemaking: (*i*) enzyme treatment of the crushed grapes in order to improve the yield and extraction, and (*ii*) enzyme treatment of the juice or wine in order to facilitate clarification.

3. Production of red wine

Two basic technologies are used today in the production of red wine: classical fermentation on skins and thermovinification. In both processes the use of enzymes presents advantages.

Classical fermentation on skins. Enzyme treatment of the crushed red grapes during fermentation on skins has the advantages:

(*i*)   More rapid, less turbulent fermentation

(*ii*)   Less formation of foam

(*iii*)   Important reduction of time through a more rapid and better extraction of colour

(*iv*)   Increased yield of free run

(*v*)   Facilitated pressing

(*vi*)   Increased total yield

(*vii*) More rapid and better clarification

(*viii*) Easier filtration

(*ix*)   Bouquet-rich and full-bodied wines with a low content of tannins

(*x*)   More rapid maturation of the wines

Fermentation will usually start earlier in the presence of enzymes. Furthermore, in general, shorter fermentation is less turbulent and has less tendency to overfoaming. There is less increase in temperature, and there are practically no halts in the fermentation process, whether due to overheating or for other reasons. The presence of the enzyme seems to promote a smooth and complete fermentation, and because false fermentations are largely eliminated, there is a considerable reduction in the amount of volatile acids and a slight increase in alcohol content.

The reduction in the time taken for fermentation on the skins is probably the most important advantage of enzyme treatment, and usually amounts to 30–50 per cent of the normal fermentation time. Besides the economic advantage of doubling the capacity of the production unit, the enzyme treatment also considerably improves the colour extraction in cases where time for fermentation on skins is already short. When special equipment, for example, 'rototanks' or continuous fermentors, is used, the production capacity can be very considerably increased by the use of enzymes.

In order to achieve these advantages it is essential to mix the contents of the fermentation tank very thoroughly twice a day. A slight rinsing of the grape skins floating on the surface is definitely not sufficient; this layer should be completely broken up and brought down into the mash again. As is well-known, the colour is found in the skins and can only be extracted when the enzyme, which is dissolved in the must, comes into contact with them.

Colour extraction by means of pectinase is initially slow. The mucous substances found beneath the skin (mainly pectins) must be decomposed during the first six to eight hours, for only then are the colouring substances available for extraction. This important fact is only a disadvantage in the production of red wine; in white wine production it improves the yield by allowing a short enzymatic maceration without a concurrent increase in colour. However, not all pectinases are equally well-suited for obtaining maximum colour yield. Suitability depends on the so-called 'protopectinase' activity (i.e. the ability to release highly esterified, unmodified pectin directly from the solids and further decompose this material).

When the correct enzyme preparation has been used in skin fermentation, the result is a striking improvement in wine quality due to the fact that more colour and aroma substances can be extracted while at the same time reducing the content of tannic substances. In addition, the maturation of these wines is considerably accelerated, so that they are sometimes ready for bottling after only one or two months. In general, the wines become much more supple, agreeable and flavoursome, as well as full-bodied and rich in bouquet.

In practice, for traditional fermentation on skins one adds for example, 2 grammes of *Ultrazym® 100* (dissolved in water) per 100 kilogrammes crushed grapes. The total amount of enzyme required for the whole batch should be added to the filled fermentation tank immediately after the grapes have been crushed. For an optimal distribution of the enzymes in the mash it is necessary to mix the crushed grapes as thoughly as possible. It is also recommended that this mixing be repeated every 12 hours.

It is essential to drain the liquids as soon as the desired colour intensity has been reached. Thus, the determining factor is not the attainment of a certain density of the must, as used to be the case. Usually this would already be too late, and abnormally high tannin contents or the appearance of other undesirable side reactions could result. It is also important to make full use of the advantages of the shorter fermentation time on skins, as this also produces the highest quality of wine (*see* Figure 4.17.1).

Thermovinification. For various reasons this technique for producing red wines has been much publicized over the past few years. It represents an alternative to the fermentation on skins and furthermore provides a means of improving the quality of the wine when mould-damaged grapes are used.

Depending on the character of the vintage, thermovinification is performed with or without flash-heating of the crushed grapes to a high temperature. The heat treatment leads to a plasmolysis of the cells, disrupting the cell membranes and thus increasing their permeability, so that the rate of extraction of cellular substances such as sugar, acid, colour and aroma components, is also increased.

Advantages of using pectolytic enzymes in thermovinification are:

(*i*)   More rapid extraction of colour and aroma components
(*ii*)   Reduction of holding time at 50 °C and thereby increase of the must separator capacity
(*iii*) Larger volume of free-run must
(*iv*)   Increased total yield
(*v*)   Easier pressing of the crushed grapes
(*vi*)   More rapid and better clarification
(*vii*) Less foam formation during the fermentation
(*viii*) Facilitated filtration of the wines

The more of the mucous substances at the cell walls that can be broken down, the more rapid is the extraction process and the lower the viscosity of the liquid used for extraction – the must. The

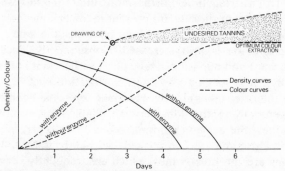

Figure 4.17.1   Enzymic improvements in red wine processing.

enzyme-induced solution of the mucous substances, mainly protopectin, increases the amount of must released by 20 per cent, and this can more easily flow from the crushed grapes, thereby increasing both the yield of free-run must and thus the total yield.

By increasing the viscosity, dissolved pectin inhibits the clarification and produces a troublesome foaming, which prevents the full utilization of the fermentation capacity. The enzymatic breakdown of pectin eliminates these problems and so facilitates filtration of the wines.

The best results have been achieved using 2–3 grammes of *Ultrazym® 100* (dissolved in water) per 100 kilogrammes crushed grapes. If the pH value is below 3.2 the higher dosage rate is recommended. The enzyme treatment (0.5–4 hours) is stopped as soon as the desired colour intensity has been reached.

Enzyme treatment of red wines and red press wines. When the alcoholic fermentation is complete, most wines have only a low pectolytic activity which originates from the fruit, because the added sulphur dioxide, the alcohol formed and the polyphenols extracted from the crushed grapes rapidly destroy the enzyme activity. For these reasons the wines often contain considerable amounts of pectin, particularly when they have been made from grapes or must which has not undergone any enzyme treatment. In such cases spontaneous clarification and filtration are extremely difficult and time-consuming.

The advantages of an enzyme treatment here are obvious. In cases when it is less certain that the problems are due to a high pectin content, the alcohol test is a simple method of assessing the level of pectin present.

In the clarification of red wines it is important to add the enzyme as early as possible, both because the higher temperature caused by the fermentation promotes the enzyme reaction, and because at this stage the fresh pectins are easier to decompose than when they have been held for a certain time. Furthermore, inhibiting substances like alcohol and sulphur dioxide are present in lower concentrations at the beginning of the process than later on, whilst the time available for sedimentation is also increased. Consequently, decantation from the yeast and deposit can take place earlier. With press wines particularly, it is striking how soon clarification occurs, while, in contrast, this often takes months when no enzyme has been used. As there is generally a rather long period available for the clarification of red wines, the enzyme amounts required are correspondingly low, 0.5–1 gramme of *Ultrazym® 100* per hectolitre. Possible exceptions are the press wines, which are frequently very rich in pectin and colour, and often require higher enzyme dosages.

4. Production of white wine

The following are among the advantages of enzyme treatment of crushed white grapes shortly before pressing.

(*i*)   Easier and more rapid pressing of the crushed grapes
(*ii*)   Prevention of splashing
(*iii*)   More free-run must of first-rate quality and less pressings
(*iv*)   Increase in total yield
(*v*)   Higher content in the must of: fermentable sugars, flavour components, acids, minerals, possibly colour compounds
(*vi*)   General improvement in the quality of the wine

Thus with this treatment there is better utilization of the winemaking equipment, less load on the machinery, a saving in time and an increase in capacity. As the grapes can be accepted for processing at a higher rate, bottlenecks in the procedures can be prevented and a better wine quality achieved.

Splashing is very troublesome, although it does not cause a significant loss of must. Its elimination, however, results in cleaner working conditions.

The increase in yield of free-run must can reach 20 per cent, and even pressings are less 'stressed'. They are less extensively oxidized and contain less turbidity-forming matter.

Regarding the content of the extracted components, increases in Oechsle readings of up to 5 degrees can be obtained. As the concentration of sugar in the grape is highest just under the skin, this means that especially with pectin-rich varieties, this sugar is often not extracted completely during pressing and therefore stays in the residue. With enzyme treatment the sugar is completely transferred into the must, resulting in a higher alochol yield. The same applies to flavour compounds and other components.

Treatment of crushed grapes. In practice, enzyme treatment requires the insertion of a maceration tank between the crusher and the press. Furthermore, destemming is essential. The enzyme dosage generally used is 1–2 grammes of *Ultrazym® 100* (dissolved in water) per 100 kilogrammes of crushed grapes. It is best to add the enzyme in the form of a 2–10 per cent solution to the funnel of the pump under the crusher. Expensive precision pumps are unnecessary; a small peristaltic pump, for example, is sufficient. If the enzyme is added to the tank the total quantity of enzyme should be added right from the beginning, and when the tank has been filled its contents should be thoroughly mixed. The holding time should be at least two to four hours, somewhat dependent, however, on the mash temperature. At temperatures below 15 °C the holding time or the enzyme dosage should be increased.

Treatment of the must before fermentation. Other advantages in treating the must before fermentation include the following:

(i)   More rapid removal of mucous substances
(ii)   Better clarification and less oxidation
(iii) Lower volume of lees
(iv)   Improved fermentation, no excessive foaming
(v)   More rapid and better clarification of the wine after fermentation
(vi)   Easier filtration
(vii) Intensification and refining of flavour, clean, authentic and lively bouquet

The more rapid removal of mucous substances ensures optimal precipitation of turbidity-causing substances before the fermentation. As the polyphenol oxidases are always linked to the turbidity-causing substances, improved clarification also means reduced oxidation.

The fermentation is much less turbulent without prolonging the fermentation time; it is often shorter and begins earlier. The foam is fine and disintegrates much more quickly; therefore, the fermentation tanks can generally be filled to a higher volume than would be the case without enzyme treatment. A gain in fermentation volume of 30–40 per cent has been observed, and the risk of losses due to overfoaming is thus considerably reduced.

Experience has shown that a wine which has been made from a must that clarifies easily, will also clarify easily after fermentation. This is even more true when enzymes have been used to eliminate the mucous substances, thereby also ensuring a trouble-free filtration. The improvement in wine quality, particularly in the bouquet, may be explained by the more efficient removal of mucous substances.

Enzymatic removal of mucous substances from the must carries few disadvantages. However, when the crushed grapes are treated with an enzyme, an increase in the content of tannins and colouring substances due to high enzyme concentration, maceration time, maceration temperature or the omission of destemming, cannot be excluded. In such cases the most suitable conditions must be found and maintained. On the other hand, enzyme treatment of the must does not cause any problems, and the low cost of this treatment makes it highly economical.

Treatment of the must in practice. The use of enzymes for the treatment of must is very simple indeed.nImmediately after the juice has been separated, an enzyme dosage of 0.5–1 g *Ultrazym®* *100* per hectolitre is added.

In practice it is most common to add the whole enzyme dose required for the batch at the beginning, just after sufficient must has been let in to cover the bottom of the tank. Additional mixing of the contents is then seldom necessary. On the other hand, a thorough mixing is necessary if the enzyme is added to the tank after it has been filled. The enzyme dosage must be adjusted according to the pectin content of the must and the desired time for removal of the mucous substances.

5. Application of a β-glucanase

The wines made from botrytized grapés very often present serious problems in clarification and filtration. It was Laborde (1907) who discovered that the origin of these problems was a polymer of glucose synthesized by *Botrytis cinerea*; he called this polysaccharide dextran. Recently, Dubourdieu (1981a) established clearly that the glucose polymer was not a dextran but rather a glucan. Using methylation analyses, he showed that the polysaccharides have a structure consisting of β 1,6-linked side chains attached to a β 1,3-linked backbone. This basic difference in structure explains the failure of the numerous dextranase treatments attempted. Different β-glucanases used in the brewing industry have also been tested without success. Failure in this case was due to the different nature of the barley β-glucan which causes filtration problems in beer, as this glucan is mostly composed of β 1,4-linked glucose units, unlike the *Botrytis* glucan, which contains almost only the β 1,3 type of linkage. This glucan is produced in all berries infected with grey mould (pourriture grise) or noble mould (pourriture noble) (Villettaz, 1981). Therefore, all wines made from botrytized grapes present filtration and clarification problems (Dubourdieu, 1978). In addition to the use of dextranases and beer β-glucanases, several treatments (Dubourdieu *et al.*, 1976) including fining agents have been used to overcome these problems, but none has been satisfactory. Recently, a specific β 1,3/β 1,6 glucanase has been developed in our laboratories. Figures 4.17.2 and 4.17.3 show the influence of pH and temperature on β-glucanase activity. The remaining enzyme activity at the pH and temperatures of the wines allows the use of this enzyme under the conditions found in practice.

First evaluations on a laboratory scale under different conditions indicate a considerable amelioration of the wine clarification as well as a 10–20-fold improvement in filterability (Dubourdieu *et al.*, 1981b). The laboratory results have been confirmed at an industrial scale even with wines presenting extreme clarification and filtration difficulties.

Let us consider the treatment of one of those wines as an example. The wine chosen was a Bordeaux sweet white wine, vintage 1980, with the following characteristics:

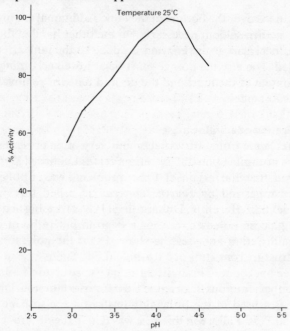

Figure 4.17.2   Percentage activity versus pH (β-glucanase).

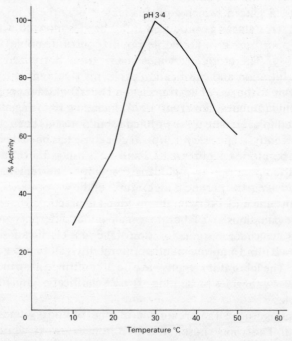

Figure 4.17.3   Percentage activity versus temperature (β-glucanase).

Alcohol:	13.7 per cent
Residual sugar:	22.6 g/l
Total acidity (as H_2SO_4):	3.14 g/l
pH	3.6
β-glucan content:	22.0 mg/l

Three tanks containing 130 hectolitres each were used for the trials. The first tank was treated 2 grammes per hectolitre with *Novozym® 116*, the second tank was treated with only one gramme per hectolitre *Novozym® 116*, while the third was kept as a control and therefore not treated at all. The enzyme treatment was performed in late August, which means that the wine was almost one year old. However, it was still highly turbid due to the continued presence of yeast and bacteria in the wine suspension. As the results in Table 4.20.1 indicate, 22 milligrammes glucan are sufficient to stabilize the cloudiness of such a wine for a very long period.

TABLE 4.17.1

Evolution of turbidities of wines during the different steps of winemaking

	Turbidity index in mg/l silica		
	Before enzyme treatment	*After enzyme treatment with Novozym® 116 (2 g/hl)*	
		And after racking (10 days)	*And after filtration*
Control	800	800	22
Enzyme-treated sample	800	400	3

In comparison, a white wine has a turbidity corresponding to 1000 milligrammes silica per litre by the end of the alcoholic fermentation. Enzyme treatment reduces this to 400 milligrammes within 10 days. Therefore, in the late summer (as in the present case), when the temperature of the wine is about 20 °C, complete clarification after enzyme treatment in a 130-hectolitre tank would take only two to three weeks.

The presence of botrytic glucan in grape juice or wine can be easily detected by the following alcohol test. To 5 millilitres of wine or grape juice in a test tube 2.5 millilitres 96 per cent alcohol are added and the solution is shaken thoroughly. If the test sample contains β-glucan a precipitation will be visible in filament form within a few seconds. In wine treated with two grammes per hectolitre *Novozym® 116* (containing 220 β-glucanase units per gramme), the alcohol test was negative after 72 hours. The filtration was performed 10 days after enzyme addition even though the sedimentation of yeast was not complete.

In the wine treated with one gramme per hectolitre *Novozym®* *116* the alcohol test was negative after 80 hours but the filtration was performed 4 weeks after the enzyme addition, to ensure that the sedimentation of the yeast was complete. Filtration was carried out under the following conditions. Using a gasquet filter unit with a four-square metre filtration surface, two grammes per litre of Kieselguhr Hyflo supercell were added to wine which had been treated with two grammes per hectolitre *Novozym®* *116*, operating with a recycling of two-thirds of the filtered wine. As can be seen from the filtration curve in Figure 4.17.4, even after 50 hectolitres had been filtered there was no reduction in filtration rate (colimation). The filtration was stopped after 50 hectolitres had passed through because the filter unit had become filled with Kieselguhr (25 kilogrammes). The high dosage of Kieselguhr was necessary because of the large amount of yeast remaining in the wine suspension.

The filtration of the control samples gives an idea of the problems which arise during the treatment of such wines. In this case the colimation point was reached after only seven hectolitres had been filtered. On the other hand, even after filtration the turbidity of the control sample was much higher than that of the treated sample (*see* Table 4.17.1). With a *Novozym®* *116* dosage of one gramme per hectolitre, however, 72 hectolitres of wine were filtered before the colimation point was reached, using only 12 kilogrammes of Kieselguhr.

After the degradation of the *Botrytis* glucan it is therefore possible to filter the wine without any problems. If sufficient time for natural sedimentation remains between the enzyme treatment stage

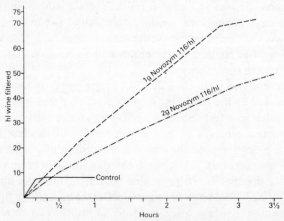

Figure 4.17.4   Filtration improvement by addition of *Novozym®* *116*.

and the filtration an even better effect is obtained using less enzyme and less Kieselguhr. The filtration of the control wine is practically impossible even if an extremely large amount of Kieselguhr (one kilogramme per hectolitre) is used. Therefore, the β-glucanase treatment solves the clarification and filtration problems of the wines made from botrytized grapes and considerably reduces their filtration costs. On the other hand, it is equally important to note that because the glucan keeps yeasts and bacteria in suspension, high sulphur dioxide levels are needed to prevent microbiological infections in such wines. The specific β-glucanase treatment would therefore allow a reduction in the amount of sulphur dioxide added to these wines.

6. Future developments

Various possibilities for new applications of enzymatic treatment in winemaking are anticipated. For instance, the use of an acid protease to hydrolyse the native proteins of the grapes would be of great interest. Of equal interest would be the use of an enzyme complex for extraction processes to treat the byproducts of juice and winemaking in order to isolate any biologically active substances. Finally, the development of enzymes better adapted to solving specific problems such as clarification, maceration, total liquefaction, filtration and colour extraction will become increasingly important.

References

Dubourdieu, D., Lefebre, A. & Ribereau-Gayon, P. *Conn. Vigne Vin* **10**, 73–92 (1976).
Dubourdieu, D. Thesis, Univ. Bordeaux II (1978).
Dubourdieu, D., Fournet, B. & Riberau-Gayon, *Carbohydrate Res.* **93**, 294–299 (1981a).
Dubourdieu, D., Villettaz, J-C., Desplanques, C. & Ribereau-Gayon, P. *Connaissance de la Vigne et du Vin* (1981b).
Laborde, J. *Cours d'Oenologie*, Mulo, Paris (1907).
Villettaz, J-C. & Amado, R. *Lebensm.-Wiss. Technol.* **14**, 176–181 (1981).

DEXTRANASE AND SUGAR PROCESSING
T. Godfrey

The annual production of sucrose from cane and beet crops has now reached almost 100 million tons, with sugar cane the dominant source. Although the basic extraction and purification process can be performed without using enzymes, there are two polysaccharides, starch and dextran, that are often troublesome contaminants of the sugar juice extracts. Their influence is primarily on the clarification and crystallization in the vacuum pans.

1. Removal of starch

Starch occurs naturally in small amounts in sugar cane, and the quantity present will depend on the growth conditions and the variety. It can be hydrolysed by the action of α-amylases that randomly attack the α-1,4 bonds to produce dextrins, which are soluble. Sugar juices are usually heated to inactivate natural invertases that are extracted from the cane, and this treatment also inactivates any natural amylases that might have been present.

The removal of starch is achieved by introducing a thermostable α-amylase of bacterial origin. In particular, the enzymes from *Bacillus licheniformis* have the ability to function in the temperature range 85–95 °C, which satisfactorily destroys cane juice invertase so that no sucrose yield is lost by inversion. The broad pH range of this type of amylase also means that it can tolerate variations in the pH of the juice from the milling plant. Small amounts of enzyme are introduced to the juice prior to the evaporation stage. For example, *Termamyl®* 120L (Novo) would be used at doses of one to two grammes per tonne of cane. Where multi-effect evaporators are used, the second stage often operates at around 90°C and the enzyme may conveniently be introduced at the inlet to this stage. Enzyme treatment to approximately halve the starch level in the juice will result in a reduction of about 90 per cent of starch from juice to raw sugar, as the largest part of the starch passes to the molasses byproduct.

Treatment with amylase is indicated if starch contents exceed 4000 parts per million based on juice solids.

2. Removal of dextran

Unlike starch, dextran is not a natural constituent of sugar juice, but is formed by bacteria of the *Leuconostoc* species, which grow in damaged cane, and in juices that are delayed before processing through the evaporators. Dextran is an α-1,6-linked glucose polymer which exhibits a high viscosity in solution. This increased viscosity becomes troublesome to both clarification and evaporation of juices and to the concentration of molasses when the level of dextran approaches 1000 parts per million of juice solids.

Industrial dextranases are obtained from strains of *Penicillium lilacinum* and degrade the polymer to smaller molecules that do not create excessive viscosity. The enzyme has a pH optimum range between 4.5 and 6.0 and a temperature optimum of 50–60°C, and functions well if added to the raw juice between extraction and clarification. Maximum effect is obtained if the juice is held for 15 minutes at 50–55°C in a tank placed between the primary and secondary heaters that precede the clarifiers. An example of enzymic treatment under these conditions would be 10–20 grammes of *Dextranase* 25L (Novo) per tonne of cane processed.

When raw sugar requires treatment on arrival at a refinery, enzyme treatment of the syrup may be similar to that described above. At the higher temperature and solids concentrations of these syrups, the enzyme dose levels or the holding times will be increased. Between 85 and 90 per cent of the dextran should be removed to obtain a satisfactory sugar.

Dextran is also formed occasionally in the processing of sugar beets, particularly when they have been exposed to freezing and thawing cycles during harvest and storage. The same treatment concepts as those described for cane sugar can be applied when this occurs.

3. Removal of raffinose

When beet sugar juice is produced by the recycle Steffen process, the presence of traces of raffinose, which is a trisaccharide of glucose, fructose and galactose, can interfere with sucrose crystallization. This molecule is concentrated during the recycle and in some cases it is treated by the application of the α-galactosidase, melibiase, which splits off the galactose molecule, leaving sucrose. Industrial sources of this enzyme are rare and the low heat tolerance of the available types is not ideally suited to the process conditions.

EDIBLE OILS
T. Godfrey

1. Introduction

For more than ten years, the practical benefit of enzyme additions in the extraction of oil from palm olives has been known. The extent to which the applications have become accepted practice is not clear, but is not thought to be very great.

With rising value for the oils and with higher demands on oil quality, coupled with several years of unfavourable climatic conditions in growing regions, there has been a noticeable increase in trials of enzyme processing for a wider range of oil sources including palm, olive, soya bean, rapeseed, sunflower seed, cottonseed, corn germ and groundnut. The use of certain enzymes shows some improvement in the yield of oil, together with a reduction in the acid development and oxidation of the oil during further processing and storage. The degree of improvement is very variable and depends on the variety of the crop being processed, the seasonal growing conditions and the conditions used for the enzyme treatment.

2. Enzyme action in conventional vegetable oil extraction

Oil is extracted from the individual oil-bearing cells of plant material by pressing and expeller systems. Depending on the initial amount of oil released, processing may have to be extended to extract the maximum quantity of oil. Since the ultimate quality of oil is impaired by acidic and oxidative reactions, extended processing may be needed to separate oil and water emulsions.

The principle of action of the industrial enzymes so far considered for improving these conditions (the cellulases, pectinases, hemicellulases and sometimes proteases), is hydrolytic depolymerization of plant components and requires an aqueous environment and adequate contact time at appropriate pH and temperature. It may not be coincidental that the two most investigated and practically resolved oil extractions are for palm and olive, where the harvested oil crop contributes some water, and processing is tolerant of further water being introduced.

While the degradation of the plant tissues is highly desirable for oil release, experience has shown that the operation of expeller and pressing systems does not show benefit if too substantial a degradation is achieved. The maceration of tissues to a homogenous pulp renders them unsuitable for these machines, although the most recently available enzymes suggest that sufficient degradation can make it practicable to separate the oil with centrifuges in a continuous process.

*Examples of established enzyme applications_*

Palm oil. The use of cellulases and pectinases at rates of 200–1000 grammes per tonne to pretreat before extraction using minimum water addition, to facilitate adequate distribution, and at 25–40°C for 2–4 hours, has demonstrated economically worthwhile yield improvements.

The main benefits of enzyme treatment are to be found by treating the oil and water mixture after separation. Here the objective is the reduction of oil losses at the clarifier stage, partly by further degradation of plant tissues and partly to reduce the molecular weight of viscosity contributing components of the aqueous extraction fraction. These components consist of varying amounts of glucans, hemicellulose, cellulose and pectins. Generally, sufficient action is obtained by use of equal amounts of cellulase and pectinase to a total of 500 to 1000 grammes per tonne of oil/water mixture at temperatures between 25 and 50°C for periods of 1–3 hours in the holding tanks preceding the clarifier system.

Olive oil. The enzymic treatment of the macerated pulp at pH 5.5–6.2 (the natural pH of the pulp) at 35–40°C over periods of 1–2 hours has been established on the industrial scale, with improvements in oil yield, reduced acidic content and lower oxidation. Cellulases, pectinases and bacterial neutral proteases have all shown some positive effect, but all 3 in combination have given 10–15 per cent yield improvements depending on olive variety. Equal quantities of the 3 enzymes, totalling 500–1000 grammes per tonne of olives, provide these gains, but overtreating can result in difficult processing.

When macerating enzymes are used alone, the results are comparable, since these enzyme preparations usually contain cellulase, hemicellulase and acid optimum protease activities. Cellulases or macerating enzymes are used at 100–500 grammes per tonne of olives according to variety, and give typically 7–10 per cent yield improvement and markedly reduced oil acidities.

Extraction of oil seeds. While most current extraction processes seek the minimum of water in the incoming seeds and then keep further water usage to very low levels, the application of industrial enzymes will largely be impractical. However, a number of industrial systems have been established that include a wet conditioning stage, permitting favourable conditions for enzymes to function. Both cellulases and macerating enzymes have been shown to offer a small increase in oil yield of two to six per cent with additional benefit in lower oil content in the final seed cake feedstuffs.

Rapeseeds have been processed by a conditioning of the seeds before the first pressing in expeller equipment. Using cellulases, 500–1000 grammes of enzyme per tonne seeds is dissolved or dispersed in the permitted water. This volume will be as large as possible, but will not exceed the subsequent processing limits of the overall plant. Heating to 50 °C and holding for 30–60 minutes is followed first by raising the temperature to 95 °C and then the first pressing. The residue is reslurried and treated in a similar way with fresh enzyme for the second pressing.

Similar treatments can be applied to cotton seed and sunflower seeds, although industrial scale operation has not been reported to date. It has also been demonstrated that a wet slurry of corn germ may be beneficially treated with one to five kilogrammes cellulase per tonne and result in useful oil yield gains. The contact time is long and requires antimicrobial treatment to prevent spoilage.

Extraction of soya bean oil. Extraction of soya bean oil has been subjected to research ftr improved yield by both acidic and alkaline wet processes. It has been established that the oil is bound more to protein than to carbohydrate and consequently proteases have proved the most promising. The use of acid or alkali optimum proteases results particularly in considerable degradation of the proteins which manifests itself as considerable emulsion formation after separation. Further protease treatment can break these emulsions, but too great a loss of valuable proteins cannot be accepted in the present process.

A more satisfactory system has been developed that utilizes proteases at 0.5–1.5 per cent on protein content with careful control of temperature and time to give maximum oil release and subsequently to break the emulsions with suitable polyelectrolyte flocculants.

Newer processing concepts. Recent work has been seen to establish a completely wet approach to vegetable oil extraction, with emphasis on the elimination of solvent treatments, low fuel costs and maximum recovery of high value proteins together with carbohydrates in solution to act as fermentation feedstocks. Several schemes have been proposed based on an initial gelatinization of starch content and degradation of carbohydrates by typical starch processing enzymes (*see* Chapter 4.13, 'Plant tissues', and Chapter 4.15, 'Starch').

Separation of proteins has been proposed both by heat coagulation in rapid flow heat exchangers or by the use of flocculating agents. Oil recovery is then to be via the application of powerful centrifuges to remove the protein and then oil-water separators. Emulsion formation must be anticipated to be a major limitation requiring innovative process engineering and careful selection of plant.

3. Fish oil extraction

As for soya bean oils, fish oils are protein-associated and industrial proteases have been used successfully to improve oil yields. Pre-treatment of macerated fish with papain and bromelain at levels between 1–5 kilogrammes per tonne fish weight using incubations of 30 minutes at 65 °C for papain and 1–2 hours with bromelain at 50–55 °C has been shown to yield more oil. When heat treatment to coagulate the proteins follows the enzyme stage, the recovery of the oil is further improved and emulsion formation is not a serious problem.

Microbial enzymes, particularly the relatively thermostable bacterial alkaline proteases, have been used to advantage by being less likely to contribute microbial contamination than plant proteases. They can be used at temperatures of 70–75 °C and so prevent any microbial growth during the treatment. Generally, 1–2 kilogrammes per tonne of fish are found effective in treatments of 30–90 minutes.

4. Oil quality

It has been established that the enzymes described for edible oil extraction are almost excluded from the oil phase and remain in the process waters. For most countries, this permits their use as processing aids rather than additives, and they do not adversely affect the quality criteria for edible oils. Where the acidity and oxidation characteristics of the oil are improved by enzyme-aided processing, there is a positive contribution to oil quality. If the newer ideas find economic favour, the elimination of the use of organic solvents will further enhance the quality of extracted oils.

GLUCOSE OXIDASE
G. Richter

1. Introduction

Glucose oxidase differs from other enzymes used in food processing. While most such enzymes are hydrolases which break down organic compounds of high molecular weight, glucose oxidase is an oxidoreductase, oxidizing D-glucose to gluconic acid. Commercial products do not consist of a single enzyme, but of an enzyme system comprising glucose oxidase plus catalase and often other components, including the coenzyme flavine-adenine-dinucleotide. The reaction can be shown thus

$$1. \quad 2C_6H_{12}O_6 + 2H_2O + 2O_2 \xrightarrow[\text{oxidase}]{\text{glucose}} 2C_6H_{12}O_7 + 2H_2O_2$$

$$2. \quad 2H_2O_2 \xrightarrow{\text{catalase}} 2H_2O + O_2$$

$$3. \quad 2C_6H_{12}O_6 + O_2 \xrightarrow{\text{MKC Glucose Oxidase}} 2C_6H_{12}O_7$$

Glucose oxidase is mainly available from microbial sources and is normally produced by the controlled fermentation of *Aspergillus niger* (submerged culture) or *Penicillium amagaskinense*.[1] Unlike most other industrial enzymes the fungal mycelium must be mechanically broken down to liberate the enzyme and make it soluble. Depending on the form required for the final product, the subsequent stages of processing can be varied accordingly. Hence, for a liquid formulation of this enzyme, after the initial mechanical breakdown of the mycelium the resultant solution is filtered and concentrated. Solid forms are filtered and then precipitated and dried. For a purified solid enzyme a futher stage of purification and lyophilization is added. Enzyme activity is measured by one or two assay methods: in the manometric (Baker Units)/titrimetric method, one Glucose Oxidase Unit (GOU) is that activity which causes the uptake of 10 mm³ of oxygen per minute under the conditions of the assay. In the colorimetric method (Boehringer Units), one Glucose Oxidase Unit is that activity which catalyses the conversion of one μmol glucose per minute under the conditions of the assay. Activity of unpurified industrial products is in the range of 750–1500 manometric units per millilitre or gramme and of purified analytical products of 60,000–120,000 manometric units per gramme.

Glucose oxidase from *Aspergillus niger* has a molecular weight of

192,000 an optimum temperature of 30–50 °C and optimum pH of 4.5–6.5. Glucose oxidase is inhibited by heavy metal salts (mercuric chloride, silver chloride and silver nitrate) and sulfhydryl chelating agents, while catalase is inhibited by urea, freezing and sunlight under aerobic conditions. Glucose oxidase has a fairly high substrate specifity for β-D-glucose, but there are also several hydrogen acceptors (e.g. molecular oxygen), certain dyes and even benzoquinone.[2,3]

Because it is a naturally occurring antioxidant, the use of glucose oxidase in the food industry is legally accepted in most countries. The safety and specifications data are given in two WHO reports.[4,5] In the USA glucose oxidase is listed under GRAS (Generally Recognized as Safe) specifications.[6]

2. Antioxidant applications

Glucose oxidase is used in the food industry primarily to prevent changes in the colour and flavour of food products – both during processing and in storage. It is well-known that the oxidation which occurs when oxygen is present causes significant deterioration of foods. Hence the addition of chemical antioxidants such as ascorbic acid and glucose oxidase. Because of its mild action glucose oxidase is widely used for this purpose. To exemplify the applications of glucose oxidase as an antioxidant several instances of its use in different products are described.

Citrus drinks. Both citrus concentrates and carbonated citric beverages contain dissolved oxygen which, once converted to peroxide by sunlight, adversely affects the flavour and hence the shelf life of the citrus products. The addition of 40–200 Glucose Oxidase Units (colorimetric) per litre to concentrates, and 20–90 Glucose Oxidase Units to carbonated soft drinks, allows the original fresh taste and colour to be maintained.[7,8] However, as the cloud stability of citrus juice appears to depend on the presence of cellulose[9] the glucose oxidase and catalase preparation must be free of, or low in cellulase.

Canned soft drinks. Oxygen can either be dissolved in the drink itself or in the 'head space' at the top of the can between the can and the fluid. The effects of glucose oxidase in soft drinks such as cola and black cherry beverages, orange soda, sassafras soda and grape soda have been described elsewhere[10,11] but can be summarized as follows: (*a*) there are no detectable flavour changes; (*b*) the tendency of sensitive colours to fade is impeded; (*c*) iron pickup is decreased by removing residual oxygen. The efficiency of the glucose oxidase and catalase system is adversely affected by high temperatures when the canning or filling of such liquids is carried out. For instance, during canning the cans may be heated to high temperatures for a very short period and this may totally inactivate the enzyme.[12] Thus, the enzyme should only be added shortly before the cans are filled and sealed – usually by automatic dosing systems.

Beer and wine. 10–70 Glucose Oxidase Units (colorimetric) per litre are sufficient to retard the oxidation of beer. Glucose oxidase and catalase can also be used to advantage in rosé and white wines, since they reduce undesired enzymic browning and changes in flavour and taste.[13]

Mayonnaise and salad dressings. These emulsions of oil in water contain oxygen. When the product is packed in glass or plastic containers and exposed to light and normal room temperatures, the oxygen present may cause off-flavour, broken emulsions and badly faded colour.[14] However, this may be avoided by the addition of 20–100 Glucose Oxidase Units (colorimetric) per kilogramme product, which will ensure excellent quality even during prolonged storage. Mayonnaise and salad dressings contain starch or modified starches, and so the actions of amylase in conjunction with glucose oxidase and catalase preparations may influence the consistency of the goods. This must be carefully monitored.

Dried foods. Instant coffee, cake mixes, milk powders, dried soups, active dried yeast, concentrated aroma components and other dried foods are very susceptible to oxidative deterioration. Again the addition of glucose oxidase and catalase is recommended to avoid any undesired changes. A special technique for adding the enzyme to these products is described below (*see* 'Future developments').

3. Glucose removal

The second most important market for glucose oxidase and catalase in the food industry is the removal of glucose from egg whites and whole eggs. Since eggs have a glucose content of approximately 0.5 per cent, browning (Maillard reaction) and development of off-flavours may occur, especially in the manufacture of dried eggs, if this glucose is not removed.

Two practical examples of desugaring of egg whites and whole eggs or yolks are given.[15] The method for desugaring egg whites is described below. The liquid egg whites are heated to approximately 32 °C. The pH is adjusted to 7.0–7.5 by adding citric acid. Approximately 1 kilogramme of citric acid per 100 kilogrammes of liquid egg whites will usually give a pH 7.0. To ensure an adequate oxygen supply for the reaction to occur, hydrogen peroxide (35 per cent) is added gently and continuously prior to the addition of glucose oxidase and also during the reaction period. About one third of the hydrogen peroxide requirement is added to saturate the egg white and this should be done as soon after pH adjustment as possible and approximately 30 minutes prior to the addition of the glucose oxidase. Approximately 3.6–4.0 litres of hydrogen peroxide per 1000 kilogrammes of liquid whites is needed for the complete reaction.

The glucose oxidase requirement is normally 100–150 grammes of *Glucose Oxidase P* (1500 Glucose Oxidase Units per gramme) or 200–250 millilitres *Glucose Oxidase L* (750 Glucose Oxidase Units per millilitre) per 1000 kilogrammes of liquid whites, while stirring gently. With this concentration and an adequate oxygen supply from the hydrogen peroxide the desugaring process should be complete in about eight hours, as indicated by the Somogyi Test Procedure. Since the reaction time is directly dependent on enzyme quantity, if the amount of enzyme used is doubled, the desugaring time may be approximately halved. After the glucose oxidase is added, the rate of hydrogen peroxide addition is halved, and for this regulation a proportionating device is recommended.

The method for desugaring whole eggs or yolks is as follows. The egg liquid is heated to 30–31 °C. The desugaring vat should be filled through the bottom to minimize foaming and no pH adjustment is needed. While filling the vat the hydrogen peroxide (35 per cent) may be added as soon as the agitator blades are covered, and the egg liquid can be agitated without undue whipping. About four kilogrammes of hydrogen peroxide (35 per cent) per 1000 kilogrammes is usually required to desugar the egg liquid in about three hours. Before addition of glucose oxidase, about one third of the total hydrogen peroxide requirement is added to the egg liquid. After the vat has filled and the egg liquid is saturated with hydrogen peroxide, 200–250 grammes of *Glucose Oxidase P* or 350–400 millilitres of *Glucose Oxidase L* per 1000 kilogrammes of egg liquid is added while gently stirring. The desugaring time should be at least three hours according to the Somogyi Test Procedure. For regulation of hydrogen peroxide addition a proportioning device is again recommended.

Other applications of this type have been suggested (i.e. in the drying of meat and potatoes) but there is no evidence of widespread. industrial practice.

4. Other applications in the food industry
There are a few other applications which have been reported. Glucose oxidase (without catalase) together with ascorbic acid can

be used for oxidizing flour. The hydrogen peroxide formed oxidizes ascorbic acid to dehydroascorbic acid which may replace other chemical oxidants used in baking.[16,17]

Glucose oxidase in conjunction with cellulase should protect against infection in fermented fruit mashes. The addition of cellulase ensures a supply of glucose as substrate for glucose oxidase. The hydrogen peroxide so formed improves the microbial stability of the mash.[18] The production of acid in cottage cheese or yoghurt in skim is caused by fermenting microorganisms. Addition of lactase together with glucose oxidase, and hence the formation of gluconic acid from lactose, has been proposed as an alternative measure.[19]

5. Measurement of glucose levels

Glucose oxidase is widely used as a diagnostic agent in medicine in the measurement of glucose levels in patients. This test is valuable in diagnosing diabetes, hypoglycemea and various adrenal and pituitary disorders. Low levels of glucose in the cerebrospinal fluid are indicative of a bacterial infection in the central nervous system, while low urinary glucose levels are a reliable index of urinary tract infection.

Glucose oxidase is also used to detect glucose in starch hydrolysates, fermentation liquids and several foods. The assay involves a two-stage reaction. After initial oxidation of glucose by glucose oxidase, the reduced chromogen and the hydrogen peroxide produced in the first reaction are converted to oxidized chromogen and water by peroxidase. The quantity of oxidized dye can be accurately measured and is a direct measure of the glucose originally present. A number of dyes can be used, such as o-dianisidine, o-toluidine, 2,6–dichlorophenolindophenol and several others available under various trade names.

The great advantage of this method is that it is specific for glucose; no other carbohydrate will react. The method is very accurate, reliable and fast, and requires very little sample volume. However, the glucose oxidase applied must be free of catalase, as this enzyme reduces the quantity of hydrogen peroxide available for the reaction with colourless chromogen. Other enzymes, such as amylase, saccharase and maltase are also undesirable and together with catalase must be separated or inactivated by means of a column or gel filtration or other methods.[20]

The complete test for glucose is available commercially, in the form of test strips impregnated with glucose oxidase peroxidase and a dye, as individual test kits or packaged for use in an automated system which includes an immobilized glucose oxidase. The biggest suppliers of the corresponding reagents and equipment are

Boehringer Mannheim, Technicon, Beckman, Miles (Ames Division), Lilly and Worthington Biochemicals.

6. Future developments

There are a number of future applications for glucose oxidase which have significant potential. For instance, the addition of glucose oxidase and amyloglucosidase to tooth pastes could grow to a market of unexpected commercial significance.

Dental caries is a result of bacteria colonizing the tooth surface and forming dental plaques. These plaques consist of bacterial cells, residues of cells and bacterial products, especially extracellular polysaccharides. Microorganisms present in the plaques convert sugars into lactic and other organic acids and the drop in pH causes demineralization in the deeper layers of the tooth.

There are, however, antibacterial factors in the saliva, such as lysozyme and an antibacterial system based on lactoperoxidase, thiocyanate and hydrogen peroxide. The hydrogen peroxide appears to be especially effective against lactobacilli and streptococci, but in reality its effect is limited by the fact that the hydrogen peroxide is produced solely by the oral flora. In order to increase the amount of hydrogen peroxide available to this antibacterial system in the saliva glucose oxidase (low in catalase) and amyloglucosidase (formation of glucose from starch) has been added to a new tooth paste and the results recorded.[21]

It was found that up to 21 hours after the application of enzymes generating hydrogen peroxide the acidity measured on the tooth surface was less. Also, the formation of plaques appeared to be significantly less in those subjects with an activated lactoperoxidase system. A new tooth paste, based on these results, is now offered in several countries under the trade name *Zendium.*

Food may be protected against oxidative deterioration by a moisture-proof sheet of plastic coated with glucose oxidase and catalase. Inactive in a dry state, the enzyme picks up moisture from the food and is reactivated when a moist food is packaged with this film. Scott[22] described the effectiveness of this system for protecting cheese loaves against formation of brown rings and for protecting prepacked luncheon meats against the abnormal greyish colouration. Other similar techniques may well be of interest and may find wider application in the future.

The possibility of oxygen removal from dried foods by glucose oxidase and catalase was discussed above. It is not feasible to add glucose oxidase directly to dried foods, since the enzyme is inactive in the dry state and the addition of moisture would cause the dried food to deteriorate. In order to use this enzyme with dried foods, it must be added together with glucose, buffer and filler in a separate pouch, consisting of a foil impermeable to water vapour but permeable to oxygen.[23] The effect of such antioxidative units in cans of powdered coffee packed under vacuum has been studied.

The heat sealed pouches, consisted of polyethylene films filled with 50–250 milligrammes glucose, 50–200 milligrammes MKC *Glucose Oxidase P* and 5 millilitres 1.0 M sodium acetate buffer at pH 6.0. These pouches were added to the cans just before they were filled and sealed. Measurement of residual oxygen in the cans has shown the benefits of the removal of very small traces of oxygen, especially after very long periods of storage (more than six months). This procedure may be applied to cake mixes, milk powder and dried soups to prevent their deterioration by oxidation and also to other consumer products such as cakes, rolls, whipped desserts etc, to prevent microbial deterioration. However, it must be noted that these pouches are rather expensive to prepare and therefore are only feasible for highly priced products.[24,25] Althogh the economic benefits of such enzyme preparations may be doubtful, they represent a development in the use of immobilized glucose oxidase and catalase.

Immobilization of glucose oxidase would provide the means for continuous and economic processing. Of the many preparations of carrier-bound enzymes reported in the literature, only seven are cited here.[26,27,28,29,30,31,32] Obviously it is impossible to oxidize high glucose concentrations completely under the conditions commonly used. This is because of the sensitivity of glucose oxidase to heat, the necessity to continually adjust the pH and the high demand for oxygen.

Therefore, other types of reactor must be considered and among these a stirred tank with oxygen supply and autotitrator (or a cascade) seem to be the most promising. In this case the particle structure of the enzyme and the resistance to mechanical destruction must be specially considered.[33] Immobilization has been carried out by fixation to the cell (with glutaraldehyde) and to many organic (collagen, cellulose, dextrans) and inorganic (metals, metal oxides, silicates, glass beads, ion exchange resins) carriers.[34]

The use of an immobilized glucose oxidase without catalase in diagnostic tests has already been described above, and this system appears to offer the greatest commercial potential.

Other suggested uses of immobilized glucose oxidase with catalase have been in the deoxygenation of beer[35,36] and fruit juices, the concentration of fructose from glucose and fructose mixtures[37,38] and in industrial gluconic acid production.[39,40,41] This latter application could become important in the future. Gluconic acid is currently produced by fermentation, and thus has all the economic disadvantages of mandatory sterile fermentation, losses of substrate by cell metabolism and a long fermentation time. However, this might be overcome by using immobilized glucose oxidase with catalase and hydrogen peroxide as the oyxgen supply.[42] The reaction velocity may be increased five to ten times when hydrogen peroxide is continuously added, but a high load on the catalase carrier is necessary to avoid a rapid and irreversible inactivation of both glucose oxidase and catalase. In general the half life of immobilized glucose oxidase with catalase preparations is quite short, except when the process is carried out at a low and in most cases unrealistically low temperature (close to 0°C).[43,44]

Interest has been renewed also by a recent publication that describes the production of hydrochinone from benzquinone by immobilized glucose oxidase.[45] Obviously the enzyme is quite specific as to the donor of the hydrogen atoms, but unspecific as to the acceptor of the same. In future this could result in new energy saving and milder processes, for instance in the chemical industries.

References

1. Kusai, K. *Annu. Rep. Sci. Works Fac. Sci.* **8,** 43 (1960).
2. Reed, G. (ed.) *Enzymes in Food Processing* 2nd edn (Academic, New York, 1975).
3. Underkofler, L.A. in *Product and Application of Enzyme Preparations in Food Manufacture*, 72–83 (Society of Chemical Industry, London, 1961).
4. 'Toxicological evaluation of some enzymes, modified starches and certain other substances', WHO Food Additive Series no. 2, Geneva (1972).
5. 'Specifications for the identity and purity of some enzymes and certain other substances', FAO Nutrition Meetings Report Series no. 50B Rome; WHO Food Additive Series no. 2, Geneva (1972).
6. *Food Chemicals Codex* 2nd edn 1st Suppl. (National Academy of Sciences, Washington, DC, 1974).
7. *MKC-Glucose Oxidase* (MKC Product Information, 1981).
8. Barton, R. R., Rennert, S. S. & Underkofler, L. A. *Fd Technol.* **11,** 683–686 (1957).
9. Reed, G. (ed.) *Enzymes in Food Processing* 2nd edn (Academic, New York, 1975).
10. Barton, R. R., Rennert, S. S. & Underkofler, L. A. *Fd Technol.* **11,** 683–686 (1957).
11. Barton, R. R., Rennert, S. S. & Underkofler, L. A. *Fd Engng*, 79, 80, 198, 199 (Dec. 1955).
12. Reed, G. (ed.) *Enzymes in Food Processing* 2nd edn (Academic, New York, 1975).
13. Ough, C. S. *Am. J. Enol. Viticult.* **26,** (1), 30–36 (1975).
14. Underkofler, L. A. in *Production and Application of Enzyme Preparations in Food Manufacture*, 72–83 (Society of Chemical Industry, London, 1961).

436   Chapter 4   Industrial applications

15. *Enzymatic Desugarisation of Eggs with Dee O®, no. 34000-3* (MKC Product Information).
16. Gams, T. *Getreide, Mehl und Brot* **30,** 113-116 (1976).
17. Silberstein, O. *Bakery Dig.* **35,** 44 (1961).
18. Bruchmann, E. E. & Kolb, E. *Lebensm.-Wiss. u. Technol.* **6,** 158-164 (1973).
19. Richardson, T. in *Food Proteins* (eds. Feeney & Whitacker), Advances in Chemistry Series 160, 232-234 (Washington, DC, 1977).
20. *Glucose Oxidase purified* (MKC Product Information, 1981).
21. Hoogendoorn, H. *The Effect of lactoperoxidase–thiocynate–hydrogen peroxide on the Metabolism of Cariogenic Microorganisms in Vitro and in the Oral Cavity* (Mouton, The Hague, 1974).
22. Scott, D. *Fd Technol.* **7,** 8, 11, 13-17 (1958).
23. Underkofler, L. A. *Manufacture and Uses of Industrial Microbial Enzymes,* Chemical Engineering Progress Symposium Series, Vol. 62 no. 69 (1966).
24. Scott, D. *Fd Technol.* **7,** 8, 11, 13-17 (1958).
25. Bärwald, G. *Oxidationsschutzmittel,* Germ. Patent DOS 25 20 792 (1976).
26. Kirstein, D. & Kühn, W. *Lebensmittelindustrie* **28,** 205-208 (1981).
27. Atkinson, B. & Lester, D. E. *Biotchn. Bioeng.* **16,** 1299-1320 (1974).
28. Messing, R. A. *Biotechn. Bioeng.* **16,** 897-908 (1974).
29. Greenfield, P. F. & Laurence, R.nL. *J. Fd Sci.* **40,** 906-910 (1975).
30. Buchholz, K. & Gödelmann, B. *Biotechn. Bioeng.* **20,** 1201-1220 (1978).
31. Buccholz, K. & Reuss, M. *Chimia* **31,** 27-30 (1977).
32. Karube, I., Hirano, L. & Suzuki, S. *Biotechn. Bioeng.* **19,** 1233-1238 (1977).
33. Hartmeyer, W. *Stärke* **33,** 97-102 (1981).
34. Kirstein, D. & Kühn, W. *Lebensmittelindustrie* **28,** 205-208 (1981).
35. Hartmeyer, W. *Studies on the Application of an Immobilized Glucose Oxidase-Catalase,* First European Congress on Biotechnology, Interlaken 1978.
36. Hartmeyer, W. *Biotechnol. Letters* **1,** 21-26 (1979).
37. Hartmeyer, W. & Tegge, G. *Stärke* **31,** 348-353 (1979).
38. Hartmeyer, W. *Stärke* **33,** 97-102 (1981).
39. Coppens, G. *Procédé pour la fabrication d'acides aldoniques par vole enzymatique,* European Pat. No. 0 014 011 (1979).
40. Richter, G. & Heinecker, H. *Stärke* **31,** 418-422 (1979).
41. Kirstein, D., Kühn, W. & Mohr, P. *Lebensmittelindustrie* **28,** 444-446 (1981).
42. Richter, G. & Heinecker, J. *Stärke* **31,** 418-422 (1979).
43. Hartmeyer, W. & Tegge, G. *Stärke* **31,** 348-353 (1979).
44. Hartmeyer, W. *Stärke* **33,** 97-102 (1981).
45. Alberti, B. N. & Klibanov, A. M. *Enzyme Microb. Technol.* **4,** 47-49 (1982).

IMMOBILIZED ENZYMES

T. Godfrey

1. Introduction

During the 1960s the frequency of scientific reports of laboratory immobilization of enzymes reached a peak that produced several hundred papers per year. By 1969 industrial pilot plant operations with both penicillin acylase (by Malcolm D. Lilly of University College, London, in association with Beecham Laboratories) and glucose isomerase (by Standard Brands, Clinton Division) had been demonstrated.

By the end of the next decade, there had been some increase in industrial scale-up and new pilot operations, leading to some six versions of glucose isomerase, two penicillin amidases and single examples of aspartase, aminoacylase, fumarase and lactase.

A further expansion has followed, so that by 1981 there were seven glucose isomerases, four penicillin amidases, three aminoacylases and lactases, two glucoamylases but still only one aspartase and one fumarase offered for commercial industrial operations by immobilized systems.

In principle, the concept of enzymes in a physically immobilized state, whereby they may be used to process large columes of substrate and not remain in the product, is highly attractive. The comparative instability of enzymes when exposed to suboptimal pH, temperature, and substrate conditions and their general intolerance of organic solvents provide a strong stimulus for the investigation of alternative processing concepts. The traditional use for industrial enzymes is in batch processes with soluble enzyme and aqueous substrates. Recovery and reuse of the enzyme catalyst is extremely difficult under these circumstances and enzyme is therefore consumed in each batch operation. In other cases, the inactivation of the processing enzyme is necessary at the conclusion of the desired reaction stage, and the changed conditions required to effect this inactivation may impose undesirable chemical or thermal effects on the products.

A further consideration would be the processing of large volumes of dilute substrates where enzyme cost contributions become significant if continuously lost.

Many of these disadvantages of the soluble enzymes, theoretically, can be overcome by the use of immobilized forms if the result is more stable, reusable and available for continuous processing. It is

437

therefore surprising, at first consideration, that so few of the industrial enzymes are available as immobilized preparations (*see* Data Indexes 3 and 4 for lists of available enzymes in all forms). Table 4.21.1 outlines the factors to be evaluated for immobilized enzyme applications.

TABLE 4.21.1
Factors for consideration when favouring immobilized over soluble enzyme systems

Reaction	Is the process uniquely possible by immobilized systems? Are there unique properties for the product by immobilized systems?
Substrate	Is the substrate suitable (e.g. low concentration, low molecular weight, free of suspended solids)?
Control	Is the rate and byproduct level more readily controlled?
Product	Will it be more pure? Is the yield higher?
Enzyme	Will it be more cost-effective? Is the stability appropriate?
Operation	Are pre- and post-treatments required? Can the process be automated?
Plant	Is the technology available and appropriate to local and site facilities?
Economy	Does it represent the best overall solution to the performance of the chosen reaction?

Interpretation of the theoretically unlimited range of potential immobilized enzymes to determine the industrial success rate shows that certain important factors are responsible for the very critical approach that has been seen from industry. These factors may be of different significance in each specific case, but they represent the pragmatic view of commercial investment and operating practice.

Firstly, the genuine practicability of performing the process on an industrial scale should be considered. Although many laboratory reports show that a conversion can be achieved by immobilized enzymes, inspection often reveals a very low conversion efficiency at an unrealistically low substrate concentration. The concept of processing an insoluble substrate with an immobilized enzyme is also hard to visualise in terms of 'collision theory' and industrial practice, although some surprises have been noted from very small scale trials with both starch and protein substrates when apparently insoluble. However, the economics would not justify attempts to scale up the idea in the face of low cost soluble enzyme technology.

The second consideration is that of novelty represented in the ultimate product. If sufficiently novel results can be demonstrated it is likely that industrial interest will be aroused. Examples of this type of immobilization occur in the use of enzymes for control systems by the use of the specificity of the enzyme in a detector electrode (*see* the paragraph 'Analytical applications of immobilized enzymes' below).

The third point to consider is the contribution to overall process economy that can be made by using immobilized enzymes in the place of soluble ones. A successful introduction will contribute in some or all of several areas, such as smaller plant size for a given throughput, lower product purification costs, elimination or minimization of residual enzyme testing in the product, overall lower process energy costs and higher productivity per enzyme activity unit.

The penetration of immobilized enzyme systems into industrial operations can be seen to be restrained by several consequential factors deriving from the previous discussion of the positive value they may contribute. These limitations include: (*a*) the comparatively low cost of soluble enzymes for many industrial processes, coupled with traditional attitudes that are slow to change; (*b*) the capital costs involved in introducing new equipment to existing process plant; (*c*) the nature and cost of the immobilizing support and the immobilizing process, including losses of activity at this stage; and (*d*) the performance of the system in both half life and productivity terms which relate to overall operational economy and plant design scale.

2. Types of immobilization

Although it is not intended to discuss the details of the large variety of immobilization methods, some comments are made here regarding the main variations, and the reader is offered further detailed reading at the end of the section.

Most industrial scale products are the result of a combination of two or more of the five bonding methods:

(*i*) adsorption onto or within a carrier

(*ii*) crosslinking of enzymes onto the surface or within a carrier

(*iii*) crosslinking of enzyme molecules without carrier

(*iv*) covalent bonding to a carrier

(*v*) encapsulation or entrapment with a carrier

By the use of natural and synthetic polymers, the bound enzyme may be offered as a granule, fibre, membrane sheet or tube reactor; the equipment to utilize each type is different and highly specific.

The variety of carrier substances is very wide and while not all have found extensive use or development for industrial practice, examples using colloidal silica, clays, glass particles, titanium dioxide, hydroxyapatite, charcoal, alginates, starches, gelatine, cellophane, polyacrylamide, nylon, cellulose, collagen, dextran and agarose can be found.

3. Changes in enzyme character on immobilization
Many of the altered characteristics that are observed when an immobilized enzyme is compared with the soluble form should be considered only as apparent changes.

Rate of reaction. Diffusion of substrate from the bulk of the reactants to the immediate environment of the enzyme will be an important factor. In the area in contact with the immobilized enzyme the substrate concentration is lower than in the overall bulk. This concentration will depend to some extent upon the rate of flow of substrate and this also influences the product concentration and its rate of removal. Consequently, within limits a faster flow of substrate over a packed bed, or faster stirring of suspended enzyme, will increase the rate of reaction.

The molecular weights of both substrate and products will influence these diffusions, and in turn the rate of reaction. The larger the molecular weight, the lower the reaction rate; this is a general rule for immobilized enzyme systems, reflected in the industrial processes currently using them, which are dominated by small molecular weight substrates such glucose or lactose.

Where the enzymic reaction is subject to product inhibition, it is commonly observed to be more inhibited when operated with immobilized enzymes. This has resulted in a variety of attempts to remove the products of reaction from the system by, for example, chemical precipitation treatments or selective membrane filtration.

If the reaction includes charged molecules the local environment of the immobilized enzyme may have a markedly different pH to that of the bulk. The inclusion of strong buffering agents in the reaction is often essential to protect the enzyme from inactivation by extremes of pH and to operate as near the optimum pH as possible.

Changed pH optima. Most carriers are charged to some extent and during immobilization of the enzyme these charges will change and modify the overall enzyme charge character. The apparent pH optimum of the enzyme can be expected to shift quite considerably, by as much as two pH units, which, of course, is a hundredfold change in local hydrogen ion concentration, as pH is a logarithmic scale. Under these circumstances, it may be necessary to establish that the carrier is itself stable and will not steadily dissolve. Glass will dissolve at alkaline pH values, quite seriously lowering the retention of the bound enzyme. In many cases, the pH of the product stream from an enzyme reactor will be quite different from the feed stream unless strong buffering is included.

Altered temperature optima. Although in a few cases reduced thermal stability has been noted, most industrial immobilized enzymes show an increased thermotolerance. Papain has shown to be particularly less heat tolerant when immobilized using a variety of carriers and binding methods.

Where the enzyme has been bound to a thermosensitive carrier, the limitations of the carrier are often compounded by a reduced thermotolerance of the enzyme itself. Many of the gel-entrapped enzymes show only up to 10 °C increased tolerance within the limits of the entrapment system.

The stability of immobilized enzymes in relation to storage conditions is extremely variable, with some aqueous suspensions having long activity storage at 4 °C and other being poorly stable, even when held dry at low temperatures.

4. Industrial applications of immobilized enzymes

Two of the main application areas have already been discussed in this book: lactase in the dairy industry (Chapter 4.6) and glucose isomerase in the starch industry (Chapter 4.15). Additional comments on possible developments have been made in the leather industry (Chapter 4.11) and in the discussion of glucose oxidase

---

TABLE 4.21.2
Typical process data for glucose isomerase

Enzyme form	Rigid granules
Reactor	Column (packed bed)
Feedstock	95% dextrose @ 40–50% w/w
Additives	Magnesium sulphate
Temperature	58–65 °C
Inlet pH	7.5–8.5
Operating life	1000–2000 hours
Productivity	2000–4000 kg/kg enzyme producing 42% fructose

(Chapter 4.20), and their potential in analytical systems has been mentioned in the chapter on analytical applications (Chapter 4.3).

There have been many reports of pilot scale evaluation of most of the various physical presentations of immobilized enzymes, but apart from analytical systems, the most accepted use is as packed beds of granular supports, or columns, of cellulose derivatives and organic gels. In all these cases, the principle is to have a flow of substrate solution passing through the bed, often by a downward feed to minimize the risk of disturbing the bed and causing channelling. Some studies have shown good performance when upward flow is used to create a fluid bed of immobilized enzyme, but control systems are more complex in this case.

Immobilized glucose isomerase. The most widely recognized and by far the largest tonnage of immobilized enzyme used is glucose isomerase. Reference to Data Indexes 2, 3 and 4 shows that several different source organisms have been adopted for industrial production and at least five producers now offer bulk product.

In each case, the conditions for application are roughly similar and demonstrate the fact that in industrial enzymology there are few alternative solutions to the problems faced by the producers in creating a viable continuous process. Early examples of this enzyme rapidly confirmed the view that the activity was not that of the primary natural enzyme; it is considered to be a xylose isomerase according to its preferred substrate. Discussion in Chapter 2 ('Practical kinetics') and Chapter 4.15 ('Starch') has shown that a remarkably high Michaelis Constant value occurs when glucose is the substrate. Consequently the feed syrup must be 40–50 per cent in relation to glucose for a satisfactory rate of reaction. This syrup has a troublesome viscosity for continuous pumping when at ambient temperatures, but becomes acceptable at around 60 °C.

Another early observation was the general need for chemical activators such as cobalt salts, which are unacceptable in food processing. It was found that by operating at a higher pH value, typically 7.5–8.5, the dependence on these activators was reduced. Calcium ions, if present, act as inhibitors and can either be removed, or countered by the addition of magnesium salts.

These imposed conditions were seen to be serious objections for the use of a soluble enzyme in a batch process, since colour formation was likely to be a major problem when holding the syrup in alkaline conditions at high temperatures. By the introduction of an immobilized enzyme, first as a stirred tank system and later as packed beds in columns, the necessary contact time for the isomerization of glucose to fructose could be reduced, the syrup product could be cooled before colour formation and the enzyme would not be present to need inactivation. It was therefore of considerable value to develop this immobilized process to solve the practical obstructions to isomerization, and this largely explains the success of the product.

Industrial evaluation of the sweetener syrup demonstrated further product qualities that have enhanced its value to the point where thousands of tonnes of immobilized enzyme are producing many millions of tonnes of syrup. Current estimates indicate up to 5 million tonnes of syrup at 42 per cent fructose will be produced in 1982 in the USA alone.

A further technical encouragement has been the development of industrial scale chromatography to separate fructose and glucose to yield enriched fructose syrups containing 55 per cent and higher fructose levels. The byproduct stream from these enrichment processes contains both glucose and residual oligosaccharides from the initial saccharification process. These interfere with recycle isomerization as they accumulate, and stimulated investigation into the possibilities of immobilized enzymes to degrade them to more glucose (*see* 'Immobilized glucoamylase' below).

Immobilized glucoamylase. Although soluble glucoamylases are readily available at commodity prices, the technical benefits of introducing a further continuous system in tandem with the glucose isomerase were a sound stimulus to developing an immobilized form. The reduction in vessel sizes for primary starch saccharification was also attractive. Several immobilized forms have been offered to industry, and tests have shown that, while they perform

TABLE 4.21.3
Typical process data for immobilized glucoamylase

Enzyme form	Rigid granules
Reactor	Column (packed bed)
Feedstock	Filtered or centrifuged liquefied starch @ 30% w/w and 15–26 DE
Temperature	53–60°C
Inlet pH	4.0–5.0
Operating life	300–500 hours (high DX syrups)
	600–1000 hours (high conversion syrups)
Productivity	500–800 kg/kg enzyme (high DX syrups)
	1000–3000 kg/kg enzyme (high conversion syrups)

quite well, there are still serious problems in achieving the high dextrose levels found with the soluble enzymes. The highest figures so far suggest that a maximum of 94 per cent glucose is reached in comparison with 95–96 per cent with the soluble systems. It is thought that diffusion resistance is the most likely explanation. The thermal stability of the immobilized glucoamylases is some 10°C lower than that of the soluble enzymes, and the economy of use is poor at this lower temperature unless the half life of the enzyme can be raised to in excess of 1000 hours.

Finally, the added cost of physical clarification of the dextrinized starch feedstock reduces the economy of the immobilized enzyme system still further. Without clarification, the enzyme packed bed is prone to blockage, and uneven distribution of the feed syrup further reduces the productivity.

Immobilized bacterial α-amylase. Many attempts have been made to establish a complete starch hydrolysis process by an all-immobilized enzyme system. The stimulus of the rapid continuous liquefaction of starch by thermostable α-amylases using jet cooking techniques, together with the highly successful immobilized glucose isomerases, has not been rewarded to date. As discussed above, the glucoamylases have not yet been found to be competitive in the immobilized form, and this leaves the dextrinization stage also unresolved.

Some covalently resin bonded bacterial amylases have been tested as fluidized beds of granulated material acting on pregelled starch slurries at around 50°C. Reaction rates are slow and yields, at best only 90 per cent of the soluble enzymes, are inadequate.

Immobilized lactase and inulinase. If the lessons learnt thus far are to be fully applied, these two enzymes represent good future prospects, since the substrates, lactose, and inulin juice from chicory or artichokes, are only small molecular weight molecules and economic processing of dilute solutions suggests some merit from immobilized enzyme processing.

Lactase acts on the milk sugar, lactose, to produce the highly soluble sugars glucose and galactose. Much has been written on the merits of performing this conversion on whole milk to alleviate the symptoms of lactose intolerance for many millions of people. What is not clear is the availability of milk and its acceptability as a food to the greater part of the population that has this intolerance.

Where cheese is produced, the utilization of whey presents many interesting opportunities. Higher unit consumption by young animals could be expected from hydrolysed whey. The removal of valuable whey proteins is a regular practice in large modern creameries, and the residual lactose stream has been viewed as a

TABLE 4.21.4

Typical process data for immobilized lactases

	Bacterial enzyme	*Fungal enzyme*
Enzyme form	Rigid granules	Porous glass beads
Reactor	Column (packed bed)	Column (packed bed)
Feedstock	Lactose, whey or whey permeate up to 20% solids	Whey or whey permeate unconcentrated
Temperature	60–65°C	30–40°C
Inlet pH	7.5–8.0	3.0–5.0
Operating life	500–800 hours	not available
Productivity	Dependent on the chosen degree of hydrolysis	

potential sweetener syrup if converted to the monosaccharides. A substantial increase in the use of soluble lactase enzymes has occurred, but the introduction of immobilized lactases is very slow. Some reasons for this slow response can be seen when whole milk is considered as a substrate. Whole milk is a complex product with colloidal protein and emulsified fats that present significant problems in the use of an immobilized enzyme. Spoilage during processing must be avoided, which suggests processing would be at either low temperatures of around 4°C or above 60°C.

At 4°C the productivity is too low to be economic, and at 60°C plus the milk develops flavours that are unacceptable. Whey and the whey permeate, from protein recovery by ultrafiltration, represent acceptable technical substrates but the economy of processing is doubtful, since there is no shortage of more effective, inexpensive and reliable sweeteners from the starch industry. Possibly the production of ethanol represents a viable route, but the need for hydrolysis is questionable, since yeasts are already used that do not require pretreatment of the lactose to produce monosaccharides. The system is also challenged by purely chemical methods that are very effective.

Inulin is not yet a readily available substrate for industrial processing, but the interest for fructose makes it very attractive to develop, since it is a fructose saccharide. The juice can be extracted from crushed or milled material to give a 13–15 per cent inulin solution. Enzymes for its hydrolysis are known and available and it remains to be seen if the fructose demand will produce a sufficient stimulus to agricultural enterprises to produce enough to create a market for the immobilized enzyme concept to develop.

TABLE 4.21.5
Typical process data for immobilized penicillin acylase

Enzyme form	Rigid granules or dextran/Sephadex
Reactor	Column
Feedstock	Penicillins or cephalosporins @ 4–15% w/w (dependent on enzyme preparation)
Temperature	35–40°C
Inlet pH	7.0–8.0 (dependent on enzyme source)
Operating life	2000–4000 hours
Productivity	1000–2000 kg/kg enzyme

Immobilized penicillin acylase. The enzymic deacylation of the side chain of penicillins produces 6-aminopenicilloic acid, which is used to manufacture a number of semisynthetic antibiotics. In its immobilized form the penicillin acylase is used to a wide extent for the production of more than 3500 tonnes of 6-aminopenicilloic acid (6-APA) per year. To achieve this conversion the enzyme preparations amount to no more than 30 tonnes per year and are usually in the form of granules suitable for fixed bed operation in columns. Stability towards a range of pH values is necessary, since the reaction product is acidic and the reactor pH will reflect this, especially in the microenvironment of the immobilized enzyme particles. Working at around 35°C, the residence times in the reactors are short and range from 20–60 minutes. The practical use of this immobilized enzyme reflects several of the major factors that have been considered necessary for successful operation of immobilized systems. The enzyme use is low, the productivity high, and the value of the product is measured both in its currency worth and uniqueness for further processing.

Other immobilized enzymes with industrial status. L-amino acid acylase has been established, particularly in Japan, for the preparation of resolved L-amino acids. The system is used especially for the production of L-methionine. The process selectively deacetylates the L-form leaving the D-amino acid easily separated for reracemization and recirculating. The enzyme is typically bound to a suitable cellulose ion-exchange carrier and operates at neutral pH and 50°C. Working life of the system is quoted to be up to 40 days in 1000-litre reactors, giving around 50 per cent yields.

Fumarase is reportedly in use as an immobilized system for the production of L-malic acid operating in the range pH 7.0–8.0 at low substrate concentrations.

Nucleases have been established for the production of 5′-nucleotides for flavour enhancement and it is reported that immobilized systems have been developed to production scale.

Invertase is frequently described as a candidate for industrial scale immobilization for the inversion of sucrose syrups. The routine use of immobilization reactors with this enzyme in industry has yet to be confirmed.

TABLE 4.21.6

Some enzymes regularly used for analytical purposes (Immobilized forms)

Continuous reactors	Enzyme electrodes	Measuring
Glucose oxidase	Glucose oxidase	Glucose
Cholinesterase		Oxygen
Hexokinase	Hexose oxidase	Glucose
Peroxidase		D-galactose
Alcohol dehydrogenase	Alcohol oxidase	Ethanol
L-Amino acid oxidase	D-Amino acid oxidase	D-alanine
Alkaline phosphatase	D-Aspartate oxidase	D-aspartate D-glutamate
Urease	Urease	Urea
Uricase	Uricase	Uric acid
Aryl sulphatase	Aldehyde oxidase	Acetaldehyde Formaldehyde
Lactate dehydrogenase	Lactate dehydrogenase	Lactate
	Lactose oxidase	Lactose
	Xanthine oxidase	Benzaldehyde
	Penicillinase	Penicillins
	Nitrite reductase	Nitrite

Analytical applications of immobilized enzymes. The sensitivity and specificity of enzyme catalysed reactions has increasingly encouraged their application to analytical procedures. Many laboratories seek to automate routine methods to achieve high throughputs and reliability, and many enzymic methods have been developed using soluble enzymes (*see* Chapter 4.3, 'Analytical applications').

The stability of enzymes in dilute solutions and their high cost at extreme purity have been overcome in many cases by immobilization. Enzymes have been successfully bound to the lumen walls of tubular systems and are commonly used as part of many commercial

TABLE 4.21.7

Industrial enzymes in use or under development as immobilized systems

Aminoacylase	Amino acids	Stirred tank/Packed bed
α-Amylase (bacterial)	Dextrinization	Stirred tank/Fluid bed
Endo/exonucleases	Nucleotides	Stirred tank/Packed bed
Ficin	Soluble proteins	Packed bed (with cofactors)
Glucoamylase	Saccharification	Packed bed
Galactosidase	Raffinose hydrol.	Stirred tank
Glucose isomerase	Fructose syrups	Packed bed/Fluid bed
Glucose oxidase/ catalase	Food preservation Soft drinks Gluconic acid	Stirred tank Packed bed Packed bed
Hydrogenases	Hydrogen by photosynthesis	Stirred tank
Invertase	Sucrose inversion	Packed bed
Lactase	Milk, whey and lactose	Tubular/Packed bed
Lipases	Fatty acids	Fluid bed
Papain	Soluble proteins	Packed bed
Pectin esterase	Juice clarification	Fluid bed
Protease (microbial)	Soluble proteins	Packed bed/Fluid bed/ Stirred tank
Rennet	Milk coagulation	Packed bed/Fluid bed/ Open surface
Steroid esterase	Steroid modification	Stirred tank
Sulphydryloxidase	Flavour control in heat treated milk	Fluid bed
Tannase	Instant tea	Packed bed
Trypsin	Soluble protein	Packed bed/Stirred tank

automatic analysis systems. They include determinations of glucose, uric acid, urea and certain specified amino acids, alcohol, acetic acid and inorganic phosphate and sulphate.

The more common application of immobilized enzymes for analyses is the enzyme electrode. This is an electrochemical detector that responds to the product or products that are generated in the immediate environment of the detector. Earlier developments were based largely on the use of glucose oxidase reactions to measure oxygen or glucose. High enzyme concentrations are necessary in relation to the substrate being measured, so that long operating lives may be expected. The electrode system is largely using the completed enzyme reaction and so depends on total exhaustion of substrate and would be variable in response if excess enzyme were not present (*see* Chapter 2, 'Practical kinetics'). Modern enzyme electrodes use platinum, which responds to the reaction of oxidase enzymes via hydrogen peroxide. The current produced at the electrode is proportional to the hydrogen peroxide concentration.

With response to oxygen, ammonia, carbon dioxide and hydrogen by a variety of electrodes, it is possible to develop immobilized enzyme systems to make a wider range of analyses than those producing peroxide. A summary of the industrial enzymes currently under evaluation for use in immobilized forms is given in Table 4.21.7.

References

Johnson, J. C., ed. *Industrial Enzymes: Recent Advances* (Noyes Data Corporation, Park Ridge, New Jersey, 1977).

Messing, R. A. *Immobilized Enzymes for Industrial Reactors* (Academic, New York, 1975).

Weetall, H. H. & Suzuki, S. *Immobilized Enzyme Technology* (Plenum Press, New York, 1975).

Wiseman, A. *Topics in Enzyme and Fermentation Technology* (Ellis and Horwood, London, 1977–9).

Wiseman, A. *Handbook of Enzyme Biotechnology* (Ellis and Horwood, London, 1975).

MEMBRANE CLEANING
T. Godfrey

Industrial filtration systems incur high operating costs when they require frequent down-time for cleaning of the filter surfaces. Many more recent filtration systems have been built to include *in situ* cleaning operations using combinations of back-flushing and the circulation of cleaning solutions, but the selection of the appropriate chemicals for these operations is still important. In many cases it is considered efficient to use a highly alkaline formula containing sodium hydroxide, for this will solubilize fats and proteins when used at relatively high temperatures. The disposal of these washing solutions can add to the effluent load, and throughout their use great care is required to ensure safe operating conditions. It is now recognized, however, that enzymatic cleaning solutions can be prepared which are of modest ionic strength and pH values of between 5 and 9 and which will remove many filter soils without hazard, while producing a very modest effluent load on discharge from the system.

Filter systems that use permanent or semi-permanent cloths, stainless steel or ceramic meshes or ultrafiltration membranes are best cleaned by application of a flow of washing solution over the soiled surface rather than by a pressure systems, since the latter will encourage penetration of the soiling substances. Back-washing from the clean side is often not practical from either an engineering or product hygiene point of view, although where possible it does provide a means of removing the gross soil very quickly.

The choice of enzyme washing system will depend upon several factors. These include the nature of the soiling materials, the acceptable pH and temperature range for the system, The available washing cycle time and the critical chemical standard applied to the process being cleaned (e.g. food products, pharmaceuticals or other high purity products).

Proteins and microbial cells form the largest sector of soiling on filter systems and are usually cleaned effectively by adopting an industrial biological detergent and choosing a concentration that is effective in the system being cleaned. Under certain circumstances, the chemistry of these products is excessive to requirement or even unacceptable because of the product being filtered in the system. It is practical to use preparations of the typical detergent enzymes offered in liquid and possibly food grade qualities with acceptable surface-active agents and pH stabilizing substances such as phosphates, to produce a specific washing solution for the system (*see* Chapters 4.7 and 4.14).

In addition to the bacterial proteases for detergents, proteases such as trypsins and papain are useful for cleaning systems used to filter cheese whey for protein recovery. Moreover, apart from proteins, the brewing industry may also encounter starches and gums from processed yeast and cereals, and so benefit from cleaning systems containing α-amylases and β-glucanases of bacterial origin. In particular, filter cloths may be rapidly washed using solutions of thermostable α-amylase and β-glucanase together with small quantities of non-foaming, non-ionic surfactants.

The filtration of fruit juices and wines will tend to load the filter with cellulose fibres and pectin products; these can be rapidly cleaned by using both cellulases and pectinases in the washing liquid.

In all cases of filter cleaning, it is necessary to ensure the removal of the washing agent, and this can take very large volumes of clean water when strong chemical systems have been employed. An advantage of enzymic washing is partly the low ionic strength of the supporting chemistry, and also the sensitivity of most enzymes to change of pH. In most cases the use of a dilute acid solution, mineral acids in general and organic acids for food systems, will produce a pH of around 3, which will inactivate residual enzymes very rapidly. Thus a recycle of a small volume of acid water, followed by a small volume rinse, will clear the system of cleaning agents.

Examples of cleaning formulations
A. For removal of microbial and protein soils:

Non-ionic surfactant	1.00 g/l
(e.g. ethoxylated nonylphenol or ethoxylated fatty alcohol)	
Proteolytic enzyme (e.g. *Alcalase 0.6L*; Novo)	2.5 g/l
Sodium tripolyphosphate	1–3 g/l
Sodium hexametaphosphate (soluble)	4–2 g/l
pH adjusted	7–10 as required

Preparation. Dissolve all chemicals except the enzyme in 20 per cent of final wash volume using warm water. Add this to the remaining water volume and check the pH and adjust as necessary. Preheat to the cleaning temperature and add the enzyme. Use the finished solution promptly and do not store, as the enzyme activity will rapidly fall due to lack of stabilizers in this formulation.

This system may be used for clothes, solid filters and membranes, in accordance with the manufacturer's advice regarding pH and temperature tolerance.

B. For removal of starches and gums:

Prepare chemical solution as in (A), but use bacterial amylase (e.g. *BAN*® 120L; Novo), 2 grammes per litre, with pH adjusted to 6–7. Use as hot as possible up to 75 °C. Do not store the washing solution as it will rapidly lose enzymic activity.

C. For removal of cellulose fibres and pectin residues:

The enzymes require lower pH, and a citric buffer is suitable.

Non-ionic surfactant (as in (A))	1 g/l
Sodium citrate	3–4 g/l
Citric acid	1–2 g/l
pH adjusted to	4.5–5.0
Pectinase (e.g. *Pectinex*® 3XL; Swiss Ferment Co.)	1–3 g/l
Cellulase (e.g. *Celluclast*® 200L; Novo)	2–4 g/l

Prepare the chemical solution as for (A) and warm to a maximum of 55 °C before adding the enzymes.

Note: When enzyme solutions are used to clean filter systems, the action is primarily that of releasing the soil from its binding site, rather than a complete solubilizing hydrolysis, and this reinforces the fact that the best results are obtained with an effective flow of cleaning solution over the face of the filter.

MINERAL OILS AND DRILLING MUDS
T. Godfrey

1. Introduction
It can be said that the presence of undegraded oils for recovery from geological formations after the passage of such enormous time spans predicts a very limited potential for microbial, and hence enzymic, attack. Recent studies of the degradation of polluting oil and petrochemicals have shown that certain organisms can metabolize these hostile substances if other supportive nutrients are provided. Based on this information, a few instances of enzymic assistance have already been noted for oil pollution treatment and the improved dispersal of industrial waste oils and cutting oils.

The drilling industry uses a variety of specilized muds as lubricants and coolants that contain many additives to provide viscosity and response to shear forces. The production of some of these additives is increasingly biotechnically based with opportunities for enzyme treatments to contribute. The tertiary recovery process involving massive hydraulic fracture of shale oil deposits employs special additives in the pumping fluids which improve the viscosity. Enzymes are now being used in the control of this viscosity.

2. The degradation of polluting oils
Up to two million tonnes of oil enters the sea each year as a result of seepage, industrial wastes and tanker operations. Microbial activity is considered to be the largest contribution to the degradation of this pollutant oil. Infection of surface oil is rapid, but biodegradation begins following emulsification by surfactants and wave activity. Where oil-in-water emulsions are formed, the microbial attack is rapid, but commonly a water-in-oil emulsion forms and biodegradation is generally limited to the outer layer. In either situation, the artificial surfactants and dispersants now used are of low toxicity to the infecting organisms, but add more carbon to an already severely overloaded system.

Hydrocarbon-oxidizing strains of *Pseudomonas*, *Mycobacterium* and *Arthrobacter* species are common in oil slicks and dispersed oil and they rapidly metabolize the normal alkane components of the oil. The aromatic and cycloalkane components are far more resistant. The degradation of both alkanes and aromatic compounds requires the addition of oxygen and the resultant fatty acids are

453

metabolized by the β-oxidation pathway. If the system was able to operate optimally, the result would be biomass, carbon dioxide and energy. However, apart from the toxic nature of the oil and the many derivatives, the temperature of the marine environment is low, usually below 5 °C. Since microorganisms are only able to multiply at about one fifth of their optimal rate at these temperatures, they become increasingly sensitive to nutrient limitations and oil component toxicity. Seawater contains little nitrogen and phosphorus. An oil spillagé produces considerable carbon and an imbalance of carbon to nitrogen and phosphorus. This stimulates microbial activity on the surface, but mineral compounds and simple organisms in the seawater may become less plentiful.

It has been found that preparations containing complex nutrients including carbohydrates and proteins and the enzymes to degrade them to suitable microbial nutrients provide excellent nutrients for oil-slick bacteria. The preparations for effluent treatment and drain cleaning described in Chapter 4.7 have proven extremely suitable under test conditions. Modification of the products can enhance their entrapment by oil slicks and water-in-oil emulsions so that dispersal is minimal.

It is also practicable to spray both detergents and emulsifiers simultaneously. While not yet publicly declared, this treatment does not appear to be over-costly. Effective degradation of surface pollution has been noted in as little as 3 days where water temperatures were 10–15 °C and in 7–10 days in water temperatures below 5 °C.

3. Industrial waste oils

Essentially, the same concepts of nutrition and the removal of sensitivity to toxic components can be related to the treatment of industrial waste oils and cutting oils. It has been observed that microbial spoilage of oils occurs most rapidly if nitrogen compounds are present, and the addition of crude proteins, together with emulsifiers and proteolytic enzymes, has been shown to accelerate the biodegradation of waste oils.

Biomass from normal effluent treatment systems can provide both a nitrogen source and a heavy inoculum of microorganisms. The addition of an appropriate amount of industrial protease, selected on the pH of the total system and related to the content of inhibiting heavy metals to be overcome, together with the effluent biomass, will initiate rapid fermentation, which in turn generates the enzyme systems of the microorganisms to attack the hydrocarbons. Suggested treatment levels range from 1–5 kilogrammes of biomass sludge and 200–1000 grammes of proteolytic enzyme per cubic metre of oil waste. The addition of emulsifiers and surfactants that are nontoxic, small levels of metal-chelating agents and pH stabilizing phosphates create a rapid degradation in the presence of adequate aeration and mixing.

4. Enzyme applications related to oil drilling and recovery fluids

Viscosity producing compounds are not extensively used at the moment. However, much research is now being directed towards choosing compounds suited for use in oil drilling. Traditionally, natural compounds such as the plant gums (e.g. gum tragacanth and guar gum) have found application in drilling fluids because of their retention of viscosity under extreme conditions coupled with a tendency to thin down under shear stress but return to normal when the shear is removed. The suspension of fine particles is maintained while these fluids are pumped, but the viscosity fails under the cutting shear to facilitate good heat transfer to the liquid system.

Massive hydraulic fracture fluids. The use of modified cellulose formulations as well as partly degraded starches and their derivatives to give good suspending viscosity has been tested for these fluids. It is necessary that the characteristics of the fluid alter when underground and at very high pressures to facilitate the flow of oil from shale deposits. The inclusion of cellulases in the original formulation in amounts that are compatible with the operating requirements, and the environment within the rock regarding heavy metal salts that will inhibit their action, permit planned failure of the viscosity and suspending character of these hydraulic fluids. Doses of industrial cellulases ranging from 0.1 to 1 kilogramme per 100 cubic metres of fluid have been noted from test reports, but the overall composition of the fluid and the operating times are not available. Further developments will depend on the relative cost and performance of these systems in comparison with the more conventional fluid additives.

Refining of microbial thixotropic additives. In the discussion of the use of enzymes and nitrogen sources in relation to the treatment of polluting oils and waste materials, it has been established that these factors are important in oil treatment. When oil is to be extracted, it is necessary that nitrogenous compounds should not contaminate the oil and stimulate microbial growth. This has produced an interest for the purification of natural polymers produced by microbial fermentation and which are used as suspending agents for drilling and hydraulic fluids.

When xanthans and other microbial polysaccharides are produced by fermentation, they are initially recovered with substantial contamination by both the producing organisms and the fermentation medium. There is a practical use for bacterial proteases to solubilize these proteinaceous substances to aid their removal by suitable washing and precipitation techniques during the purification of the polysaccharides. Treatment with bacterial alkaline proteases in the pH range 7.5–10, or with neutral proteases between pH 5.5–7.5, using temperatures up to 60°C, will require doses of 0.5–2 per cent of protein content according to the reaction time available. Incubations of 30 minutes to 1 hour at the higher doses extending to several hours for the lower doses will be suitable.

YEAST EXTRACT
M. Kelly

1. Introduction

The yeast *Saccharomyces cerevisiae* is grown on a very large scale for industrial use in the baking and alcoholic beverage industries. Molasses provides carbohydrate, essential vitamins and minerals and is the preferred growth medium. In these two substantial applications the yeast, after its active period of converting glucose to carbon dioxide and ethanol, is subsequently killed by the baking or distilling process. However, in the brewing of beer the live yeast (*Saccharomyces cerevisiae* or *Saccharomyces uvarum*), which has multiplied up to sixfold during the fermentation, is separated from the beer. Although a portion is required for addition to the next batch of fermentation wort, three-quarters or more is surplus. This surplus may still be disposed of by direct dumping, but in many countries this is illegal, and in any case most brewers seek profitable outlets for this byproduct. Some goes into animal feed while, because of its high content of vitamin B, some is made into tablet form for pharmaceutical use. In the UK, some, in particular that from the Scottish breweries, is used in the production of malt whisky, but in general a large proportion of the surplus yeast is used to produce yeast extract, a product which has been of commercial value since the late 19th century.

Yeast extract is also produced from yeast especially grown on molasses and can also be made from other species of yeast such as *Kluyveromyces fragilis* grown on whey, or *Candida utilis* grown on carbohydrate wastes. The key factors for a manufacturer in the choice of starting material are primarily the price and availability of the yeast itself. Brewer's yeast as a byproduct costs less than the molasses-grown material. Secondly, choice may be influenced by the type and cost of the process needed to give acceptable yields and extract quality. For example, Brewer's yeast from hopped beer will require additional and costly debittering treatment. The quality of the final product is of very great importance and Brewer's yeast tends to produce higher yields and darker extract of a different flavour from that obtained from Baker's type yeasts.

Accurate figures for total world production are not available but estimates for various countries, for example 4000 tonnes per year in the UK, 3000 tonnes in France, 1000 tonnes in Australia and

200 tonnes in Peru, have led to a suggested world annual production figure at approximately 25,000 tonnes.

2. Yeast extract

This section will describe the product and the principal steps in its production, and will be followed by a discussion of production methods which use enzymic hydrolysis catalysed by the yeast's own enzymes, in particular the proteases, acting alone or supplemented with papain. The technology involved is simple but this traditional and proven way is still used by manufacterers throughout the world. Advances both in the large scale production of microbial enzymes and in knowledge of optimum conditions for their use now permit their successful application to the commercial production of yeast extract, as will be discussed in the final section.

The simplest definition of a yeast extract is 'a concentrate of the soluble material obtained from yeast following treatment'. It is possible to break down yeast by acid hydrolysis, or to disrupt the cells mechanically, and then to concentrate the released cell contents. Neither process involves significant enzymic hydrolysis. This process of yeast extraction has the objective of degrading the macromolecular structure of the yeast, in particular the proteins, into the maximum amount of soluble material at a commercially acceptable cost.

The composition of a typical yeast extract from Brewer's yeast is given in Table 14.24.1, with figures in parentheses giving the values for a molasses-grown yeast for comparison. The extract has high levels of the B group vitamins and all the essential amino acids (though only low levels of methionine and tryptophan). However, the nucleic acid content limits its use in an individual's diet to no more than 20 grammes per day. Thus, yeast extract is not comparable with treated single-cell protein in nutritional terms and values.

Table 4.24.1 shows that extracts from two yeast types exhibit only minor differences in composition between different batches from the same yeast type. The high salt levels are a consequence of the manufacturing process (*see* below), but modifications to the process can produce low salt extract, with correspondingly higher values for protein, carbohydrate and other components in the final material at around 75 per cent solids. For special applications a dry extract can be made by either spray or vacuum drying.

What is not shown by Table 4.24.1 is the flavour quality for which yeast extract finds its major applications in the food industries. The extract is particularly valuable for its contribution of a 'meaty' flavour to a wide range of products including snacks, soups, meat

TABLE 4.24.1

Average composition of Brewer's yeast extract

General		
	Moisture	27% (28)
	Protein (N× 6.25)	44 (37)
	Sodium chloride	10 (11)
	Ash (excluding NaCl)	13 (7)
	Fat	Traces
	Carbohydrate	6 (17)
	(by difference)	

A 5% solution should be a clear, light brown colour with pH about 5.5

Vitamin B content (mg per g extract)

Thiamine	20–70	(10–20)
Riboflavin	55–100	(50–100)
Pyridoxine	12–16	(10–16)
Niacin	250–700	(300–500)

Amino acids (of total protein (N×6.25) after hydrolysis)

Essential		Non-essential	
Isoleucine	4.7 (4.9)	Alanine	7.2 (6.5)
Leucine	7.0 (6.2)	Arginine	1.7 (3.8)
Methione	1.5 (1.4)	Aspartic	10.2 (14.0)
Phenylalanine	3.8 (4.5)	Cystine	Trace
Threonine	0.5 (7.1)	Glutamic	11.5 (19.6)
Tryptophan	1.7 (0.6)	Glycine	5.5 (4.4)
Valine	6.0 (5.6)	Histidine	2.3 (1.8)
		Proline	5.0 (3.5)
		Serine	4.5 (4.6)
		Tyrosine	3.0 (2.5)

Values in parentheses are for Baker's yeast extract of lower protein and higher carbohydrate content.

pies, sausage, fish products and deliveries of hydrolysed vegetable protein. In addition, after modification, by the addition of spices and vegetable proteins, yeast extract forms the basis of savoury spreads. Thus, although of some nutritional value with certain special applications, for example in the production of microbiological culture media, the chief role of yeast extract is for flavour in the food industries. This is largely due to the low cost for the intensity of flavour generated. Nevertheless it is still suggested that the cost limits wider use, and that even less costly flavour materials can be made by combining yeast extract with modified whey solids.

The flavour of yeast extract is influenced by many factors; these include the type of yeast, the presence of impurities, process conditions, salt levels, the extent of enzyme hydrolysis and the effects of any contaminating bacteria. The effect of impurities is demonstrated by the observation that very well washed Baker's yeast gives an extract with a blander flavour than that of unwashed molasses-grown yeasts. Extract prepared from whey-grown *Kluyveromyces fragilis* is bland, but its flavour becomes more meaty if molasses residues are added to the yeast before the start of extraction.

3. Production process outline

Although the details of the production process are confidential to each manufacturer, for example with regard to temperature, pH or the duration of each step, most follow the general pattern of 'plasmolysis, autolysis, pasteurization, clarification and extract concentration'. As an example, the conditions suitable for preparing an extract of a Baker's type yeast are outlined in Table 14.24.2.

TABLE 4.24.2
Process conditions for yeast extract production

---

(1) Adjust pH of cream to about 5.
    Raise temperatures over 5–8 hours to 55 °C. Hold at 55 °C for 24 hours (plasmolysis and autolysis).

(2) Raise temperature to about 70 °C and hold for 15 hours (completes autolysis and starts pasteurization).

(3) Centrifuge to remove cell debris. Counter current washing for maximum yield of extract.

(4) Heat the extract to 70–75 °C for 2–5 hours (completes pasteurization). If evaporator capacity is limiting, extract may be held at this temperature for longer.

(5) First evaporation under partial vacuum to about 30 per cent solids.

(6) Polish filtration to remove any precipitate during concentration 1.

(7) Second evaporation to 70–75 per cent solids using vacuum. Ensure temperature does not exceed 55 °C.

Baker's yeast	28% dry matter and 600 g/l yeast cream
	Sodium chloride 3.5 g/l
	Papain (commercial) 0.4 g/l

Note: Variations in the temperatures, holding times, pH, proteases and the level of salt used will affect yield and flavour.

Step 1: Plasmolysis. A simple method for the initiation of cell disruption. Favourable conditions include raised temperature – sufficient to kill the yeast but not to inactivate its enzymes – and the addition of chemicals, in particular salt or organic solvents such as ethyl acetate or isopropanol. These added plasmolysing agents, because they have some bacteriostatic or bactericidal effect, perform the additional role of reducing the growth of any contaminating bacteria which may reduce the lead to a high viscosity in the extract, hamper clarification and impair the flavour of the final product. Salt as a condiment contributes to the flavour. However, extract made by heat alone or just an organic plasmolysing agent, although it will be more bland, will also be low in salt, so that it could be considered for use in special applications, such as food for convalescents and in infant and baby foods.

Suitable conditions for a temperature-alone plasmolysis are pH 5.5 and 40–48 hours at 58 °C, for yields of around 65 per cent. Similar results are obtained if isopropanol is used at 0.5 per cent v/v, but with a lower temperature for the first 5 hours.

Step 2: Autolysis. Self-digestion of the yeast cell contents. During this stage the enzyme-catalysed hydrolysis is achieved commercially by relying on the yeast's own enzymes, possibly augmenting them with an added protease which is most frequently papain at about 0.04 per cent w/w of the yeast cream. The added papain is useful for increasing solubilization in long autolysis processes only. Using the process outlined in Table 4.24.2, the presence of papain increases the final yield by only about five per cent, and this is only detected relatively late in the autolysis. Short autolysis times do not give maximum yield of extract when relying on the yeast's own proteases, but under such conditions the addition of papain also does not increase the yield. Higher levels of papain do not produce improvements, and for this reason, preparations from both Japan and France do not contain added enzyme, and have plasmolysis and autolysis times of only about 20 hours. One French process uses 55 °C and pH 5.5 for a total time for the two stages of just 20 hours. However, the yields are only 50–60 per cent solubilization, as compared with the 70 per cent expected for the process outlined in Table 4.24.2. The yeast for this French process is specially grown to have a high protease level and takes advantage of the fact that the short process time limits the growth of contaminating bacteria.

By contrast, in the UK the use of papain increases the final yield in extract processes using either baking or brewing yeast, and a similar approach is taken in Australia where a mixture of the two yeast types may be used in a process of 30 hours at 60 °C.

Step 3: Pasteurization. This serves to kill vegetative cells of bacteria, and the temperature used is sufficiently high to inactivate papain and yeast proteases. In the process typified by Table 4.24.2 no further solubilization occurs during the pasteurization, but interaction of small molecules, in particular amino acids and sugars, leads to flavour development.

Step 4: Clarification. Clarification removes insoluble cell wall debris, glucans, mannans and some protein which, despite the presence of mannanases and glucanases in the yeast, is only slightly degraded during autolysis. Some solubilization of this debris can be achieved with commercial protease preparations such as *Pronase®*, but such secondary treatment is generally uneconomical. Therefore, insoluble material is discarded as waste or can be incorporated into animal feed, or may find use as a flavour carrier.

A second clarification step may follow the first concentration to produce a clear extract by removing the haze formed during concentration. This haze is partly composed of less soluble oligopeptides precipitated at the concentration stage.

Step 5: Concentration. This is usually in two stages, with a final filtration interposed. During the extraction concentration further development of flavour occurs, influenced by the increasing concentration of the components and by the temperature. To avoid the development of burnt flavours during the final evaporation, a temperature not exceeding 55 °C is achieved by the application of vacuum techniques.

4. Application of proteases

Yeast enzymes and papain. Although yeast contains a wide variety of degradative enzymes that include lipases, nucleases, mannanases and glucanases, the most studied and probably most important ones in yeast extract production are the proteases. These have been investigated by various workers, and their roles in extract formation considered. Maddox and Hough (1969, 1970) found evidence for four enzymes which they designated A, B, C and D, and reported that B, C and D attacked proteins to yield polypeptides, whereas the action of protease A was to form free amino acids. The optimum temperatures and pH values for these enzymes are given in Table 14.24.3.

TABLE 4.24.3

Endogenous yeast proteases

	pH optimum	Temperature optimum (°C)
Protease A	8–9	35–40
Protease B	5–7	45–55
Protease C	5–7	45–55
Protease D	3–4	60

pH and temperature optima according to Maddox and Hough (1969).

The process conditions used by most extract manufacturers are therefore not optimal for these enzymes; in particular, the pH is too acid and the temperature too high for enzyme A. Nevertheless, they represent a compromise for producing a simple and cost-effective operation. By using a more acid pH, growth of any contaminating microorganisms is reduced and the risk of protease-induced bitter polypeptides decreased. However, a process such as that described in Table 4.24.2 would produce little improvement in yield by modifying temperature or pH to promote protease activity, since the remaining debris is mostly cell walls. Nevertheless, modifications of process conditions to optimize the activity of the yeast proteases might reduce the total autolysis time.

The role of added proteases is to increase both the rate of solubilization and the final yield. As already noted, papain, although much studied, seems to be of no advantage in short autolysis reactions. Its role in yeast autolysis may be to act on material exposed by the endogenous yeast enzymes, or upon relatively insoluble protein materials not susceptible to hydrolysis by the yeast enzymes. Papain acts on proteins to yield free amino acids, and some peptide bonds are broken at faster rates than others. These favoured bonds include those involving lysine, arginine and in particular phenylalanine. Only about 50 per cent of the protein nitrogen of yeast extract exists as free amino acids, and considerable hydrolysis (although not a higher yield of soluble protein) would be obtained by prolonging the protease action. Any alteration in the ratio of peptides to amino acides would also alter the final flavour of the extract product. Alkaline proteases may produce bitter oligopeptides, but even these can be avoided, or hydrolysed out by the application of peptidases.

Microbial proteases and future prospects. Yeast extract manufacturers have varied their operating conditions to suit local conditions of the market and flavour preferences; they have not for the most part undertaken elaborate analysis of, for example, the hour by hour composition of autolysate, to determine the changing chain length and amino acid compositions of the polypeptides, oligopeptides and sugars. Work done to investigate the solubilization of, for example, fish protein could, in principle, be applied with the objective of obtaining higher yields in shorter process times. Also, any modification of the process or the use of newer, more potent enzymes must be a cost-effective change.

Enzyme mixtures have been obtained from various microorganisms that attack the intact yeast and could be used to initiate autolysis. Some investigations of these enzymes have been made and the enzymes can be obtained commercially. Such enzymes demand careful selection if their use for large scale operations is to be commercially economic. They contain β-glucanase and protease activity, but may require supplementation with other enzymes such as papain for the highest yield. When used at, for example, 0.05 per cent by weight of yeast cream together with 0.04 per cent papain, the yields from Baker's yeast may be raised to 75–80 per cent by reactions at 45 °C and pH 5.0 lasting 48 hours. Since cell walls can be totally solubilized by such enzyme mixtures, it would be possible to obtain an even higher percentage of soluble compounds, although the final product would differ in flavour from present yeast extracts. Alternatively, a mannannase–glucanase mixture with limited protease content could be used to digest cell wall debris after extract production had been completed by conventional methods.

Another and more speculative improvement to current practice would be the use of 'debittering enzymes'. At present the bitter compounds, isohumulones, are removed from Brewer's yeast by various processes including prewashing the intact yeast at pH 9, ion exchange chromatography and by a type of exclusion chromatography. All these processes represent significant additional cost to the process, and enzymic degradation of these 'bitter substances' would simplify the processing of this yeast type. Such an enzyme system might require cofactors and would not be a simple hydrolase. Brewing developments such as post-fermentation hopping could provide yeast that needed no debittering treatment.

There is still much surplus Brewer's yeast not used for extract production, and in some countries it would be easy to increase the supply of molasses-grown types. Extract production will only increase in response to demand, and this will depend on the price of the various flavouring additives available, together with the changing pattern of opinion and legislation about such additives. Possibly, the nutritional value of the yeast extracts, and particularly those with low salt levels, has been under-emphasized to date.

The current processes are still empirical and do not take full advantage of advances in enzyme technology, particularly the increased microbial sources, so that yield and cost-effective process changes vcan be anticipated as the knowledge is absorbed into the industry.

References and further reading
Baker, J. C. et al. Cereal Chem. **30** (1953).
Berry D. E. The Physiology of Whisky Fermentation
Edozien, J. C. et al. Nature **228**, 180 (1970).
Hata, T. et al. Agr. biol. Chem. **31** (1967).
Hevia, P. et al. J. Agr. Fd Chem. **24** (1976).
Holzer, H. Biochim. biophys. Acta **384**, (1) (1976).

Knorr, D. *et al. J. Fd Sci.* **44** (1979).

Kraft Foods Ltd Br. Pat. 1 578 440 1976.

Maddox, I. S. & Hough, J. S. in *Eur. Brewers Conv. Proc.* 1969.

Maddox, I. S. & Hough, J. S. *Process Biochem.* **5,** (5) 50–52 (1970).

Onoue, Y. & Riddle, V. M. *J. Fish Res. Can.* **30** (1973).

Perlman, R. L. & Lorand (eds) *Meth. Enzym.* **20,** 226 (1970).

Phaff, H. J. in *Food Proteins: Improvement through Chemical and Enzymatic Modification* (eds Feeney & Whitaker, J. R.) (American Chemical Society, 1977).

Reed, G. & Peppler, H. J. in *Yeast Technology* (AVI, Westport, Connecticut, 1973).

Riviere, J. *Industrial Applications of Microbiology* (Mason & Cie, Paris, 1975) (Transl. Moses, M. C. & Smith J. E., 1977).

Seeley, R. D. *MBAA Tech. Q.* **14,** (1) 35–39 (1977).

Smith, J. T. *Brewers Guardian* **108** (1977).

Stauffer Chemical Company. Flavouring agent. Br. Pat. 1 561 202 1977.

Whitaker, J. R. in *Food Proteins: Improvements through Chemical and Enzymic Modification* (eds Feeney & Whitaker, J. R.) (American Chemical Society, 1977).

Chapter 5

COMPARISON OF KEY CHARACTERISTICS OF INDUSTRIAL ENZYMES BY TYPE AND SOURCE
T. Godfrey

1. Introduction
This section describes many main features of 85 industrial enzymes currently in commercial production. Information is given concerning their action, reaction optima, sensitivity to inhibitors and activators, and, where available, their stability in practical reaction conditions.

For each enzyme described there are two graphic representations giving the analytical profiles for pH and temperature, together with additional plots of special features of industrial significance where available.

The precise characteristics of an industrial enzyme preparation are likely to alter to some extent if the medium of production, or the fermentation method or conditions, are altered by the producer. This can sometimes be a response to technical improvements or to economic considerations influencing the choice of substrate components. In most cases the primary activity of the preparation will be standardized so that only the side activities will vary. However, the continued strain improvement policy of most producers of enzymes can result in small changes in the major characteristics of the main activity. Consequently, although prepared from data published by the manufacturer or supplier, the information given in this section should not be taken as absolute, and for this reason the manufacturer or supplier of a particular preparation has not been specified. Indeed, it is apparent that many variants of apparently identical enzymes have different operating parameters, and these are described in this section where they are available.

2. Evaluation of analytical pH and temperature curves
For the practical necessity of standardizing their products, and to supply reproducible information, manufacturers and suppliers choose to issue these curves based on analytical methods. In almost all cases the ratio of enzyme to substrate is enormously greater in these assays than would be adopted in industrial applications of the enzymes, and because of this, narrower optima are presented in the plots than would occur in practical applications. Most enzymes are significantly less sensitive to changes of pH and temperature, and more tolerant of higher temperatures when protected by substantial substrate concentrations. Consequently, industrial operations may often be carried out over longer time periods and at higher temperatures than the curves given here would suggest.

Many of the curves for pH have a horizontal bar below the curve, and this serves to indicate the stability of the enzyme towards wider ranges of pH. Within the limits of pH so defined, it is likely that the enzyme will demonstrate better than 80 per cent of its maximum activity, when used at or below the maximum performance temperature. Even wider tolerance of pH is observed when temperatures below the optimum are used. For the same reasons, temperature tolerance will be greatest at the optimum pH for many enzymes.

Activators. As most industrial enzymes function without additional cofactors and almost all of them are simple hydrolases, there are few specific activators mentioned in this section. The amylases are generally most active and stable when in the presence of adequate free calcium ions. Some proteases are aided by the presence of cysteine and reducing agents. The lactases demonstrate some enhanced action when potassium ions are plentiful, but this condition is not a practical commercial consideration in view of the dominance of sodium ions that would have to be overcome in the typical substrates of milk and whey that these enzymes are required to act upon.

Inhibitors. Most enzymes are seriously inhibited or inactivated by the presence of salts of the heavy metals (mercury, silver, iron, tin, copper, lead and zinc). In some cases very low levels of zinc can result in activation, but it is generally considered important to avoid heavy metal salts in enzyme processing.

Many enzymes are sensitive to free halogens which thus limits the use of modern antimicrobial additives. The same applies in the case of phenolic compounds, and any known toxic material should be considered a likely inhibitor. Oxidizing agents, chelating agents and active polyelectrolytes, and cationic surfactants exhibit varying degrees of inhibition towards enzymes and it is advisable to introduce prior testing of the proposed system whenever these substances are known or suspected to be present.

Purity. Most industrial enzymes contain some side activities to a greater or lesser extent. Enquiry of the manufacturer or supplier is recommended if certain activities are contraindicated in the proposed system or unexpected results are observed during practical processing operations. There are many cases where the side activity makes a positive contribution to the overall outcome of an enzyme process, but it is very difficult for industrial producers to establish the specific activity responsible, and even more unlikely that a nominated side activity can be supplied in standardised amount or proportion related to the main activity.

Note: The enzymes described in this section are arranged in the same way as set out in Data Index 1.

3. Enzyme data

Amylases

Action: Endohydrolysis of α-1,4-D-glucoside bonds of polysaccharides having more than three adjacent α-1,4 linked D-glucose units. EC 3.2.1.1.
On amylose: random hydrolysis, ultimately yielding maltose and maltotriose.
On amylopectin: restricted by α-1,6 branch points. Products are glucose, maltose and α limit dextrins.

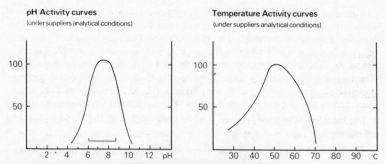

pH Activity curves
(under suppliers analytical conditions)

Temperature Activity curves
(under suppliers analytical conditions)

(1) Pancreatic amylase (bovine/porcine): Neutral pH range, thermolabile above 50 °C. Industrial preparations generally contain lipase and protease. Calcium stabilizes the enzyme towards heat (above 150 parts per million as calcium). For reactions of more than 30 minutes the temperature should be reduced to 40–45 °C. Maximum activity is between pH 6 and 7.

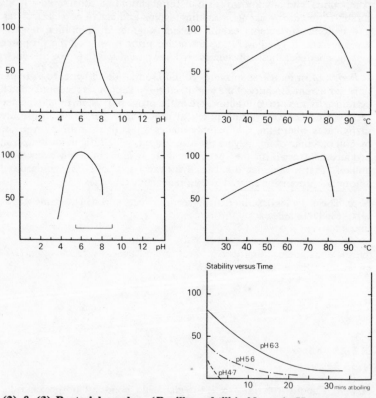

Stability versus Time

(2) & (3) Bacterial amylase (*Bacillus subtilis*): Neutral pH range; some stability in mildly alkaline conditions. In the presence of stabilizing calcium ions (150 parts per million) the activity is greater than 80 per cent of maximum for 60 minutes (pH 6.5–7.5), with excess substrate at 75–80 °C. Many strain variations exist with small differences of pH and temperature

pH Activity curves
(under suppliers analytical conditions)

Temperature Activity curves
(under suppliers analytical conditions)

characteristics. Short reaction times permit powerful liquefaction at 85–90 °C.

Inhibitors: chelating agents removing calcium, for example, EDTA, nitrilotriacetic acid, oxalate, polyphosphate. Heavy metal ions. Free chlorine.

Inactivation: the stability curves at boiling indicate that under industrial operating conditions the pH should be reduced and the temperature raised to achieve the minimum holding time and the maximum inactivation where this is desired. The pH of 4.5–5.0 adopted for subsequent saccharification of starch slurries is appropriate. Where heating is not selected, the pH should be reduced to below 4, as, for example, in the preparation of starch pastes for textile and paper operations. Inactivation at 70–85 °C is then very rapid.

Other activities: commonly the bacterial amylases have β-glucanase and sometimes proteolytic side activity. These activities are both generally less heat stable and are destroyed rapidly at starch processing temperatures.

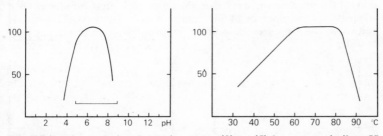

(4): Whilst the varieties shown in curves (2) to (5) have very similar pH characteristics, enzyme (4) has a more rapid response to temperature, reaching a maximum activity at 55–60 °C and retaining this to 80 °C before thermal inactivation commences. The thermal stability is further enhanced under substrate protecting conditions.

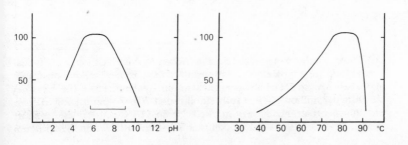

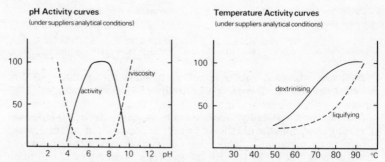

pH Activity curves (under suppliers analytical conditions)

Temperature Activity curves (under suppliers analytical conditions)

(5): Enzyme (5) is reportedly produced from *Bacillus subtilis* through surface fermentation and it exhibits a much narrower temperature profile than the other representatives of this group.

Starch conversion: the curves illustrating the relationship of activity to viscosity reduction, and the differing rates of dextrinization and liquefaction demonstrate the utility of pH selection and heating rates in the various applications of amylases. Both curves are based on typical industrial practice with starch slurries in the range 25–35 per cent solids.

A rapid drop in viscosity can be achieved between pH 5 and 8.5 in the presence of calcium and near the optimum operating temperature. Where subsequent processing requires that retrogradation (a form of recrystallization) of the starch must be avoided when the temperature is reduced, then adequate time at operating temperatures is essential as liquefaction lags behind the rise in reducing power resulting from the production of oligosaccharides.

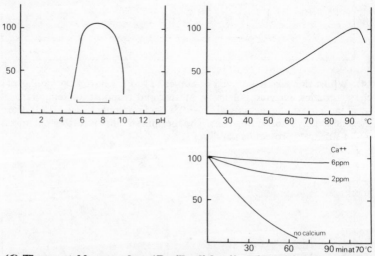

(6) Thermostable α-amylase (*Bacillus licheniformis*): Although analytical assays indicate a typical pH relationship, these enzymes are demonstrably more sensitive to pH when operated at minimal calcium levels (i.e. less than 20 parts per million). When fully stabilized at 70 parts per million calcium the thermal tolerance is very marked. Up to 110°C is tolerated for short reaction times. When used for continuous starch liquefaction the optimum performance lies between 95 and 110°C, depending upon calcium levels, starch concentration (18–40 per cent dry solids) and pH (5.8–8.0). Typically, with 30 per cent starch at pH 7 the liquefaction may be operated at 103–105°C for 7–11 minutes.

Inactivation: due to the thermal tolerance, these enzymes must be treated by lowering the pH to 3.5–4.5 and maintaining the high temperature for 4–30 minutes to achieve full inactivation.

Inhibitors: as for the conventional amylases but with a greater sensitivity when operating near the limits of pH tolerance, at low calcium levels, or at extreme temperatures.

Calcium dependence: under controlled conditions with high substrate levels at optimum pH, these enzymes require very low calcium levels for maximum activity. In practical use, the starch slurry will provide more than sufficient calcium to permit working stability in most cases. Where very low pH slurry is used (e.g. pH 4.5–5.5) the calcium dependence is higher and should be supplemented to give 70–100 parts per million free calcium ions in the reacting slurry.

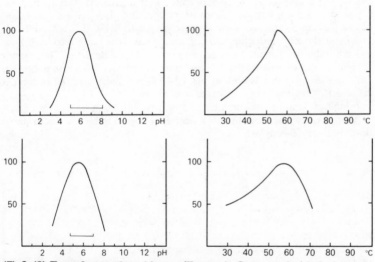

(7) & (8) Fungal α-amylase (*Aspergillus* spp.): Compared with the bacterial amylases, the products of these enzymes are often of smaller molecular weight, with maltose forming a major end-product.

The pH curve is typical, but under substrate protection an activity of at least 60 per cent of maximum can be obtained at pH levels down to 4. Calcium ions greatly improve the reaction at low pH values. Generally, thermal tolerance is limited to 55–60°C for prolonged maltose producing reactions, but short reactions at higher temperatures are valuable in, for example, baking flour applications.

Inactivation: this is easily achieved by brief exposure to higher temperatures, and can be assisted by a shift of pH to either end of the stability range.

The reaction products: these depend greatly on the substrate. The choice of liquefaction enzyme(s) for starch processing to syrups will influence the level of maltose that can be produced by fungal amylase action. The generation of glucose from acid liquefaction methods, together with a wider spectrum of oligosaccharides, also limits the control of the final sugar spectrum after fungal amylase action.

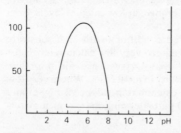   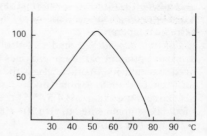

(9) Cereal β-amylase (barley and wheat malt): *Action:* maltogenic hydrolysis of α-1,4-D-glucoside bonds of starches and glycogen with inversion of the C-1 configuration of the glucose from α to β. EC 3.2.1.2. Action is exo from the non-reducing ends of the chains. It is unable to hydrolyse α-1,6 bonds, and the products are maltose and limit dextrins. These latter represent approximately 40–45 per cent of product from the action on amylopectin. Activity is high at pH 4.5–7.0 and the temperature limit is around 55 °C even in high substrate conditions.

Inhibitors: heavy metals and their salts.

Note: the enzyme from the soya bean is reported to be more acid stable than the cereal β-amylase.

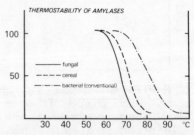

Thermostability/inactivation of amylases: the processing control of the application of the various amylases can be determined by reference to the thermal sensitivity of the selected variant. In addition, the influence of operating pH and the amount of stabilizing calcium present should be taken into account. The balance of heating rate to accelerated activity as the temperature approaches thermal inactivation becomes most significant when bacterial enzymes are used, as rapid heating is essential for maximum control.

pH Activity curves
(under suppliers analytical conditions)

Temperature Activity curves
(under suppliers analytical conditions)

(10)–(16)   Glucoamylase   (amyloglucosidase)   fungal   sources:
Action: exo-α-1,4-D-glucosidase. EC 3.2.1.3. Hydrolysis of terminal glucose residues successively from the non-reducing ends of chains. Most preparations also act on α-1,6-D-glucoside branch points and some also on α-1,3-D-glucoside bonds.

Reaction products: glucose and small amounts of oligosaccharides. Routine production of hydrolysates with 93 per cent dextrose is performed, and under carefully controlled liquefaction treatments the subsequent saccharification can reach 96 per cent dextrose.

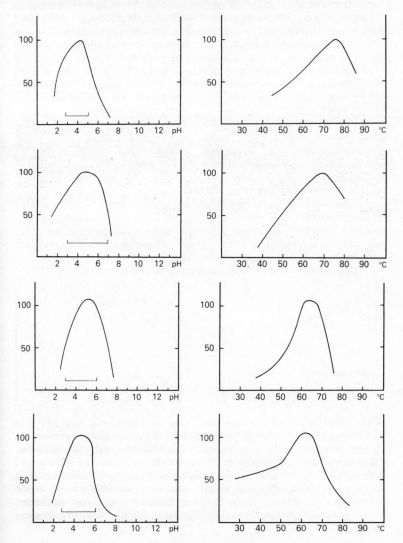

474   Chapter 5

(10)–(13) Glucoamylase (*Aspergillus niger var.*): All four enzymes represented here demonstrate small variations of optimum pH and temperature. In the range pH 4–5.5 all are able to perform well at temperatures of 55–65°C. The stability towards lower pH values that are often encountered in alcoholic fermentations varies with the variety from which the enzyme is produced. Practical testing is the only sure way to establish the one most suited to the pH operating. Some of the variants are able to operate at greater than 50 per cent of the optimum at pH values of 3.4–3.6.

Optimum temperature depends on substrate concentration and pH, and variants can be selected for higher than average operating temperatures.

Application differences: in general, saccharification for sugar syrups and dextrose production will be performed at optimum pH and close to the optimum temperature. Long reaction times are often encountered and the higher temperature assists in maintaining microbial control. Where the application is for fermentation saccharification, the time course is even more extended, the temperature is around 15–25°C and the pH below 4.5. Under these conditions the acid stability of the variants is demonstrated to the fullest extent.

The rates of reaction fall rapidly with decreasing substrate size, being greatest on liquefied starches, and falling to only some 25 per cent of the maximum activity when hydrolysing maltose. Thus, the declaration of activity, based on an analytical hydrolysis of maltose, will underestimate the performance of the enzyme when acting on the long chain oligosaccharides produced by starch liquefaction.

Reaction limitations: at high glucose concentrations, as at the end of saccharification for glucose syrup production, there is a tendency for the reverse reaction to synthesize oligosaccharides. Choice of enzyme dose, substrate concentration and reaction time require balancing to take this into account.

Some preparations contain amylase and protease activity.

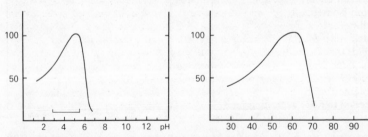

(14) Glucoamylase (*Aspergillus awamori*): Very similar to the *Aspergillus* spp. enzymes in terms of pH and temperature profile. Distinguished by a significant reaction rate on unhydrolysed starches. A greater activity at low pH values is also exhibited by this enzyme type.

Some preparations are very low in side activity and can be considered for sugar syrup saccharification, others are frequently found to be active proteolytically and this limits their applications for starch conversion except for some specialized baking treatments and the large fermentation industry.

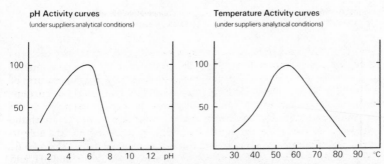

pH Activity curves
(under suppliers analytical conditions)

Temperature Activity curves
(under suppliers analytical conditions)

(15) Glucoamylase (*Rhizopus niveus* and *Rhizopus delemar*): Another varient range showing similar characteristics to the *Aspergillus* types, They are generally derived from surface cultures and usually have high concentrations of side activities. Proteases of neutral and acid pH optima, cellulases and other glucanases, together with pentosanase and amylase activities, are to be found. The general lower thermal tolerance predicts their use in short reactions at optimum temperature; they are more particularly suited to the saccharification of mashes for fermentation.

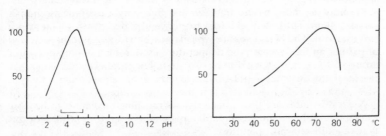

(16) Glucoamylase – immobilized (*Aspergillus niger*): This enzyme has undergone limited industrial development to date. Some interest has centred on the processing of raffinates from the enrichment of fructose from isomerized corn and other syrups. Continuous operation shows that temperatures around 55 °C are suited to longest runs. Due to the enhanced back-reaction at high glucose concentration, the feed syrup is kept at lower solids than for conventional saccharifications with soluble enzyme, a typical level for this enzyme being around 20–25 per cent solids. Maximum glucose levels obtained tend to be some two to three per cent lower than obtained with the soluble enzyme. For maximum operating life, care is needed to offer a clarified feed syrup to the enzyme reactor bed.

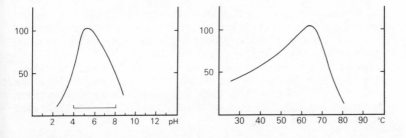

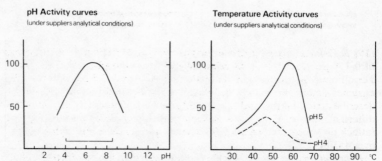

pH Activity curves
(under suppliers analytical conditions)

Temperature Activity curves
(under suppliers analytical conditions)

(17) & (18) Pullulanase – debranching enzyme (*Klebsiella aerogenes*):
Action: endohydrolysis of α-1,6-D-glucosidic bonds of pullulan, amylopectin and glycogen together with limit dextins. EC 3.2.1.41. Recent interest in this enzyme has been in its application for the optimal fermentation of cereals for ethanol production, and the economics of starch-based sugar syrup production has stimulated its development for industrial use. Maltose is the only reaction product, with maltosyl maltose being the smallest substrate acted on. The α-1,6 branch points of liquefied amylopectins from starches are rapidly degraded to yield high levels of linear fragments which are rapidly degraded by the amylases and glucoamylases of the saccharification system. The potential is for a higher yield of dextrose from the starch raw material. The pH curves are typical of carbohydrases but the selection of pH has an influence on the temperature used. Maximum temperatures around 60 °C require the pH to be close to 6. For parallel action with other saccharifying enzymes at pH 4 to 5, the enzyme has a maximum activity at 45 °C, and is rapidly inactivated at higher temperatures.

Note: it is indicated that a bacterial debranching enzyme is to become available that has a more compatible pH and temperature profile for this type of multi-enzyme operation.

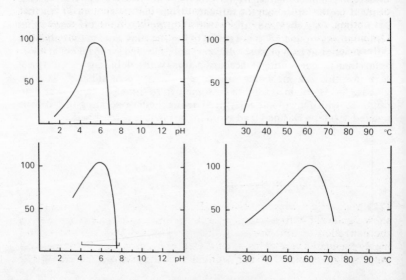

(19) & (20) Cellulase (*trichoderma viride/reseei*): *Action:* endohydrolysis of β-1,4 glucosidic bonds of cellulose and cereal β D-glucans. EC 3.2.1.4. Terminology is confused by restriction of the name Cellulase to describe enzymes acting on 'native' cellulose by some workers, and widened to describe action on this and substituted celluloses by others. Variants show differing ratios of C_1 type (attack on 'native' cellulose) and the C_x type (attack on substituted cellulose) activities and temperature optima range over 45–60°C.

Reaction product: this is largely cellobiose, which inhibits further cellulose hydrolysis. Application of cellobiase (*see* enzyme (42)) can increase the total hydrolysis and the glucose yield. Although maximum activity occurs in the narrow range of pH 4–6, considerable stability is seen up to pH 8. For long reaction times the thermal tolerance is best around 40°C, but care is needed to control microbial activity that will utilize the glucose produced in the reaction. For fermentation of the reaction products this is used to advantage, as the removal of glucose will increase the thermodynamic drive for cellulose hydrolysis.

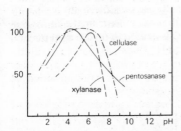

(21) Macerating enzyme complex (*Trichoderma viride*): The evident heterogeneity of these preparations is used to advantage in the degradation of plant tissues. Although the three main activities present have different pH optima, they show good tolerance and stability in the pH range 3–7. Optimum action is obtained by either operating a pH gradient through the differing optima, or adopting a compromise pH value of around 4.5–5.0.

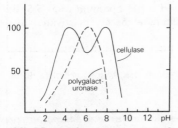

(22) Macerating enzyme complex (*Rhizopus* spp.): These preparations show significant polygalacturonase activity and so are especially suited to the extraction of fruit tissues. Colour is extracted from red grapes and highly pigmented vegetables, and the yield of juices and oils from fruits and pulps is also enhanced. A sharp optimum for the maximum effect of both

pH Activity curves
(under suppliers analytical conditions)   Temperature Activity curves
                                                          (under suppliers analytical conditions)

cellulase and polygalacturonase activity occurs at pH 4. The cellulase is typical and its reaction is improved in the presence of cellobiase. The polygalacturonase activity is inhibited by polyphenolic compounds to an extent that depends on the concentration and the pH and temperature of exposure.

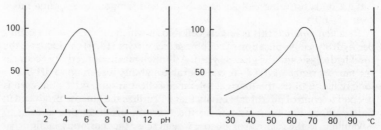

(23) Cellulase (*Penicillium funiculosum*): A typical fungal cellulase with the usual characteristics described. EC 3.2.1.4. It is reported that preparations from this source organism are markedly less sensitive to the inhibitory action of the reaction product cellobiose than are those from the *Trichoderma* species. Thermal stability above 50°C is limited to reactions at pH 5.0, and inactivation by heat is rapid at higher pH values.

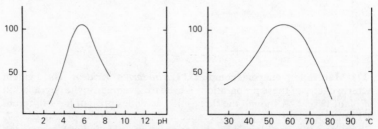

(24) Dextranase (*Penicillium lilacinum* and *Penicillium funiculosum*): *Action:* endohydrolysis of α-1,6 D-glucoside bonds of dextrans. EC 3.2.1.11. Used largely for the reduction of viscosity in cane juices infected by dextran-forming bacteria. It is used at a pH optimum of 5–5.5. In practice it is used at 45–60°C to avoid the rapid thermal inactivation that occurs above 65°C.

Care is generally taken to ensure that preparations are free of invertase activity.

Reaction products: isomaltose and isomaltotriose, which are both soluble in water.

(25) & (26) Invertases (*Saccharomyces diastaticus* and *Saccharomyces cerevisiae*): *Action:* hydrolysis of terminal non-reducing β-D-fructofuranoside residues. EC 3.2.1.26. Their main substrate is sucrose, but they can also hydrolyse raffinose to yield fructose and melibiose, and can catalyse fructotransferase reactions. Preparations often contain β-glucosidase activity but the end-products are the same as when acting on sucrose. They also contain strongly bound mannans. The pH optimum is 4.5, with more than 80 per cent of maximum between 3.5 and 5.5. They are maximally active between 50 and 60°C, and at high substrate concentration up to 70°C can be used. The reaction is maximal at 5–10 per cent sucrose, falling to 25 per cent at 70 per cent sucrose. (No relevant curves.)

(27)–(33) Lactases (fungal, bacterial and yeast): *Action:* hydrolysis of terminal non-reducing α-D-galactose residues of β-galactosides. EC 3.2.1.23. Galactosyl transfer reactions also readily occur. They show low specificity to the aglycone and high reaction rates occur with the *o*-nitrophenyl derivatives used as an assay basis by many workers. Their activity as defined by this reaction is not easily related to performance on typical milk or whey substrates.

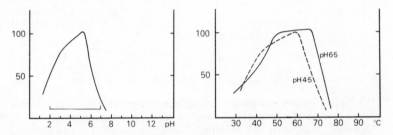

(27) Fungal lactase (*Aspergillus oryzae*): This has a wide pH stability but a narrow optimum of between pH 4.5 and 6.0. Temperature activity is markedly dependent upon pH, with good stability up to 65°C at pH 6.5.

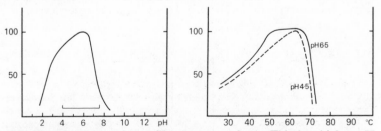

(28) Fungal lactase (*Aspergillus oryzae varient*): This has higher pH optimum and better pH tolerance at maximum temperature than enzyme (27).

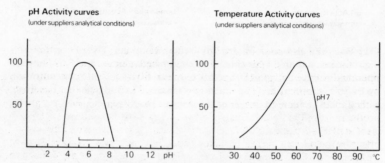

pH Activity curves
(under suppliers analytical conditions)

Temperature Activity curves
(under suppliers analytical conditions)

(29) Fungal lactase – immobilized (*Aspergillus niger*): Good operating characteristics at pH 7 and 60–65 °C enable continuous processing above the thermal death point of most bacteria found in cheese whey. Some cost is involved in maintaining the pH, as most substrate materials will be acidic.

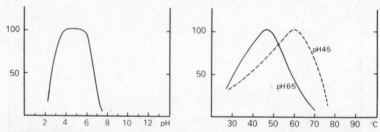

(30) Fungal lactase – soluble (*Aspergillus niger*): This has a typical pH curve but markedly different response to temperature at the two pH values illustrated, when compared with the *Aspergillus oryzae* enzymes (27) and (28). Its thermostability is better at acid pH than at neutral values, and the overall maximum temperature is only 55 °C at pH 4.5. When pushed to the lowest practical pH, these preparations allow control of microbial spoilage in whey processing.

Note: fungal lactases do not require stabilizing ions and are not inhibited by chelating agents. The reaction products are generally inhibitory and galactose is the most significant of these. Transfer reactions can generate a wide range of oligosccharides in short time courses. These tend to be rehydrolysed when reactions are carried out to high conversion levels.

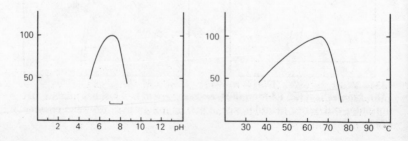

pH Activity curves
(under suppliers analytical conditions)

Temperature Activity curves
(under suppliers analytical conditions)

(31) Bacterial lactase – immobilized (*Bacillus* spp.): Noteworthy for its high thermostability, this recently introduced commercial preparation operates in the neutral pH range at between 60 and 70°C for continuous hydrolysis. Optimal activity occurs at pH 6.5, but this does not coincide with the best long-term stability, which is found at pH 7.5. For maximum productivity of the immobilized enzyme in a packed-bed reactor, operation is set at pH 7–7.5 and 60°C. During the reaction the pH falls and this can be offset by including a buffer in the feed substrate. A typical feed would have a pH of around 8.0 and the product would emerge at around pH 7.5. Up to 20 per cent lactose in the feed will give rapid conversion without microbial contamination under these conditions.

Activation: magnesium and potassium ions.

Inactivation: heavy metal ions and the antagonism of magnesium by calcium.

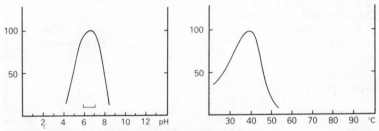

(32) Yeast lactase (*Saccharomyces lactis*): This enzyme has a narrow pH range between 6 and 7 for full activity. It shows marked thermal sensitivity, with a maximum under optimum conditions of 40°C. For prolonged reaction times 30–35°C is more generally adopted.

Activation: potassium ions, and cysteine and sodium sulphide reducing agents.

Inhibition: raw milk and reaction products.

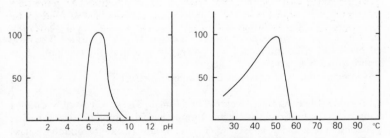

(33) Yeast lactase (*Kluyveromyces fragilis*): Very similar to *Saccharomyces lactis* but with some increased thermal tolerance when operating with substrate at 10–20 per cent lactose and pH 6.5. Thermal inactivation is rapid above 48°C.

482   Chapter 5

(34) & (35) Inulinase: *Action:* endohydrolysis of β-2,1-D-fructoside bonds in inulin. EC 3.2.1.7.

Reaction products: almost pure fructose from inulin with some side reaction and transferase activity if other fructans are present.

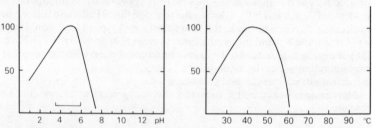

(34) Yeast inulinase (*Candida* spp.): This shows above 80 per cent activity in the pH range 3.5–6.0 and at temperatures of 35–50°C at a pH of 4.5. Rapid thermal death occurs above 55°C.

Reaction products: include up to 10 per cent oligosaccharides after a 24-hour reaction, achieving 85–90 per cent overall hydrolysis.

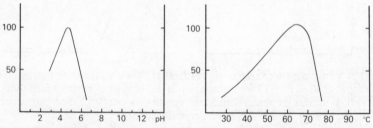

(35) Fungal inulinase (*Aspergillus* spp.): This enzyme type shows a similar pH profile to that for the *Candida* enzyme but has a markedly higher thermal tolerance and maximum activity temperature. Practical operational hydrolysis is at 60°C, with substrate at 10–15 per cent inulin.

Evidence of multi-enzyme action in these preparations enhances the hydrolytic action (on the basis of amount of enzyme required). When used for 24 hours under optimum conditions the products represent 96–99 per cent hydrolysis of substrate with only traces of oligosaccharides. The ratio of exo- and endo-inulinases may vary, but preparations available show approximately equal levels of each.

Inhibitors: heavy metals, but calcium has no effect.

(36)–(41) β-Glucanases: Bacterial and fungal. These consist of a complex range of preparations, often exhibiting several activities together. The predominant activity is EC 3.2.1.6, a broad-specificity endohydrolase acting on β-1,3 and/or β-1,4 bonds in β-D-glucans. Other glucanases present may include specific β-1,3-glucanhydrolase (EC 3.2.1.39), specific β-1,2-glucanohydrolase (EC 3.2.1.71) and the exo β-1,3- and β-1,4-glucanohydrolases (EC 3.2.1.73 and 3.2.1.74).

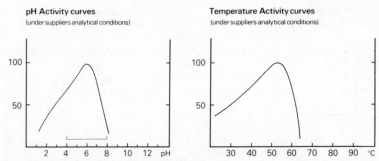

pH Activity curves
(under suppliers analytical conditions)

Temperature Activity curves
(under suppliers analytical conditions)

(36) Bacterial β-glucanase (*Bacillus subtilis*): These are typical bacterial enzyme preparations showing a pH optimum between 6 and 7.5 and thermal sensitivity above 60°C. This variant is rapidly inactivated at 60°C and has an optimum at 50°C.

Many bacterial glucanases have amylase and protease side-activities.

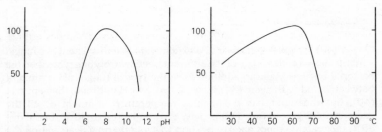

(37) Bacterial β-glucanase (*Bacillus subtilis*): This is a strain selected variant with slightly higher pH range and thermal stability. The main applications for these enzymes are in brewing and the allied malt and cereal extract production industries. The reduction in viscosity of soluble glucans is the main objective.

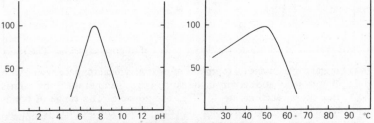

(38) Bacterial β-glucanase (*Bacillus subtilis*): This is a different variant with a sharp and slightly higher pH optimum at 7.5.

The general thermostability of bacterial glucanases, compared with the enzymes present naturally in the cereals being processed, allows for control of the filtration of the extracted materials whilst heating the mash above the thermal death point of the cereal enzymes.

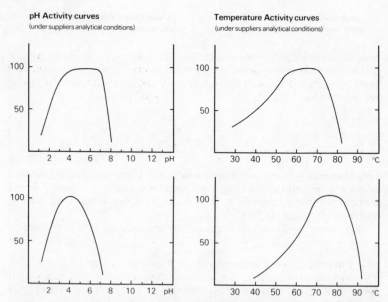

pH Activity curves
(under suppliers analytical conditions)

Temperature Activity curves
(under suppliers analytical conditions)

(39) & (40) Fungal β-glucanase (*Penicillium emersonii*): These two fungal variants show a wider range of the different bond hydrolysing types than the bacterial enzymes, and in addition have the typical fungal pH optima of between 4 and 5. Enzyme (39) shows a broader pH optimum than enzyme (40), but the latter has the highest temperature optimum of all the commercial β-glucanases. Full activity is observed up to 80°C, but above 70°C the reaction times need to be short to avoid thermal deactivation.

Both enzymes function well at pH values as low as 3 and so can be used to hydrolyse cereal gums in both mash and fermentation stages of the brewing and distilling processes.

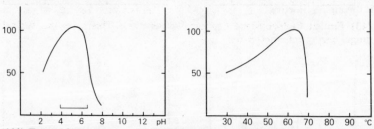

(41) Fungal β-glucanase (*Aspergillus niger*): A typical fungal enzyme with pH optimum at 5.0 and activity in the range 4 to 6. Temperature optimum is 60°C but short reactions up to 70°C are possible. The range of bonds hydrolysed is broad but not identical to that of the *Penicillium* types, so that maximum results can often be obtained by combining several types of glucanase.

(42)–(48) β-glucosidases (fungal, yeast, bacterial and plant types): Complex activities are generally present with the main activity stated to be β-glucosidase. Both exo- and endo-actions occur as defined by activity

pH Activity curves
(under suppliers analytical conditions)

Temperature Activity curves
(under suppliers analytical conditions)

towards specified substrates such as laminarin and lichenin. Hydrolysis of β-linked D-glucose polymers is very varied according to enzyme specifications present and to the bonding character of the substrate. EC types represented are 3.2.1.21, 3.2.1.58, 3.2.1.74 and 3.2.1.75.

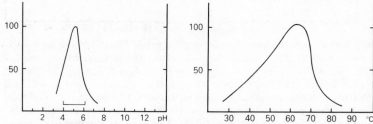

(42) Fungal β-glucosidase, cellobiase (*Aspergillus niger*): EC 3.2.1.21. The most industrially interesting member of this group, with the important function of hydrolysing cellobiose, the product of cellulose hydrolysis that is inhibitory to the cellulase enzymes.

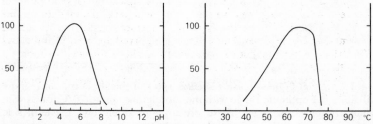

(43) Fungal β-glucosidase (*Aspergillus oryzae*): This type has typical fungal enzyme characteristics.

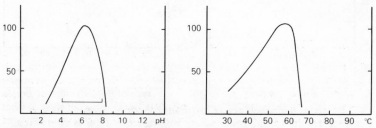

(44) Fungal β-glucosidase (*Trichoderma viride*): This variant and the *Aspergillus* type (43) are very similar.

Note: both enzymes (43) and (44) can exhibit special actions on different substrates, and will also hydrolyse galactans, arabans and xylans to some extent.

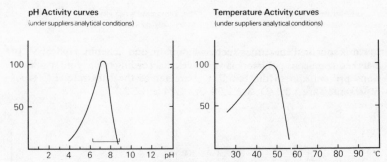

(45) Yeast β-glucosidase (*Saccharomyces* spp.): The pH optimum is very narrow, around pH 7, and the temperature optimum is typical of yeasts at 45 °C. These enzymes have very easily controlled action and find some application in the modification of sensitive molecules having β-linked glucose or similar components. Acid action would disturb the remainder of the molecule and cannot always be used when seeking to replace the glucose part, or remove it completely. This type of reaction is useful in pharmaceutical chemistry.

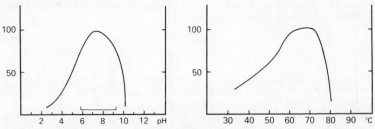

(46) Bacterial β-glucosidase (*Bacillus* spp.): Typical neutral and mildly alkaline pH optimum range with thermotolerance up to 70°C found with many bacterial enzymes. These preparations often have the least side activities amongst the available β-glucosidases.

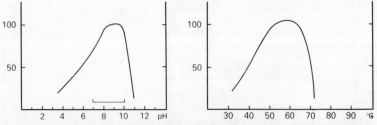

(47) Bacterial β-glucosidase (*Clostridium thermocellum*): There is little evidence for industrial application, although this type is commercially available. The alkaline pH optimum and thermolability make this an atypical clostridial enzyme. Some applications in pharmaceutical chemistry similar to those mentioned for enzyme (45) can be carried out at the higher pH values.

This is the most active type to hydrolyse β-1,3 bonds.

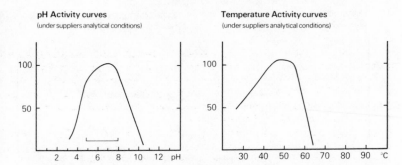

(48) Plant β-glucosidase (*Prunus amygdalus*, sweet almond): Moderate pH and temperature range. Often used as an assay tool for the determination of glycosides.

(49) – (52) Glucose isomerases, immobilized: The EC classification of these enzymes was 5.3.1.18 but this has now been deleted. Most industrial commercial preparations are classified as EC 5.3.1.5, xylose isomerase having the ability to convert D-glucose to D-fructose. Kinetic studies show that these enzymes require high glucose substrate concentrations to give adequate commercial activity. The K_m values are very high and the limit of conversion is around 50 per cent, the industrial operation usually accepting 42–45 per cent fructose in the product stream. High fructose levels are obtained by subsequent enrichment procedures, usually chromatographic.

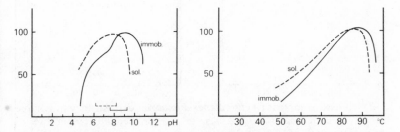

(49) Bacterial glucose isomerase (*Bacillus coagulans*): Comparison of the pattern of soluble and immobilized preparations shows the pH shift to be more marked on immobilization. There is also a small improvement in the temperature tolerance. Purity of feed syrups is important as, in addition to heavy metal inhibition, the enzyme is activated by magnesium and this is antagonized by calcium. Cobalt is not required as an extra activator for this enzyme source. The immobilized enzyme is very sensitive to oxygen, which is removed from the feed stream by heating. Operating conditions are 58–62 °C at pH of feed 8–8.3 for up to 2000 hours.

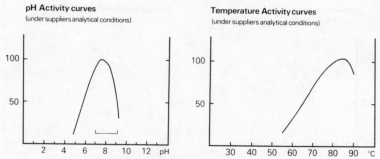

pH Activity curves
(under suppliers analytical conditions)

Temperature Activity curves
(under suppliers analytical conditions)

(50) Glucose isomerase (*Streptomyces albus*): The main feature is a somewhat lower pH optimum than the *Bacillus* type. Magnesium and cobalt cofactors are required. Operation is at 60–75 °C and pH 6–7.

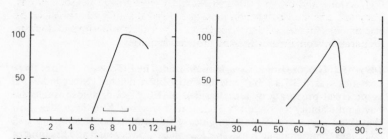

(51) Glucose isomerase (*Arthrobacter* spp.): This enzyme is similar to enzyme (49) from *Bacillus coagulans* but with a sharper thermal limit. Cobalt and magnesium cofactors are required with optimum operation at pH 8 and higher.

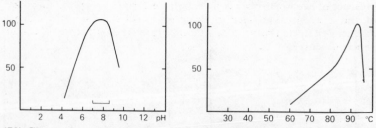

(52) Glucose isomerase (*Actinoplanes missouriensis*): An enzyme notable for its high thermal tolerance despite a recommended temperature of operation of 60–65 °C. Cobalt or magnesium cofactors allow operation at neutral pH and an operating life of 1500 hours.

(53)–(58) Lipases (animal, yeast and fungal): Triacyl glycerol lipases EC 3.1.1.3. All appear to be activated to some extent by calcium ions. Consequently, chelating agents such as EDTA are inhibitors. Heavy metal ions also inhibit. Many preparations contain esterases EC 3.1.1.1 and EC 3.1.1.2. Diisopropyl fluorophosphate will not inhibit true lipases but strongly inhibits esterases.

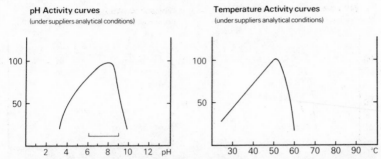

pH Activity curves
(under suppliers analytical conditions)

Temperature Activity curves
(under suppliers analytical conditions)

(53) Pancreatic lipase (porcine and bovine): Preferred substrates for this enzyme type are triglycerides, with decreasing action on di- and monoglycerides. They have a high preference for action on the 1 position, yielding diglycerides, and prefer fatty acid chains of more than 12 carbon atoms. They are activated by sodium ions, and thermal stability is enhanced by calcium ions for the porcine type. The bovine type is insensitive to calcium. Preparations contain amylase and protease activities.

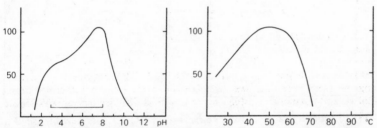

(54) Yeast lipase (*Candida cylindracea*): Some evidence of the common presence of two isoenzymes is given by the pH curve. The powerful action of this enzyme is retained in the presence of lipid solvents (except methanol) if treated below 40 °C. Its action on olive oil is three times that of pancreatic enzymes.

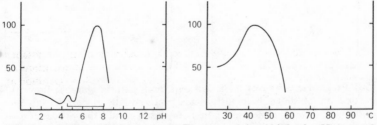

(55) Fungal lipase (*Rhizopus* spp.): The unexplained biphasic pH curve of this enzyme type may be due to isoenzymes or the influence of an acid environment. When standardized by action on tributyrin, this type shows more than 80 per cent activity on olive oil, and rape and cotton seed oils, but only 60 per cent on soya oil. It is tolerant of pH in the range 4.5–8.0 when used at or below 35 °C.

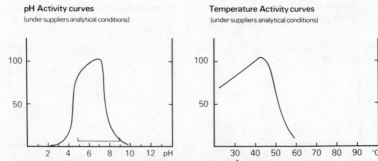

pH Activity curves
(under suppliers analytical conditions)

Temperature Activity curves
(under suppliers analytical conditions)

(56) Fungal lipase (*Aspergillus niger*): This enzyme type is very tolerant of acid conditions down to pH 3.5 when the substrate is protected and calcium ions are present. Its normal thermal limit of 45 °C can be extended to 60 °C when working with complex substrates such as seed oils. It shows a preference for fatty acids shorter than 12 carbon atoms, in contrast to the pancreatic enzymes.

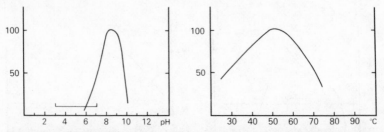

(57) Phospholipase A (porcine pancreatic): EC 3.1.1.4. Lecithinase. An enzyme stereo specific for the 2 position, with a lysophosphatide as the product. It has an absolute requirement for calcium ions. Optimum practical conditions for action on lecithins (soya, egg yolk) are pH 8 and 40°C.

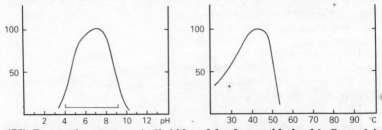

(58) Pregastric esterases (calf, kid and lamb parotid glands): Powerful production of short-chain fatty acids preferentially released from any position. These enzymes are commonly used to aid the production of 'Italian cheeses'. They have a broad pH range and low thermal tolerance.

(59) & (60) Catalase: EC 1.11.1.6. Hydrogen peroxide oxidoreductase. This enzyme type is inactivated by high concentrations of substrate (above about 0.5 M). Alkali and heavy metal salts inhibit. It is generally used in association with glucose oxidase to remove glucose or peroxides from foods and food systems.

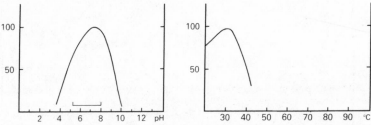

(59) Bovine catalase (liver extract): This enzyme type has a pH optimum at 7, with good activity between 5 and 8, and very low thermotolerance.

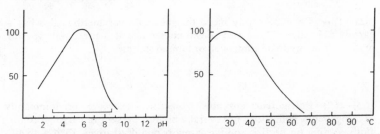

(60) Fungal catalase (*Aspergillus niger*): This enzyme has a wider pH tolerance, showing activity in the range 3 to 9 and an optimum at pH 6.5. It is very stable to low pH at temperatures below 30°C. Thermotolerance up to 60°C is good for short reaction times.

(61) & (62) Glucose oxidase (*Aspergillus niger*): EC 1.1.3.4. β-D-glucose oxidase 1-oxidoreductase. Preparations usually also contain catalase. D-glucose is converted to gluconic acid through the intermediate lactone, with hydrogen peroxide produced in the absence of catalase. Such enzymes are partly dependent on the dissolved oxygen concentration in the reaction system for maximum activity and are therefore heat sensitive partly due to loss of oxygen at elevated temperatures. They often contain side activities of invertase, amylase, cellulase and glucanase.

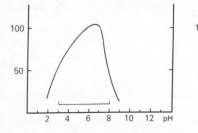
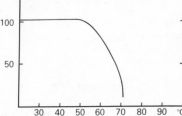

pH Activity curves
(under suppliers analytical conditions)

Temperature Activity curves
(under suppliers analytical conditions)

(61): The pH and temperatue curves indicate a normal fungal enzyme character with good operating performance up to 50 °C in the pH range 4.5 to 6.5.

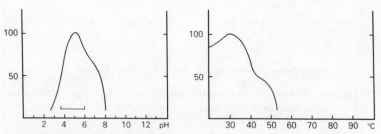

(62): This is a significantly more thermostable variant that also has a narrow pH range. These two sensitivities make it a preferred type for reactions that are to be controlled or limited sharply.

(63)–(67) Pectinolytic enzymes (fungal): These are predominantly heterogeneous collections of several activities present in varying proportions. Acting on pectins and the many pectin derivatives, their activities include: (*i*) Esterase. EC 3.1.1.11. acting to de-esterify pectins to pectic acid by removal of methoxyl residues. Polymethylgalacturonase esterase. (*ii*) Depolymerases, polymethyl galacturonases with either endo- or exo-activity. EC 3.2.1.15 endo; EC 3.2.1.67 exo; EC 3.2.1.82 exo-polydigalacturonase attacks alternate bonds; EC 4.2.2.2 endo-polygalacturonate lyase; EC 4.2.2.6 oligogalacturonate lyase (transelimination); EC 4.2.2.9 exo-polygalacturonate lyase; EC 4.2.2.10 polymethyl galacturonate lyase. Both endo- and exo-transeliminations.

The oligogalacturonases are only present in trace amounts, if at all. The pectinesterase and polygalacturonase activities are the most reliable activity descriptions and provide the basis for evidence of pectin hydrolysis.

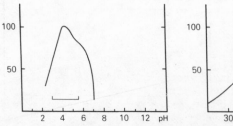
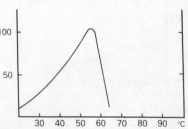

pH Activity curves
(under suppliers analytical conditions)

Temperature Activity curves
(under suppliers analytical conditions)

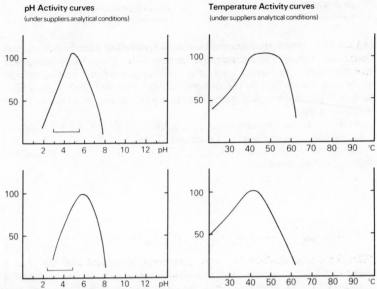

(63) – (65) (*Aspergillus niger* variants): Such enzymes show broadly similar pH and temperature characteristics but have varying enzyme components that make them more or less suited to specific applications.

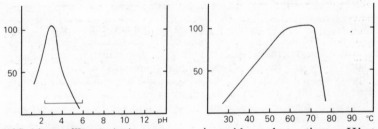

(66) (*Aspergillus* spp.): An extreme variant with very low optimum pH but good stability from pH 2 to 6. It also shows greater thermostability than many *Aspergillus* enzymes.

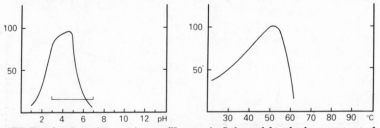

(67) Pectin glycosidase (*Aspergillus* spp.): Selected for the low content of the de-esterifying activities, this enzyme is used to produce citrus juices with high cloud content. It is also used for the production of fruit and vegetable pulps. It will not clarify fruit juices.

pH Activity curves
(under suppliers analytical conditions)

Temperature Activity curves
(under suppliers analytical conditions)

(68) & (69) Hemicellulase/pentosanase (*Aspergillus niger*): Act to produce hydrolysis of arabans, galactans, manans and xylans. They have some
traditional uses in baking and flour modification as anti-staling agents.
More recent uses include tea and coffee extraction and process modification, and pharmacological extractives and modification of components of
oil-drilling fluids.

Activities present include EC 3.2.1.55, EC 3.2.1.88, EC 3.2.1.89 and EC
3.2.1.90 acting on arabans and galactans; EC 3.2.1.25, EC 3.2.1.77 and EC
3.2.1.78 acting on manans; EC 3.2.1.8, EC 3.2.1.32, EC 3.2.1.37 and EC
3.2.1.72 acting on xylans.

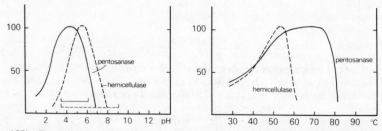

(68): By selection of substrate the two components can be seen to have
some different stability and operating optima, the so-called 'hemicellulase'
activity being generally less acid and thermotolerant than the pentosanase.
Claims for action in hydrolysing the plant gums used widely in the food
industry are not easily substantiated (guar and locust bean gums). Much
depends on the extraction and purification of the gum and the declared
objective of the hydrolysis. Some reduction in viscosity of less pure gum
products is observed.

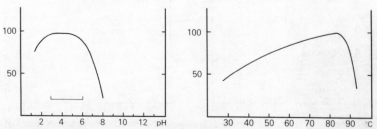

(69) Galactomananase (*Aspergillus niger*): This enzyme shows a more
marked thermotolerance than enzyme (68), together with a broader pH
range.

Proteases: Additional descriptive and comparative data will be given at the end of this section.

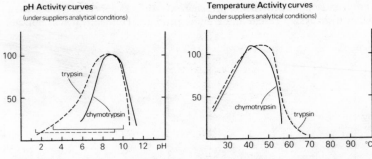

pH Activity curves (under suppliers analytical conditions)

Temperature Activity curves (under suppliers analytical conditions)

(70) Chymotrypsin and trypsin (porcine pancreas, also bovine) Mixed preparations of these two enzymes in varying proportions. EC 3.4.21.1 and EC 3.4.21.4. The pH and temperature characteristics show a wide range of action despite the alkaline optimum.

Activators: calcium ions raise the thermotolerance.

Inhibitors: heavy metals ions, organophosphorous compounds and many natural inhibitors of cereals and legumes and egg white.

Note: the bovine extracts have about 10°C greater heat stability.

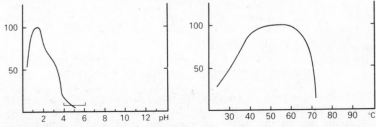

(71) Pepsin (porcine stomach): EC 3.4.23.1. The powerful action of this enzyme at acid pH values from 1.5 to 4 permits a high degree of hydrolysis of denatured proteins that have been acid precipitated. Due to attack on a narrow range of amino acid residues, the release of bitter peptides is often very marked. Usual reactions are at pH 5.0–5.5, at 50–60°C.

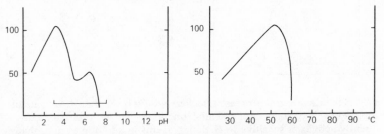

(72) Fungal acid protease (*Aspergillus niger* var. macrosporus): EC 3.4.23.6. A very effective protease acting in the range pH 3–6 but not very thermotolerant even at high substrate concentrations. It is most active on

pH Activity curves
(under suppliers analytical conditions)

Temperature Activity curves
(under suppliers analytical conditions)

casein and animal proteins, expressing only some 25 per cent of that activity when hydrolysing cereal and legume proteins. However, it will not clot milk.

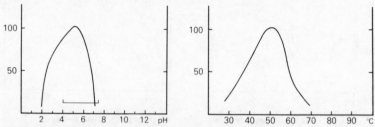

(73) Fungal acid protease (*Rhizopus* spp.): Probably still of the type classified as EC 3.4.23.6, this enzyme is considered to act like pepsin. The specificity is broad and it will clot milk but not hydrolyse synthetic substrates of small molecular weight.

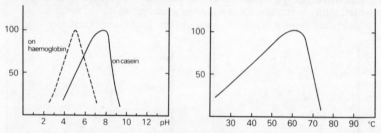

(74) Fungal acid protease (*Aspergillus* spp.): This is a variant with a more substrate specific action with regard to pH, the casein hydrolysis being essentially neutral and that of haemoglobin at pH 4 to 5. This variant performs well at up to 60 °C at pH 4 when acting on plant proteins. It is not better than 20 per cent active on egg white.

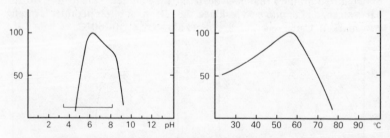

(75) Bacterial neutral protease (*Bacillus subtilis*): EC 3.4.24.4. A metalloprotease, it contains one essential zinc atom per molecule and consequently activity is sensitive to chelating agents in addition to phosphates and metal ions. The specificity of action is quite narrow, with a preference for bond pairs containing aromatic amino acids. The activity is stable in the pH range 5 to 8 at up to 55 °C, but falls rapidly above this temperature. It does not appear to be sensitive to inhibitors of alkaline proteases that are found in raw barley.

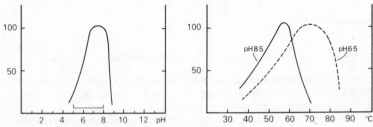

(76) Bacterial neutral protease (*Bacillus subtilis*): A neutral to mildly alkaline variant that shows marked increase in thermotolerance at the low end of the pH stability range. This cannot be demonstrated on all substrates and should be tested carefully when selecting for optimum action on a nominated substrate.

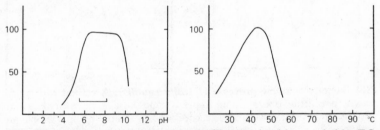

(77) Fungal neutral protease (*Aspergillus* spp.): Also probably EC 3.4.24.4. Its action is notably towards hydrophobic amino acid residues. This can be used to advantage in minimizing bitterness in protein hydrolysates. (Hydrophobic peptides are considered to be the major contributors to bitterness of food protein hydrolysates.) The extremely low thermotolerance is not improved by high substrate concentration.

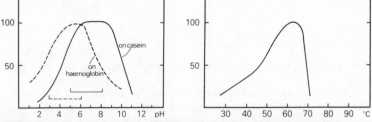

(78) Fungal neutral protease (*Rhizopus* spp.): Another enzyme with potent action on hydrophobic peptides. The pH optimum varies widely with different substrates, and the thermotolerance is good up to 60°C over reaction periods of up to 4 hours.

498   Chapter 5

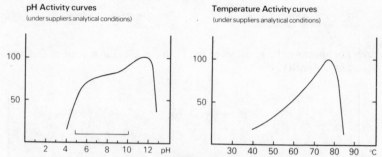

(79) Bacterial alkaline protease (*Bacillus subtilis*): Alkaline variants of EC 3.4.24.4. The pH and thermal stability of this type make them suited to the detergent industry. Sensitivity to chelating agents such as EDTA and to phosphates is low, while tolerance to oxidizing agents such as perborates is good. Full activity at pH 9 and 60°C can be maintained for at least 60 minutes. Cereal and legume inhibitors will affect these enzyme types.

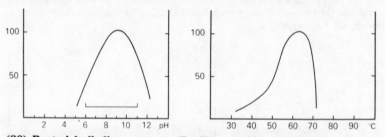

(80) Bacterial alkaline protease (*Bacillus subtilis*): A variant with a pH activity permitting operation up to pH 10, and stability between pH 9 and 10. Thermotolerance up to 55°C is good; short (30 minute) operations can accept up to 85°C. It is sensitive to cereal and legume inhibitors so that action on these protein sources is low. It is also inhibited by egg white.

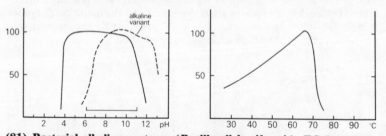

(81) Bacterial alkaline protease (*Bacillus licheniformis*): EC 3.4.21.14. Microbial serine type proteases with powerful action in the extreme alkaline range. They also hydrolyse peptides vigorously, and include the subtilisins and subtilopeptidases. pH optimum is in the range 9 to 11, and stable in the range 5 to 12 over varying time courses.

Activation: calcium ions make some contribution to thermotolerance.

Inhibitors: organophosphorous compounds, heavy metals and natural inhibitors from cereals, legumes and white of egg.

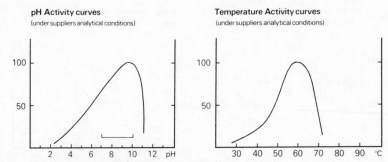

(82) Fungal alkaline protease (*Aspergillus oryzae*): Not a true serine type protease, but has a broad activity spectrum acting on many proteins. Does not hydrolyse peptides. Usually the preparations contain amylases.

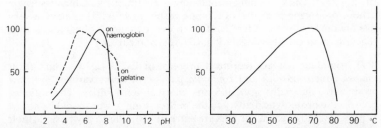

(83) Plant protease, papain (*Carica papaya*): EC 3.4.22.2. The plant proteases are thiol type and sensitive to heavy metals and oxidizing agents. They are stabilized by addition of reducing agents and cysteine, and are not inhibited by chelating agents. They show a very variable pH optimum, which depends on substrate. Their temperature optimum is 60–70°C, but they can be used up to 75°C for short reaction times. They are widely used as broad spectrum proteases with action on peptides such that bitterness is rarely found in food protein hydrolysates.

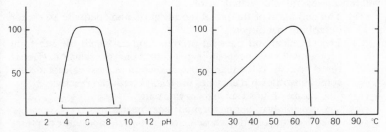

(84) Plant protease, ficin (*Ficus glabrata*): EC 3.4.22.3. Shows preferential hydrolysis of bonds containing aromatic amino acids. pH optimum in the range 5–8 varies with substrate. It has a similar sensitivity to inhibitors as does papain.

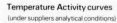

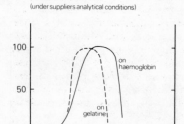

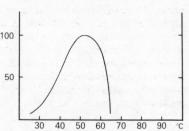

(85) Plant protease, bromelain (pineapple, *Ananas comosus*): EC 3.4.22.4. Another thiol protease but also a glycoprotein. It acts on basic and aromatic amino acids for preference. Its pH optimum varies with substrate in the range 5–8, and it has a low thermotolerance.

(86) Animal rennet, chymosin (young calves, lambs and kids – unweaned): EC 3.4.23.4. These are milk coagulating preparations with highly specific action, producing hydrolysis of kappa casein of milk. They have a very narrow range of pH and temperature in use. Calcium is obligatory for the reaction to proceed to coagulation. (No relevant curves.)

(87) Microbial rennets (carboxyl proteases): EC 3.4.23.6 from *Mucor miehei*, *Mucor pusillus* and *Endothia parasitica*. Many variants are now produced industrially with differing thermal sensitivities. The ratio of nonspecific proteolytic activity to the milk clotting power should be as low as possible. Most recent variants are extremely similar to chymosin in their characteristics. (No relevant curves.)

4. Additional data on proteases

From the comparative data and graphic representation of the pH and temperature characteristics (based on analytical methods, not working conditions) it can be seen that of all the enzyme groups used by industry, the proteases are the most difficult to specify. The choice of pH, the substrate and its concentration will all influence the type of reactivity that will occur. Temperature will only influence the rate and persistence of the reaction (*see* Chapter 2 for kinetic and thermodynamic interpretations).

The selection of an enzyme for a particular protein modification target will rely on several interrelated factors:

- (*i*) The definition of the target modification and adequate assays for describing that target.
- (*ii*) The chemical and physical pretreatment to which the substrate protein will have been subjected before enzyme reaction. (Denatured proteins can enhance enzyme action in most cases, but with some substrates and enzymes the result is reduced reactivity.)
- (*iii*) The choice of substrate and enzyme pairs.
- (*iv*) The possibility of opting for multi-enzyme action.
- (*v*) The pH of the reaction chosen, and the time/temperature relationship selected.

Making the correct choices in these areas will require empirical testing if the best economic route to the target is to be found. There are some guidelines that can be used in the selection of the appropriate enzyme and these are illustrated in the tables that follow.

Enzyme specificity. Based on a very pure, highly defined protein substrate, the specificity of a variety of proteases can be compared. Each is applied at its optimum pH and at a temperature well within the known limits. The amino acid bond pairs and the degree of vigour of the hydrolysis of these pairs can be determined. Table 5.1 sets out some of these observations for the hydrolysis of the oxidized B chain of insulin, but interpretation towards action on other substrates remains intuitative until more information is obtained from direct testing. However, some guide to the likely outcome can be inferred from the table.

TABLE 5.1
Specificity of proteolytic action on oxidized B chain of insulin

Enzyme	No. of bonds vigorously broken	Main bonds attacked	pH optimum range
Chymosin	2	Glu-Ala; Leu-Val	3.5–6.5
Microbial coagulants	2	Glu-Ala; Leu-Val	3.5–6.5
Trypsin	2	Lys-Ala; Arg-Gly	6–9
Ficin	4	Glu-Ala; Tyr-Leu; Phe-Tyr	5–7.5
Pepsin	5	Leu-Val; Phe-Tyr	1.8–3.0
Fungal (alkaline)	5	Leu-Tyr; Phe-Tyr	7–9
Bacterial (neutral)	6	His-Leu; Ser-His; Ala-Leu; Gly-Phe; Arg-Gly	6.5–7.5
Bacterial (alkaline)	7	Gln-His; Ser-His; Leu-Tyr	7.5–9.5
Fungal (acid)	9	His-Leu; Gly-Phe; Phe-Phe	2.5–4.0
Papain	9	Asn-Gln; Glu-Ala; Leu-Val; Phe-Tyr	5.0–7.0

General reactivity. The general performance of the major groups of proteases on a variety of common substrates has been compared by many workers. Table 5.2 shows a compilation of these observations. Again, the interpretation is subjective, since the only criterion for the comparisons was increased soluble nitrogen after acid treatment of the reaction mixtures. This does not form a guide to functional changes occurring with the same enzyme/substrate pair under other conditions.

More detailed comparisons – acid optimum fungal proteases. These proteases form a narrow and reproducible group and have been extensively

TABLE 5.2
General comparative reactivity of proteases
(at their optimum pH)

| | Substrate | | | |
Enzyme	Haemo-globin	Gelatine	Casein	Soya protein
Reference (papain)	1	1	1	1
Ficin	0.5	1.1	0.2	0.3
Pepsin	0.65	0.7	0.95	0.1
Trypsin	0.4	0.45	0.45	0.6
Bacterial (neutral)	1.35	1.4	1.3	1.4
Bacterial (alkaline)	1.6	1.3	1.65	1.7
Fungal (acid)	0.65	0.2	0.8	0.25
Fungal (neutral)	0.2	0.65	0.95	0.7

studied. Table 5.3 shows their general reactivity on a variety of substrates in the pH range 3–5. Similar data for the other groups have not yet emerged, but interest in protein hydrolysis is growing rapidly and such information should soon be available.

TABLE 5.3
Comparison of the acid optimum fungal proteases on various substrates in the range pH 3 to 5

Substrate	Activity value	Substrate	Activity value
Casein	100	Maize gluten	35
Blood	80	Soya protein	25
Wheat gluten	55	Egg albumen	18

DATA INDEX 1

ENZYME GROUPINGS

Enzyme groupings	Data Index reference
Amylases – General	A
Bacterial	B
Fungal	C
Plant	D
Amyloglucosidases (debranching)	E
Analytical/Pharmaceutical/Research	F
Cellusases	G
Dextranase and invertase	H
Galactosidases and inulases	I
Glucanases	J
Isomerases	K
Lipases and esterases	L
Oxidases and catalases	M
Pectinases	N
Pentosanases	O
Proteases – General	P
Animal	Q
Microbial	R
Plant	S
Rennets	T
Immobilized – General	U
Mixed activity preparations	V

DATA INDEX 2
ENZYME SUPPLIER COMPANIES
(* denotes data in Data Index 3)

No.	Company	Country	Comments	Enzyme types (see Data Index 1)
1*	A.B.M.C. Food Division Poleacre Lane Woodley Stockport Cheshire SK6 1PQ	UK	Bulk industrial	A, B, C, E, J, P, R, S
2	Akzo Chemie Statwinstraat 48 PO Box 247	Holland		Range not specified
3	Aktieboluget Montoil Fack S 100 55 Stockholm	Sweden		Range not specified
4*	Alltech Inc. 271 Gold Rush Road Lexington Kentucky 40503	USA	Specialists for alcohol production	A, B, E
5*	Alpha Color S.P.A. Via Valtellina 48 20159 Milan	Italy	Textile applications	B
6	Amano Pharmaceutical Co. Ltd 1–2 1–chome Nishiki Naka-Ku Nagoya	Japan	Bulk industrial	A, B, C, E, G, L, P, R, S

7*	B.D.H. Chemicals Ltd Broome Road Poole Dorset BH12 4NN	UK	Agents for Merck	H, S
8*	Biddle Sawyer and Co. Ltd PO Box 170 3 Lovat Lane London EC3P 3EX	UK	Brewing suppliers	S
9*	Biochemie GmbH A–6250 Kundl Tirol	Austria	Pharmaceutical specialities	F
10*	Biocon Ltd Kilnagleary Carrigaline Co. Cork	Eire	Bulk industrial	A, B, C, D, E, G, H, J, N, P, R, S, V
11*	Biozyme Laboratories Ltd Unit 6 Gilchrist–Thomas Estate Blaenavon Gwent NP4 9RL	UK	Analytical and research Wide range Agents in USA, Italy, Switzerland and France	F
12	Boehringer Mannheim GmbH PO Box 51 D–6800 Mannheim 31	W. Germany	Analytical and research Wide range Worldwide agents	F
13	Calbiochem PO Box 54282 Los Angeles California 90054	USA	Analytical and research Wide range Worldwide agents	F

No.	Company	Country	Comments	Enzyme types (see Data Index 1)
14*	Chr. Hansen Laboratories A/S 3 Sankt Annae Plads DK–1250 Copenhagen	Denmark	Dairy specialities	L, T, V
15*	Corning Biosystems Corning Glass Works Corning New York 14830	USA	Immobilized enzyme systems Agents for Rohm	A, B, I, N, P, R, U
16*	Dairyland Food Laboratories Inc. Progress Avenue Waukesha Wisconsin 53187	USA	Dairy specialities	L, M, T, V
17	Daiwa Kasei KK 3–11 Vehonmachi–5–chome Tennoji-ku Osaka	Japan	Bulk industrial	A, B, C, P
18	Diamalt Friedrichstrasse 18 D–8000 München 40 Postfach 400 469	W. Germany	Textile and leather industries	A, P, Q, R
19	Diamond Shamrock 620 Progress Avenue Waukesha Wisconsin 53186	USA		Range not specified
20	Enzyme Development Corporation 2 Penn Plaza New York New York 10121	USA	Bulk industrial Research and development	A, B, C, E, G, H, I, J, K, L, M, N, O, P, Q, R, T, U

#	Company	Country	Industry	Range
21*	Fermco Biochemics Inc. 2638 Delta Lane Elk Grove Village Illinois 60007	USA	Beverage industry Wine and juice industry Analytical	A, B, D, E, F, G, H, K, M, N
22*	Genzyme Biochemicals Ltd Springfield Mill Maidstone Kent ME14 2LE	UK	Analytical/diagnostic Division of Genzyme Corp. Connecticut, USA	F
23*	Gist Brocades NV PO Box 1 Wateringseweg 1 Delft 2600MA	Holland	Bulk industrial	A, B, C, E, G, H, I, J, K, L, M, N, P, R, S, T, V
24*	Glaxo Operations UK Ltd Ulverston Cumbria LA12 9DR	UK	Bulk industrial	C, E, J
25*	Godo Shusei Co. Ltd 6–2–10, Ginza, Chuo-ku Tokyo	Japan	Bulk industrial	A, B, I, K, R
26	Grindestedvaerket A/S 38 Edwin Rahrs Vej 8220 Braband	Denmark	Bulk industrial Wine/juice industries	J, M, N
27	Hankyu Kyoei Bussan Co. Ltd 5/6 chome Tenjin Bashi-suji Oyodo-Ku Osaka	Japan	Bulk industrial	Range not specified

No.	Company	Country	Comments	Enzyme types (see Data Index 1)
28	Hayashibara Shoji Co. Ltd 198 Shimoishii Okayama City	Japan	Bulk industrial	Range not specified
29	Henkel KG a.A D-4000 Düsseldorf 1 Postfach 1100	W. Germany	Bulk industrial	Range not specified
30	Hoechst, D-6000 Frankfurt/M80 Postfach 800320	W. Germany	Bulk industrial Textile industry	A, B, C
31*	Hughes and Hughes (Enz.) Ltd Elms Industrial Estate Church Road Harold Wood Romford Essex RM3 0HR	UK	Analytical and research	F, G, L, M, S
32	Koch-Light Laboratories Ltd Colnbrook Buckinghamshire SL3 0BZ	UK	Analytical/research	F, U
33*	Kingsbridge Industrial Inc. PO Box 24 200 Tapei	Taiwan	Bulk (bromelain)	S
34	Kyowa Hakko Kogyo Co. Ltd Ohtemachi Building 6-1, Ohtemachi 1-chome Chiyoda-ku Tokyo	Japan	Bulk industrial	G, T

No.	Company	Country	Type	Range
35	Laboratories Sanders SA 47–51 rue Henri Wafelaerts Brussels 6	Belgium		Range not specified
36	Dr. Madis Laboratories Inc. 375 Huyler Street S. Hackensack New Jersey 07606	USA	Bulk industrial	P
37*	Miles Kali–Chemie GmbH 3 Hannover–Kleefeld Hans–Buckler Allee 20 Postfach 690307	W. Germany	Bulk industrial Analytical and research	A, B, C, E, F, G, J, K L, M, N, P, Q, R, S, T, U
38*	Miles Laboratories Inc. Elkhart Indiana 46514	USA	Bulk industrial Analytical and research	A, B, C, E, F, G, J, K, L, M, N, O, P, Q, R, S, T, U, V
39	Mitsui and Co. Ltd PO Box 822 Tokyo Central	Japan	Bulk industrial	Range not specified
40*	Munton and Fison Ltd Cedars Factory Stowmarket Suffolk IP14 2AG	UK	Malt extract specialities	D, S, V
41*	Murphy and Son Ltd Wheathampstead St Albans Hertfordshire	UK	Bulk (papain)	S
42	Naarden International NV Postbus 2 Naarden–Bussum	Holland	Bulk industrial	Range not specified

No.	Company	Country	Comments	Enzyme types (see Data Index 1)
43	Nagase and Co. Ltd Konishi Building 2,2–chome Honcho Nihonbashi Chuo-ku Tokyo	Japan	Bulk industrial Analytical/pharmaceutical	A, B, D, E, F, G, K, P, R
44*	Novo Industri A/S Novo Alle DK–2880 Bagsvaerd	Denmark	Bulk industrial Analytical/research/pharmaceutical	A, B, C, E, F, G, H, I, J, K, L, N, O, P, Q, R, T, U, V
45	Okasa Industria e Comercia de Diastase Ltda Rue Araraquara 41–Diadema–SP Brazil CEP 09900 Caixa Postal 352	Brazil	Bulk industrial	Range not specified (A)
46	Oriental Yeast Co. Ltd Enzyme Development Centre 4–1 Minamisuita 4–chome Suita, Osaka 564	Japan		Range not specified
47	P. L. Biochemicals Inc. 1037 West McKinley Avenue Milwaukee Wisconsin 53205	USA	Analytical/research	Wide range
48	Premier Malt Products Inc. 1137 North 8th Street Milwaukee Wisconsin 53201	USA		D, S

No.	Company	Country	Type	Codes
49	Pfizer Inc. World Headquarters 235 East 42nd Street New York New York 10017	USA	Bulk industrial	A, B, C, G, O, P, Q, R, T
50	Pharmachim State Economic Trust 15 Iliensko Chaussee Sofia	Bulgaria		Range not specified
51	Powell and Scholefield Ltd 38 Queensland Street Liverpool L7 3JG	UK	Brewing Industry	A, B, C, J, P, S
52*	Rohm GmbH Kirschenallee Postfach 4242 D–6100 Darmstadt 1	W. Germany	Bulk industrial	A, B, C, G, N, O, P, Q, R, S, V
53	Rohm and Haas Co. Independence Mall West Philadelphia Pennsylvania 19105	USA	Bulk industrial	A, B, C, G, N, P, Q, R, S
54	Sigma Chemical Company Inc. St Louis Morrisville Missouri	USA	Analytical/research	F
55	Société Rapidase 15 rue des Comtesses 59113 Seclin	France	Gist–Brocades associate company	See no. 23

No.	Company	Country	Comments	Enzyme types (see Data Index 1)
56*	John and E. Sturge Ltd Denison Road Selby North Yorkshire YO8 8EF	UK	Bulk industrial	C, D, E, F, G, I, L, M, N, O, V
57*	Sturge Enzymes Henley and Co. Inc. 750 Third Avenue New York New York 10017	USA	Analytical See also no. 56	A, B, D, F, H
58	Sumitomo Shoji Kaisha Ltd 2 Nishikicho Building 24–1 Kandanishikicho 3–chome Chiyoda-ku Tokyo	Japan	Bulk industrial	A, B, C, E, F, G, I, N, O, P, R, S
59	Swiss Ferment Co. Vogesenstrasse 132 4056 Basel 13	Switzerland	Bulk industrial Novo Industri subsidiary	A, C, H, L, N, O, V Also see no. 44
60	Tanabe Seiyaku Co. Ltd via Siber Hegner Benelux BV Postbus 414 Rotterdam Westersingel 107	Japan and Holland	Bulk industrial Analytical/pharmaceutical	A, F, L, N, P
61*	Ubichem Ltd 281 Hithermoor Road Stanwell Moor Staines TW19 6AZ	UK	Industrial/analytical	A, G, P, S, V

62	Vifor SA 48 route de Drize 1227 Carouge Geneva	Switzerland		Range not specified
63	W.B.E. Ltd Sandyford Industrial Estate Foxrock Dublin 18	Eire	Bulk industrial	Mixed systems
64	Worthington Biochemical Co. New Jersey 07728	USA	Analytical/research	Range not specified
65	Windsor Laboratories Ltd Bedford Avenue Slough Berkshire	UK	Analytical/research	Wide range
66	Yakult Biochemical Co. Ltd 8–21 Jingikan-Machi Nishinomiya-shi Hyogo	Japan	Industrial/pharmaceutical	B, C, F, G, N, O, T

DATA INDEX 3

PRODUCT DATA SUPPLIED BY CERTAIN OF THE COMPANIES LISTED IN DATA INDEX 2

Editors' notes: All the companies listed in Data Index 2, as well as several more whose addresses proved unreliable, were invited to supply information describing their products in relation to the headings of the columns in this table. Only 27 of the total invited actually responded and their data are given here. Each supplier is identified by the number given in Data Index 2. The enzymes are then described in their alphabetical groupings as Index 1. Thus it is possible to scan quickly for the supplier of a particular enzyme type and then to use the code letter to extract specific data on those examples described in the index.

Supplier no. (Index 2) Enzyme type (Index 1)	Trade or general name	Enzyme type and source	Product form	Optimum pH (or range)	Optimum temperature (°C) (or range)	Activity (own units)	Application and special features
1 B	Nervanase	Bacterial α-amylase *Bacillus subtilis*	Powder/liquid	5.8	75–90	Range 300–10,000 SKB/g	Starch conversion
	Bacterase		Powder			CF 1–100 SKB/g CF 9–900 SKB/g	
	Nervanase T	*Bacillus licheniformis*	Liquid	5.8	90–105	1700 SKB/g	Thermostable
C	Amylozyme	Fungal α-amylase *Aspergillus oryzae*	Liquid	5.5	60–65	30,000 SKB/g	Maltogenic saccharification
			Powder			Range 450–6000 SKB/g	

	Name	Source	Form	pH	Temp	Activity	Application
D	High diastase β-amylase	Barley malt	Powder	5.5	57–65	1500°L/g	Maltogenic saccharification
	β-Amylase	Barley extract	Powder/liquid			1500°L/g	
E	Ambazyme	Fungal amyglucosidase *Aspergillus niger* var.	Liquid/powder	4.7	63–70	50–120 AG/g	Starch syrups, dextrose, food processing industries
	Pulluzyme	Pullulanase *Klebsiella aerogenes*	Liquid/powder	5.2	55–65	450–1500 PU/g	Debranching for starch processing
J	β-Glucanase	Bacterial β-glucanase *Bacillus subtilis*	Liquid	7	55–70	200 BGU/g	Brewing and food processing industries
	β-Glucanase	Fungal β-glucanase *Penicillium emersonii*	Liquid/powder	4	30–90	200 and 750 BGU/g	Brewing and food processing industries
R	Proteinase	Bacterial neutral protease *Bacillus subtilis*	Liquid/powder	7	57–70	0.1; 14 and 45 KTys/g	Brewing and protein hydrolysis
	Gelatase		Powder			7 KTys/g	Silver recovery
	Proteinase T	Bacterial neutral protease *Bacillus coagulans*	Powder	7.5	73–85	14 KTys/g	Thermotolerant
	Proteinase D	Bacterial alkaline protease *Bacillus licheniformis*	Powder	10.5	60–66	150–350 KDU/g	Detergent industry

516 Data Index 3

Supplier no. Enzyme type	Trade or general name	Enzyme type and source	Product form	Optimum pH (or range)	Optimum temperature (°C) (or range)	Activity (own units)	Application and special features
1 R	Panazyme 77	Fungal neutral protease *Aspergillus oryzae*	Powder	7	50–60	3.3 KTys/g	Baking industry
S	Scintillase	Plant protease *Carica papaya*	Liquid/powder	6.5	55–70	30–100 WG	Brewing and general protein hydrolysis
4 B	Allcoholase I	Bacterial α-amylase *Bacillus subtilis*	Liquid/powder	6.5	77	200 Units	Starch conversion
	Allcoholase High T		Liquid		90	80 Units	Thermotolerant; starch liquefaction
E	Allcoholase II	Fungal amyloglucosidase *Rhizopus niveus*	Powder	3.5–4.0	30–50	400 Units	Cold tolerant for alcohol fermentation
5 B	Sbozzimante SPC	Bacterial α-amylase	Liquid	6.5	70	60,000 Units	Textile industry
7 H	Invertin	Invertase-β-h-fructosidase Yeast	Liquid	4.5–5.0	55	Unquoted	Confectionery industry

S	Auxillase	Plant protease, papain *Carica papaya*	Liquid	—	—	12,000 E/g	Brewing (chillproof)
8 S	Papinase	Plant protease, papain *Carica papaya*	Liquid/powder/granule	4.5	65–70	100–1300 Tyr/ml or mg	Brewing (chillproof)
9 F		Penicillin amidase *Basidiomycetes* spp.	Immobilized	6–8	25–32	150 U/g	Immobilized for production of 6-aminopenicillanic acid and assay of penicillin V
10 B	Canalpha	Bacterial α-amylase	Liquid/powder	5.5–6.5	75–85	Unquoted	Starch liquefaction
	Hitempase		Liquid	5.5–6.5	95–105	Unquoted	Thermotolerant for starch liquefaction
	Biotempase		Liquid/powder	5.5–6.5	75–85	Unquoted	Brewing industry
	Biosize		Liquid/powder	5.5–6.5	75–85	Unquoted	Textile industry
C	Bioferm	Fungal α-amylase	Liquid	4–5	40–50	Unquoted	Dextrogenic saccharification
	Biobake		Liquid/powder	4–5	40–50	Unquoted	Flour milling saccharification
D	Malt diastase	Plant α- and β-amylases Barley malt	Powder	5–6	50–55	Unquoted	Maltogenic saccharification

Supplier no. Enzyme type	Trade or general name	Enzyme type and source	Product form	Optimum pH (or range)	Optimum temperature (°C) (or range)	Activity (own units)	Application and special features
10 D	β-Amylase	Plant β-amylase Soya bean	Powder	5–6	55–65	Unquoted	Maltogenic saccharification
E	Amylo	Fungal amyloglucosidase	Liquid/powder	4–5	55–65	Unquoted	Alcoholic saccharification
	Agidex		Liquid	4–5	60–65	Unquoted	Alcoholic saccharification
G	Biocellulase A	Fungal cellulase Aspergillus niger	Powder	4–6	40–50	Unquoted	Cellulose hydrolysis
	Biocellulase T	Trichoderma viride	Powder	4–6	45–55	Unquoted	Cellulose hydrolysis
H	Bioinvert	β-h-fructosidase Yeast invertase	Liquid/powder	4–5	55–60	Unquoted	Confectionery industry
J	Bioglucanase	β-glucanase	Liquid/powder	4–6	65–75	Unquoted	Brewing; thermotolerant
N	Biopectinase	Unspecified pectinase	Liquid/powder	3–5	45–55	Unquoted	Juice/wine industries
R	Acid protease	Fungal	Liquid/powder	2.5–4.0	40–50	Unquoted	Protein hydrolysis
	Fungal protease	Fungal-neutral	Liquid/powder	5–7	50–60	Unquoted	Baking, brewing, protein hydrolysis
	Alkaline protease	Bacterial	Prill	9.5–10	50–60	Unquoted	Detergent industry

S	Profix		Liquid/powder	4–6	60–70	Unquoted	Brewing, food, pharmaceutical industries
		Plant protease, papain					
V	Biase	Mixed α-amylase, β-glucanase	Liquid/powder	5–7	60–75	Unquoted	Brewing industry
	Gasolase	Unspecified	Powder	4–6	45–55	Unquoted	Alcoholic fermentation
	Oloclast	Mixed amyloglucosidase, hemicellulase, cellulase, protease	Powder	4–6	45–55	Unquoted	Alcoholic fermentation
	Progan	Mixed protease, β-glucanase	Liquid/powder	4–6	65–75	Unquoted	Brewing industry
	Promalt	Mixed amylase, β-glucanase, protease, cellulase	Liquid/powder	4–6	60–70	Unquoted	Brewing industry

11   F   Biozyme Laboratories Ltd offer a large range of diagnostic and pharmaceutical enzymes numbering 48 in 1980. Full details and catalogues are obtainable from their UK address and also from their agents:

France
Laboratoires Eurobio
20 Bd Saint Germain
75005 Paris, France

Italy
Derivati Biologica
International Spa
Cassina de'Pecchi (M1)
SSN 11 Padana Superiore Km
160 Milan, Italy

Switzerland
Biogenzia Lemania SA
27 Avenue de Morges
CH-1004 Lausanne, Switzerland
....
Dr Rudolf Strueli AG
Utoquai 29. CH 8032,
Zürich, Switzerland

USA
Accurate Chemical &
Scientific Corp.
28 Tec Street
Hicksville
New York 11801, USA

Supplier no. Enzyme type	Trade or general name	Enzyme type and source	Product form	Optimum pH (or range)	Optimum temperature (°C) (or range)	Activity (own units)	Application and special features
14 L	Lipases	Pregastric esterase of lamb, calf, kid	Powder	5.7–6.6	30–37	Unquoted	Dairy industry
T	Rennets	Carboxyl (acid) proteases Chymosin of calf, cow or pig	Liquid/powder	6–6.7	30–35	Unquoted	Dairy industry
15 B	Rhozyme GC2x 86L H39	Bacterial α-amylase Bacillus subtilis	Liquid/powder	5–7	80	12,500– 210,000 FM	Food processing, textile industries
N	Pectinol	Fungal pectinase Aspergillus niger	Liquid/powder	3–5.5	60	Range 500– 10,000 APU	Food grades
	Pectinol DL	Anthocyanase Aspergillus niger	Liquid	3–4	50	Unquoted	Decolorizing of red grape
R	Rhozyme 41	Fungal protease Aspergillus oryzae	Powder	5.5–8.5	50	54,000 HU	Food grade protease
	Rhozyme P-11	Aspergillus flavus	Powder	5.5–8.5	60	10,000 EE	Food processing industry

	Rhozyme P-53 PF P-64	Bacterial protease Bacillus subtilis	Powder/liquid	6–7.5 8–9.5	60 60	1200 EE 12,000 EE	Food industry Food processing industry
16 L	Italase C	Pregastric esterase, calf	Powder	5–5.5	5–60	Unquoted	Dairy – butter flavour
	Capalase K	Kid	Powder	5–5.5	5–60	Unquoted	Piccante flavour
	Capalase L	Lamb	Powder	5–5.5	5–60	Unquoted	Peccorino flavour
	Capalase KL	Mixed kid and lamb	Powder	5–5.5	5–60	Unquoted	Dairy industry
M	Catalase L	Bovine liver	Liquid	6.5–7.5	5–45	Unquoted	Dairy; thermolabile
T	Emporase	Fungal coagulant Mucor pusillus lindt	Powder/liquid	3–7	5–50	Single/double	Dairy industry
	American rennet	Carboxyl (acid) protease Calf chymosin	Powder/liquid	2–7	5–50	Single/double	Dairy industry
	Beef rennet	Bovine chymosin	Powder/liquid	2–7	5–60	Single/double	Dairy industry
	DFL pepsin	Acid protease Hog stomach	Liquid	2	—	Unquoted	Dairy industry
V	Quikset	Mixed veal and porcine rennet	Liquid	5.5	5–60	Unquoted	Dairy industry
	Regalase	Mixed microbial and porcine rennet	Liquid	6.3	5–50	Unquoted	Dairy industry
21 B	Spezyme	Bacterial α-amylase Bacillus subtilis	Powder	—	—	Unquoted	—

Supplier no. Enzyme type	Trade or general name	Enzyme type and source	Product form	Optimum pH (or range)	Optimum temperature (°C) (or range)	Activity (own units)	Application and special features
21 D		Plant β-amylase, barley	Liquid	—	—	Unquoted	—
E	Spezyme	Fungal amyloglucosidase	Powder	—	—	Unquoted	—
G	Cellulase	Fungal cellulase Aspergillus niger	Powder	—	—	Unquoted	Wine industry
F	Lysozyme	N-acetylmuramoyl hydrolase Hen egg albumen	Powder	—	—	Unquoted	Available as the hydrochloride
H	Fermvertase	Invertase-β-h-fructosidase Saccharomyces cerevisiae	Liquid	—	—	0.3–3.0 K by AOAC	
K	Speyzyme	Glucose isomerase Streptomyces spp.	Beads	—	—	Unquoted	Immobilized and dimensionally stable
M		Peroxidase Horseradish	Powder	—	—	1000–1500 U/mg	Diagnostics

	Product	Enzyme / source	Form	pH	Temp.	Activity	Application
	Fermcolase	Fungal catalase *Aspergillus niger*	Liquid	—	—	1000 U/ml FCC method	
N	Extractase L and P	Fungal pectinase *Aspergillus niger*	Liquid/powder	—	—	5X and 25X	Apple and wine industries
V	Fermcozyme/ Ovazyme	Mixed fungal glucose oxidase and catalase *Aspergillus niger*	Liquid/powder	—	—	750–150,000 FCC method	Diagnostics, soft drinks
	Fermcozyme 952 DM	Mixed fungal glucose oxidase and plant peroxidase	Liquid	—	—	750–5000 U/ml 1100 OD units	Diagnostics

22 F Genzyme Biochemicals Ltd. offer a wide range of diagnostic and research enzymes numbering 28 in 1981. Full details and catalogues are obtainable from their UK address and also from

Genzyme Corporation
1 Bishop Street
Norwalk
Connecticut, USA

	Product	Enzyme / source	Form	pH	Temp.	Activity	Application
23 B	Amylase THC	Bacterial α-amylase *Bacillus subtilis*	Powder	7–8	60–70	Unquoted	Detergent industry
	Bactamyl-Rapidase		Liquid/powder	—	70	Unquoted	Textile industry
	Dexlo		Liquid/powder	6.5	85–90	Unquoted	Starch industry
	Maxamyl		Liquid	6.5	85–90	6000 BAU/g	Starch industry
	Brewer's Amyliq		Liquid	6.5	70	Unquoted	Brewing industry

Supplier no. Enzyme type	Trade or general name	Enzyme type and source	Product form	Optimum pH (or range)	Optimum temperature (°C) (or range)	Activity (own units)	Application and special features
23 C	Amylase P	Fungal α-amylase *Aspergillus oryzae*	Powder	—	—	500–500,000 PS$_{50}$/g	Baking industry
	Brewer's Fermex		Powder	—	—	Unquoted	Brewing industry
	Mylase		Liquid	5.2	50–55	220,000 PS$_{50}$/g	Starch industry
	Hazyme	*Aspergillus niger*	Liquid	3.8	60–65	>1,000 BRAC/g	Juice and wine industries
E	Brewer's Diase	Fungal amyloglucosidase *Aspergillus niger*	Powder	4–4.2	50	16,000 IGA/g	Brewing industry
	Amigase		Liquid	4.2	60	16,700 IGA/ml	Starch industry
I	Maxilact	Yeast lactase, β-galactosidase *Saccharomyces lactis*	Liquid	6.4–6.8	35–40	5000 NLU/ml	Dairy industry
J	Filtrase AM	Bacterial glucanase *Bacillus subtilis*	Powder	6.8	44	Unquoted	Brewing industry
	Filtrase B		Powder	6.8	44	50,000 PC/g	Brewing industry
K	Maxazyme	Microbial glucose isomerase *Actinoplanes missouriensis*	Particles	7.5	60	900 MGIU/kg	Immobilized for starch process

L	Piccantase	Fungal lipase–esterase *Mucor miehei*	Powder	7.5	45–50	A-215 BGE/g B-20 BGE/g	Dairy industry
M	Maxazyme GO	Fungal glucose oxidase *Aspergillus niger*	Liquid	4.5	50	1500 Sarret U/ml	Juice and wine industries
N	Klerzyme Rapizyme Rapidases C, CX, CXP	Fungal pectinase *Aspergillus niger*	Liquid/powder	4–4.5	—	Unquoted	Juice and wine industries
	Rapidase CPE	Fungal pectin esterase *Aspergillus niger*	Liquid	3.8–4.3	—	Unquoted	Juice and wine industries
R	Maxatase	Bacterial alkaline protease *Bacillus subtilis*	Prills	9–10	50–60	Unquoted	Detergent industry
S	Collupulin Brewer's Chill	Plant protease, papain *Carica papaya*	Liquid/powder	—	—	Unquoted	Brewing industry
T	Fromase	Microbial rennet *Mucor miehei*	Liquid/powder	—	—	Range	Dairy industry
V	Maxatase M	Mixed bacterial protease and amylase *Bacillus subtilis*	Prills	9–10	50–70	Unquoted	Detergent industry

Supplier no. Enzyme type	Trade or general name	Enzyme type and source	Product form	Optimum pH (or range)	Optimum temperature (°C) (or range)	Activity (own units)	Application and special features
23 V	Brewer's Mylase SR	Mixed α-amylase and amyloglucosidase	Powder	5.2	50–52	25,000 PS$_{50}$/g	Brewing industry
	Battinase	Mixed proteases	Powder	—	—	Unquoted	Leather industry
	Pamylase	Mixed fungal α-amylase and protease *Aspergillus oryzae*	Powder	—	—	Unquoted	Baking industry
24 E	Agidex	Fungal amyloglucosidase *Aspergillus niger*	Liquid	4.5	60–65	3000–8000 Glaxo Units/ml	Starch conversion
J	Barley β-glucanase	Fungal glucanase *Penicillium emersonii*	Liquid	4.0	65–95	200 Glaxo Units/ml	Brewing, distilling, feedstuffs
25 B	Godo-BXA	Bacterial α-amylase	Powder	5.6	80	20,000 G/g	Calcium for heat stability; starch industry
I	Godo-YNL	Yeast lactase *Kluyveromyces lactis*	Liquid	6.5	50	50,000 G/g	High purity; dairy industry
K	Godo-AGI	Microbial glucose isomerase *Streptomyces* spp.	Granule	8.3	63	220 GI/g	Immobilized for continuous use

			Form	pH	Temp	Activity	Application
R	Godo-BNP	Bacterial neutral protease *Bacillus polymyxa*	Powder	7–8.5	52	100,000 PU/g	Calcium for heat stability
	Dispase		Ampoule	7.5	52	10,000 PU/g	Animal tissue culture
	Godo-BAP	Bacterial alkaline protease *Bacillus licheniformis*	Powder	10–10.5	60	300,000 DU/g	Food processing and detergent industries
31 F	Urease	Plant urease type A *Canavalia ensiformis*	Powder	7	50	250–350 U/g	Analytical, diagnostics
G	Cellulase	Fungal cellulase *Basidiomycetes* spp.	Powder	4.3	50	13,000 and 50,000 U/g	
L	Lipase	Fungal lipase–esterase *Rhizopus arrhizus*	Powder	7–7.5	37	20,000 U/g	
M	Ovazyme/Fermcozyme	Fungal glucose oxidase *Aspergillus niger*	Liquid	4.5–7	15–60	750 and 1500 Scott U/ml	*See* no. 21
	Fermcolase	Fungal catalase *Aspergillus niger*	Liquid	5	25	1000 Baker units	Stable to hydrogen peroxide
33 S	Bromelain	Plant protease, stem extracts *Bromeliaceae* spp.	Powder	5–7	62	Unquoted	Food processing, medical

Supplier no. Enzyme type	Trade or general name	Enzyme type and source	Product form	Optimum pH (or range)	Optimum temperature (°C) (or range)	Activity (own units)	Application and special features
37 B	MKC α-amylase, Amylase HT Series	Bacterial α-amylase Bacillus subtilis var.	Powder	5.0–7.5	Up to 90	Concentrate 2,100,000 MWU/g 700,000 MWU/g 30,000 MWU/g	Baking, starch, distilling brewing, animal feeds, textiles, paper, pharmaceutical and fermentation industries
	MKC α-Amylase L	Bacillus subtilis var.	Liquid	5.0–7.5	Up to 90	420,000 MWU/ml	Baking, animal feeds, brewing, pharmaceutical and fermentation industries
	Optiamyl–L	Bacillus subtilis var.	Liquid	5.0–7.5	Up to 90	420,000 MWU/ml	Starch industry
	Optimash–L	Bacillus subtilis var.	Liquid	5.0–7.5	Up to 90	420,000 MWU/ml	Distilling industry
	Optisize–L	Bacillus subtilis var.	Liquid	5.0–7.5	Up to 90	420,000 MWU/ml	Textiles
	Amylase–LP	Bacillus subtilis var.	Liquid	5.0–7.5	Up to 90	420,000 MWU/ml	Paper industry
	MKC Amylase LT	Bacillus licheniformis var.	Liquid	5–8	90–105	420,000 MWU/ml 210,000 MWU/ml	Animal feed, brewing, textiles, paper, pharmaceutical and fermentation industries
	Opitherm–L	Bacillus licheniformis var.	Liquid	5–8	90–105	210,000 MWU/ml	Starch industry

	Product	Source	Form	pH	Temp.	Activity	Application
	Optimash–pH	Bacillus licheniformis var.	Liquid	5–8	90–105	210,000 MWU/ml	Distilling industry
C	MKC Fungal Amylase	Fungal α-amylase Aspergillus oryzae var.	Powder	4.0–6.6	Up to 60	40,000 SKB/g 5000 SKB/g	Baking, fruit juice, distilling brewing, starch, processing
	Clarase	Aspergillus oryzae var.	Liquid			40,000 SKB/ml	
	Amylase LF 20	Aspergillus oryzae var.	Liquid	2.5–5.5	Up to 60	20,000 MWU/ml	Fruit juice processing
	MKC Dextrinase A	Formulation	Powder	5–5.3	55	Standardized powder	Starch industry, acid syrup conversion to 62–65 DE glucose/maltose syrups
E	MKC Glucoamylase	Fungal amyloglucosidase Aspergillus niger var.	Liquid	3.8–4.5	55–60	200 GAU/ml 150 GAU/ml 160 GAU/g 100 GAU/ml	Baking, starch, distilling, brewing, textiles, paper, animal feeds, pharmaceutical and fermentation industries
	Optidex–L	Fungal amyloglucosidase Aspergillus niger var.	Liquid	3.8–4.5	55–60	200 GAU/ml 150 GAU/ml 100 GAU/ml	Starch processing
	Optisprit–L	Fungal amyloglucosidase	Liquid	3.8–4.5	55–60	200 GAU/ml 150 GAU/ml 100 GAU/ml	Distilling industry
F	MKC Glucose Oxidase Pur	Fungal glucose oxidase Aspergillus niger var.	Powder	3.5–7.0	70	60,000 GOU/g (photometric) concentrate (per million titre units)	Diagnostics, low and high catalase products

Supplier no. Enzyme type	Trade or general name	Enzyme type and source	Product form	Optimum pH (or range)	Optimum temperature (°C) (or range)	Activity (own units)	Application and special features
37 G	MKC Cellulase Cellulase P4000 P20000	Fungal cellulase *Aspergillus niger* var.	Powder	3.0–6.0	50–60	Concentrate 4000 CU/g 20,000 CU/g	Food industry, baking, fruit juice, essential oils and spices, brewing and animal feed industry, sewage
	MKC Cellulase TV	Fungal cellulase *Trichoderma viride* var.	Powder	3.0–7.0	Up to 50	Concentrate (not Std)	All cellulose materials, animal feeds, sewage
I	MKC Lactase	Fungal lactase *Aspergillus oryzae* var.	Powder	2.5–7.0 4.5–5.0	Up to 55	14.000 FCCLU/g	Dairy, food processing, lactose hydrolysis and pharmaceutical industries
K	Optisweet P	Bacterial glucose isomerase *Streptomyces* sp.	Powder	6–8	Up to 75	3000 TGIU/g minimum	Starch, sweeteners, isoglucose (HFS) production, sugar industry
	Optisweet TS	Bacterial glucose isomerase *Streptomyces olivaceus* var.	Immobilized	7.0–8.5 7.8	55–65 60	100 MIGICU/g	Starch, sweeteners, isoglucose (HFS) production, sugar industry
	Optisweet 22	Bacterial glucose isomerase *Streptomyces* sp.	Immobilized	7.5	60–65	High productivity (unquoted)	Starch, sweeteners, isoglucose (HFS) production, sugar industry

	Name	Source/type	Form	pH	Temperature	Standardization/Activity	Application
M	MKC Glucose oxidase	Fungal glucose oxidase *Aspergillus niger* var.	Powder/liquid	3.5–7.5	70	1500 GOU/g (T) 2100 GOU/g (P) 1500 GOU/ml(T) 2100 GOU/ml (P) 750 GOU/ml (T) 1050 GOU/ml (P)	Food industry, analytical
N	PV 8 'Pro Vino'	Fungal pectinase	Powder	4–5	5–65	Standardized	Wine production
	Opticlar–P	Fungal pectinase *Aspergillus niger* var.	Powder	2.5–6	50–60	Standardized	Food, juice and wine industries
	Clarex–L	Fungal pectinase *Aspergillus niger* var.	Liquid	3.5–5.0	50	15,000 AJDU/ml	Food, juice and wine industries
	Sparkl–HPG	Fungal pectinase *Aspergillus niger* var.	Liquid	2.5–6.0	5–60	10,000 AJDU/ml Minimum 80 PGU/ml	Food, juice and wine industries
R	Optimase	Bacterial alkaline protease *Bacillus subtilis* var.	Powder/granulate	6–11 9–10	60–70	330,000 DU/g 330,000 DU/g 440,000 DU/g 100,000 DU/g	Detergent and leather industries
	MKC Protease P330	Bacterial alkaline protease	Powder	6–11	Up to 70	330,000 DU/g	Food processing, animal feeds, pharmaceutical industries
	MKC Protease–Conc	Bacterial neutral protease *Bacillus subtilis* var.	Powder	6.5–8.0	55	Concentrate	Food and meat processing, brewing, protein modification, leather, animal feeds, pharmaceutical and photographic industries
	MKC HT Proteolytic 200	*Bacillus subtilis* var.	Powder	5.0–9.5	Up to 60	200 NU/g	

Supplier no. Enzyme type	Trade or general name	Enzyme type and source	Product form	Optimum pH (or range)	Optimum temperature (°C) (or range)	Activity (own units)	Application and special features
37 R	Fungal Protease P	Fungal protease *Aspergillus oryzae*	Powder	4.0–9.5 6–9	Up to 60	Concentrate 31,000 HU/g	Baking, food and meat processing, brewing, protein modification, leather, animal feeds, pharmaceutical and photographic industries
	MKC Acid Fungal Protease P	Acid fungal protease *Aspergillus oryzae* var.	Powder	2.0–6.0 2.5–3.0	Up to 55 50	1000 SAPU/g	Brewing, protein hydrolysis, meat and fish processing industries
	MKC Gelatinase	Protease system	Powder	6.0–9.5	Up to 55	84 VU/g	Photographic industry

Miles Kali-Chemie manufacture enzymes under licence from Miles Laboratories Inc. Their products are listed above, under their MKC trade names, and where the products are the same, the Miles Takamine trade name is shown in the key below for reference. The other Miles enzymes are listed separately below.

A: HT Amylase = MKC Amylase HT; C: Clarone = Clarone, Milezyme LC 20,000 = Amylase LF 20, Dextrinase A = MKC Dextrinase A; F: Dee-O = MKC Glucose Oxidase; G: Takamine Cellulase = MKC Cellulase, Takamine Cellulase TV = MKC Cellulase TV; M: Dee-O = MKC Glucose Oxidase; R: Protease P330 = MKC Protease P330, Protease Conc = MKC Protease Conc; HT Proteolytic 200 = MKC HT Proteolytic 200, Fungal Protease P = MKC Fungal Protease P, Acid Fungal Protease = MKC Acid Fungal Protease, Gelatinase = MKC Gelatinase.

38		Name	Source	Form	pH	Temp (°C)	Activity	Applications
	B	Tenase	*Bacillus subtilis* var.	Liquid	5.0–7.5	Up to 90	420,000 MWU/ml	Baking, starch, distilling, brewing, animal feeds, textiles, paper, pharmaceutical and fermentation industries
		TakaTherm	*Bacillus licheniformis* var.	Liquid	5–8	90–105	420,000 MWU/ml 210,000 MWU/ml	Brewing, starch, distilling, animal feed, textiles, paper, pharmaceutical and fermentation
	C	Takamine fungal amylase	Fungal α-amylase *Aspergillus oryzae* var.	Powder Liquid	4.0–6.6	Up to 60	40,000 SKB/g 5000 SKB/g 40,000 SKB/ml	Baking
		Takamyl	Fungal α-amylase *Aspergillus oryzae* var. (low protease)	Powder	4.0–6.6	Up to 60	40,000 SKB/g	Baking, fruit juice, distilling, brewing, starch processing
	E	Diazyme	Fungal amyloglucosidase *Aspergillus niger* var.	Liquid	3.8–4.5	55–60	200 GAU/ml 150 GAU/ml 100 GAU/ml 160 GAU/g	Baking, starch, distilling, brewing, animal feeds, textiles, paper, pharmaceutical and fermentation industries

F Miles Laboratories Inc. offer a large range of research and diagnostic enzymes, numbering 130 in the 1982 catalogue. Full details are obtainable from all Miles subsidiary companies.

| | I | Takamine fungal lactase | Fungal lactase *Aspergillus oryzae* var. | Powder | 2.5–7.0 4.5–5.0 | Up to 55 | 14,000 FCCLU/g | Food processing, lactose hydrolysis and pharmaceutical industries |

Supplier no. Enzyme type	Trade or general name	Enzyme type and source	Product form	Optimum pH (or range)	Optimum temperature (°C) (or range)	Activity (own units)	Application and special features
38 J	β-Glucanase L300	Bacterial β-glucanase	Liquid	5.5–7.5	Up to 70	—	Brewing – processing aid
K	TakaSweet	Bacterial glucose isomerase Steptomyces olivaceus var.	Immobilized	7.0–8.5 7.8	55–65 60	100 MIGICU/g	Starch, sweeteners, isoglucose (HFS) production, sugar industries
L	Takamine pancreatic lipase	Porcine pancreatic lipase	Powder	6.0–9.5	Up to 50	250 MLU/g	Good processing, fat modification and pharmaceutical industries
	Takamine lipase powder	Lipase – edible glandular tissue	Powder	5.5–9.5 6.2	Up to 50	Standardized	Dairy and food processing
M	Takamine Catalase L	Bovine catalase	Liquid	6.0–8.0	Up to 60	—	Dairy, food processing, textiles, sterilization, oxidation and plastics industries
N	Clarex–L	Fungal pectinase Aspergillus niger var.	Liquid	3.5–5.0	50	15,000 AJDU/ml	Food, juice and wine industries
	Sparkl–HPG	Fungal pectinase Aspergillus niger var.	Liquid	2.5–6.0	2–60	10,000 AJDU/gl Minimum 80 PGU/ml	Food, juice and wine industries

	Product	Source	Form	pH	Temperature	Activity	Applications
O	Hemicellulase	Fungal hemicellulase *Aspergillus niger* var.	Powder	3.0–6.0 3.5–4.5	50–60	100,000 HCU/g 2500 HCU/g	Baking, fruit juice, pharmaceutical and coffee industries, reduces viscosity and solubilizes gums used in oil well drilling, textile desizing
S	Takamine bromelain	Protease, pineapple	Powder	4.0–9.0	Up to 60	1100 BTU/g 110 BTU/g 22 BTU/g	Baking, fish processing, food processing, brewing, animal feeds, leather, textile, meat tenderizers and pharmaceuticals
	Takamine papain 30,000 FCC	Protease, papaya, latex	Powder	3.0–9.0	Up to 90	30,000 NFPU/mg	Baking, fish processing, food processing, brewing, animal feeds, leather, textile, meat tenderizers and pharmaceuticals
T	Marzyme	Microbial rennet *Mucor miehei*	Liquid	6.4–6.6	28–37	1–45 clotting ratio	Dairy industry
	Marzyme 11	Microbial rennet *Mucor miehei*	Liquid	6.4–6.6	28–37	1–45 clotting ratio	Thermolabile, dairy industry
	Animal rennet	Protease, bovine, calf stomach	Liquid	6.4–6.6	28–37	1–1500 clotting ratio	Dairy industry
	Chymo SET	Chymobin, bovine	Liquid	6.4–6.6	28–37	Standardized	Dairy, food processing
V	Brew n Zyme	Formulation	Powder	5.0–7.5	Up to 90	Standardized	Brewing, malt replacement
	Tender–Meat Tenderizers	Proteases	Powder/liquid	4.5–7.5	Up to 60	Standardized	Meat tenderizers, food processing

Supplier no. Enzyme type	Trade or general name	Enzyme type and source	Product form	Optimum pH (or range)	Optimum temperature (°C) (or range)	Activity (own units)	Application and special features
38 V	Tendrin	Proteases	Liquid	4.5–7.5	Up to 60	Standardized	
	Takamine Pancreatin 4 NF	Porcine enzymes	Powder	5.0–9.0	Up to 50	Standardized	Food processing, leather, pharmaceuticals
	Milezyme 3X/8X	Carbohydrase/protease	Prills	5.0–10.0	Up to 70	2.0×10^6/MWU/g 350,000 MDU/g	Detergents, pre-soaks and formulations
	KSTUV Sanizyme	Formulation Formulation	Powder Powder	5.0–9.5 5.0–9.5	Up to 55 Up to 55	Standardized Standardized	Sewage/drain/waste processing, compost mixtures
	Talase	Formulation	Powder	6.8–7.0	50	Standardized	Textile industry
	Takabate	Formulation	Powder	7.0–9.0	50	Standardized	Leather industry
	Takaskreen	Formulation	Powder	6.0–9.5	55	Standardized	Silk screens – cleans without damage to silk
40 D	Superzyme	Plant α- and β-amylase Barley malt	Liquid	5.5–6.5	65	200° Inst. Brew 40–65 DU	Range of enzymic malt extracts for food processing
41 S	NDB3	Plant protease, papain Carica papaya	Liquid	—	—	Unquoted	Brewing industry

44		Name	Source	Form	pH	Temp	Activity	Applications
	B	Aquazym	Bacterial α-amylase *Bacillus subtilis*	Liquid	6–7	70	120 KNU/g	Textile industry
		BAN		Liquid/powder/granule	6–7	70	120–1000 KNU/g	Alcohol, baking, brewing, food processing, paper, starch and textile industries
		Termamyl	*Bacillus licheniformis*	Liquid/granule	6–7.5	90–105	60 and 120 KNAAU/g	Alcohol, brewing, detergent, food, paper, starch, sugar and textile industries
	C	Fungamyl	Fungal α-amylase *Aspergillus oryzae*	Liquid/powder	5	55	180–1600 FAU/g	Alcohol, baking, brewing and food starch industries
	E	AMG SAN	Fungal amyloglucosidase *Aspergillus niger*	Liquid	4–5	60–75	150 and 200 AGU/ml	Alcohol, brewing, food, juice, starch and wine industries
	F	Uricase S	Bacterial uricase *Bacillus fastidiosus*	Freeze-dried in vials	8.5–9.0	40–42	20 IU/vial	Analytical
		Subtilisin A	Bacterial alkaline protease Subtilisin Carlsberg *Bacillus licheniformis*	Crystalline Freeze-dried	8–9.5	60	25 AU/g	Analytical, especially forensic
		Novozym 217	Penicillin-v-acylase	Immobilized	6.5–7.5	35	30–60 PVU/g	Production of semi-synthetic penicillins
		Plasmin	Animal lysofibrin Porcine blood	Freeze-dried	7.5	37	3 Novo U/mg	Pharmaceutical, therapeutic

Supplier no. Enzyme type	Trade or general name	Enzyme type and source	Product form	Optimum pH (or range)	Optimum temperature (°C) (or range)	Activity (own units)	Application and special features
44 F	Crystalline Trypsin Novo	Animal pancreatic Bovine	Freeze-dried	7.5–8.5	37	3000 K	Analytical, medical, tissue culture
		Porcine	Freeze-dried	8	50	2500–4000 K	Analytical, medical, tissue culture
G	Celluclast	Fungal cellulase *Trichoderma reesei*	Liquid	5.5	55–60	100–200 C$_1$/g 2000–4000 c$_x$/g	Alcohol, brewing, baking, food, juice, starch and wine industries
H	Dextranase	Fungal dextranase *Penicillium* spp.	Liquid	4.5–5.5	55–60	25 KDU/g	Food processing, sugar
	Novozym 230	Fungal inulinase *Aspergillus* spp.	Liquid	4.5	60	3000 U/g	Sweeteners
I	Lactozym	Yeast β-galactosidase *Klebsiella fragilis*	Liquid	6.5–7.0	40–45	3000 LAU/g	Dairy, food processing
	Novozym 231	Bacterial β-galactosidase *Bacillus* spp.	Immobilized	7.0–7.3	60	Unquoted	Thermostable for dairy and food industries
J	Cellobiase Novozym 188	Fungal specific β-glucanase *Aspergillus niger*	Liquid	4.5–5.0	60	250 CBU/g	Used with cellulases and in food processing

	Enzyme / Source	Form	pH	Temp	Activity	Application
Cereflo	Bacterial β-glucanase *Bacillus subtilis*	Liquid	6.5–7.5	55–60	200 BGU/g	Brewing, food processing
Finizym	Fungal β-glucanase *Aspergillus niger*	Liquid	5	60	200 FBGU/g	Brewing, food processing
K Sweetzyme	Bacterial glucose isomerase *Bacillus coagulans*	Immobilized	8	60–65	300 GINU/g or 200 IGIC/g	Starch, sweeteners
L Novozym 206	Fungal lipase–esterase *Aspergillus niger*	Liquid	5–7	35–40	750 LU/g	Dairy, food processing, leather and wool industries
Novozym 244	Animal phospholipase A_2 Porcine pancreas	Powder	8	40–50	100 IU/mg	Food processing industry
N Pectinex	Fungal pectinase *Aspergillus niger*	Liquid	5–6	40	750-2.25 mio MOE/g	Alcohol, food, juice and wine industries
Ultrazym		Powder	5–6	40	1.5–3.75 mio MOE/g	Food, juice and wine industries
O Gamanase	Fungal hemicellulase (galactomannoglucanase) *Aspergillus niger*	Liquid	3–6	70–80	1.5 MVHCU/g	Food processing, tissue extraction
Q PEM	Animal protease Pancreatic extract	Powder	7–8	37	Range	Food and general protein hydrolysis
PTN	Porcine pancreas	Powder	7–8	37	1.5 and 3.0 AU/g	Food, leather, pharmaceutical industries

Supplier no. Enzyme type	Trade or general name	Enzyme type and source	Product form	Optimum pH (or range)	Optimum temperature (°C) (or range)	Activity (own units)	Application and special features
44 R	Alacalase	Bacterial alkaline protease *Bacillus licheniformis*	Liquid / Granule/slurry	9	60	0.6 AU/g / 1.5 and 2.0 AU/g	Food, dairy, leather and detergent industries
	Esperase		Liquid/granule/ slurry	9–11	60	8 KNPU/g / 4 KNPU/g / 8 KNPU/g	Protein hydrolysis / Detergent industry / Detergent industry
	Savinase	*Bacillus* spp.	Slurry	10–11	55	8 KNPU/g	Detergent industry
	NUE No. 1	*Bacillus* spp.	Powder	7–9	25–50	4 KNPU/g	Leather industry
	Neutrase	Bacterial neutral protease *Bacillus subtilis*	Liquid Powder/granule	5.5–6.5	55	0.5 AU/g / 1.5 AU/g	Alcohol, baking, brewing, food, leather and protein hydrolysis
T	Rennilase	Fungal rennet *Mucor miehei*	Liquid/powder	6.5	32	Range 11–150 KRU/ml	Dairy industry
	Rennilase TL		Liquid			50 KRU/ml	Thermolabile for dairy industry
V	Ceremix	Mixed bacterial α-amylase, β-glucanase and protease	Liquid	5.7–6.5	50–70	Standardized	Brewing industry

	Product	Enzyme / source	Form	pH	Temp	Activity	Research and analytical
	Novozym 234	Fungal system for cell lysis *Trichoderma harzianum*	Powder	4.5	50	—	Research and analytical
52 A	Rohalase PA	Animal amylase Pancreatic extract	Powder	7	50	50 Units	Starch industry
B	Rohalase A	Bacterial amylase	Powder/liquid	7	70	3000–18,000 SKB	Brewing, detergent, food, juice, starch and textile industries
C	Rohalase M	Fungal α-amylase	Powder	5.5	50	16,000–48,000 SKB/g	Baking, brewing, food, paper, juice and waste industries
	Veron AV, AC F25		Powder	—	—	500–4500 SKB/g	Baking industry
E	Rohalase HT	Fungal amyloglucosidase	Powder	4.5	60	400 GAU/g	Brewing, juice and starch industries
G	Cellulase C	Fungal cellulase	Powder	4.5–5.0	40	Unquoted	Food and paper industries
I	Rohalase I 10X	Yeast invertase	Liquid	4.5	50–65	3.0 K	Confectionery industry
N	Pectinol D	Fungal pectinase	Liquid/powder	—	—	20 and 100 PA	Juice and wine industries
	Pectinol B		Powder			1250 PGU	Juice, vegetable and wine industries
	Pectinol VR	Pectinolytic	Powder	—	—	Unquoted	Red wines
	Rohament P	Pectin glycosidase *Aspergillus* spp.	Powder	4	45	2500 PGU	Fruit and vegetables

Supplier no. Enzyme type	Trade or general name	Enzyme type and source	Product form	Optimum pH (or range)	Optimum temperature (°C) (or range)	Activity (own units)	Application and special features
52 N	Rohapect C	Fungal pectinase	Powder	2.2	25	Unquoted	Acid stable for citrus juices
O	Veron HE	Pentosanase	Powder	—	—	800 Units	Baking
Q	Pancreatic proteinase A	Animal protease Porcine pancreas	Powder	7–8	50	350 PU	Protein hydrolysis
R	Bacterial proteinase N	Bacterial neutral protease	Powder	7	50	140 and 280 P Units	Brewing, protein hydrolysis
	Veron P, PS	Fungal protease	Powder	—	—	11 P Units	Baking
	Fungal proteinase P	Fungal neutral protease Aspergillus spp.	Powder	5.5–8.0	50	350 P Units	Protein hydrolysis
	Fungal proteinase S	Fungal acid protease Aspergillus spp.	Powder	4–6	50	2550 Hb Units	Protein hydrolysis
S	Corolase	Plant protease, papain	Powder/liquid	3–9	60–70	800–8000 Hb Units	Baking, brewing and food processing industries
56 C	Starzyme HM	Fungal α-amylase Aspergillus oryzae	Powder	4.0–6.2	50	17,500 SKB/g	Baking, sugar syrups

	Name	Source / Description	Form	pH	Temp	Activity	Application
	Starzyme LM					4500 SKB/g	Speciality syrups
D	Malt diastase	Plant α- and β-amylase Barley malt	Powder	4–8	50	1500° Litner	
E	Starzyme AG	Fungal amyloglucosidase *Aspergillus niger*	Powder	3–6	60	100 AGU/ml	
G	Cellulase CA	Fungal cellulase *Aspergillus niger*	Powder	2.5–6.5	60	10,000 CMCU/g	
	Cellulase CP	*Penicillium funiculosum*	Powder	3.5–6.5	60	10,000 CMCU/g	Active on crystalline cellulose
	Cellulase CT	*Trichoderma reesei*	Powder	3.5–7.5	60	10,000 CMCU/g	Active at neutral pH and on crystalline cellulose
I	Hydrolact	Yeast β-galactosidase *Saccharomyces fragilis*	Liquid/powder	6–7.5	35	5000–25,000 EU/g	Dairy industry
L	Lipase A	Fungal lipase–esterase *Aspergillus niger*	Powder	3–8	40	2300 U/g	Natural oil and fat emulsions
M	Catalase LA3	Fungal catalase *Aspergillus niger*	Liquid	3–10	35	430 Baker U/ml	Acid stable form
	Glucox	Fungal glucose oxidase *Aspergillus niger*	Liquid/powder	2.5–8.0	40	750–70,000 SU/g	Analytical, food processing industry

Supplier no. Enzyme type		Trade or general name	Enzyme type and source	Product form	Optimum pH (or range)	Optimum temperature (°C) (or range)	Activity (own units)	Application and special features
56	N	Panzym	Fungal pectinase *Aspergillus niger*	Powder	3–6.5	50	Unquoted	Juice and wine industries
	V	Panzym Combi	Mixed activities *Aspergillus niger*	Powder	3–6.5	50	Unquoted	Juice and wine industries
57	F	α-Amylase	*Bacillus subtilis*	Powder	—	—	Unquoted	Analytical
		β-Amylase	Barley	Powder	—	—	Unquoted	Free of α-amylase
		Malt diastase	Barley malt	Powder	—	—	1000°Lintner	Analytical
		Invertase	Yeast	Scales	—	—	10 K/g	Analytical
61	S	Bromelain	Plant protease	Powder	5–7	60	Range	Protein hydrolysis
	V	Actizyme	Mixed enzymes and bacteria	Pellets	6–8	35–38	Unquoted	Waste treatment

ALPHABETICAL LISTING OF INDUSTRIALLY
AVAILABLE ENZYME SOURCES

Many of the enzymes listed here are available from the suppliers listed in Data Index 3. Where not available from those in Data Index 3, enquiry should be made to others as listed in Data Index 2. Only enzymes known to be in industrial scale production are included in this Index.

Enzyme type (Common synonyms)	IUB No. (EC numbering 1978)	Natural source
Amylase – alpha Diastase	3.2.1.1	Malted cereals Animal pancreas *Aspergillus oryzae* *Aspergillus niger* *Bacillus subtilis* *Bacillus licheniformis* *Endomyces* spp.
Amylase – beta Diastase	3.2.1.2	Barley, wheat, soya Malted cereals *Bacillus megaterium* *Bacillus cereus*
Amylase – gamma Amyloglucosidase Glucoamylase	3.2.1.3	*Aspergillus niger* *Aspergillus oryzae* *Aspergillus phoenicis* *Rhizopus delemar* *Rhizopus niveus*
Amyloglucosidase		*See* Amylase – gamma
Catalase	1.11.1.6	*Aspergillus niger* *Penicillium* spp. *Micrococcus lysodeikticus*
Cellobiase	3.2.1.21	*Aspergillus niger* *Aspergillus oryzae* *Aspergillus phoenicis* *Aspergillus wentii* *Trichoderma viride (reesei)* *Mucor miehei* *Saccharomyces cerevisiae*
Cellulase	3.2.1.4	*Aspergillus niger* *Aspergillus sojae* *Penicillium* spp. *Penicillium funiculosum* *Trichoderma viride (reesei)*
Diastase		*See* Amylases
Dextranase	3.2.1.11	*Penicillium funiculosum* *Penicillium lilacinum* *Klebsiella aerogenes*

Enzyme type (Common synonyms)	IUB No. (EC numbering 1978)	Natural source
Esterase amino acid fatty acid		*See* Proteases *See* Lipases
Galactomannanase		*See* Hemicellulase
Galactosidase – alpha Melibiase	3.2.1.22	*Aspergillus niger* *Saccharomyces cerevisiae*
Galactosidase – beta Lactase	3.2.1.23	*Aspergillus niger* *Aspergillus oryzae* *Bacillus* spp. Saccharomyces fragilis Saccharomyces lactis *Kluyveromyces* spp.
Glucanase – alpha		*See* Amylases
Glucanase – beta	3.2.1.6	*Aspergillus niger* *Aspergillus oryzae* *Bacillus circulans* *Penicillium emersonii* *Saccharomyces cerevisiae* *See also* Cellobiase and Cellulase
Glucoamylase		*See* Amylase – gamma
Glucose isomerase Xylose isomerase	5.3.1.5	*Actinomyces missouriensis* *Bacillus coagulans* *Streptomyces albus* *Streptomyces olivaceous* *Streptomyces olivochromogenes*
Glucose oxidase	1.1.3.4	*Aspergillus niger* *Penicillium glaucum* *Penicillium notatum*
Hemicellulase Pentosanase Galactomannanase	3.2.1.78	*Aspergillus niger* *Aspergillus oryzae* *Aspergillus saitoi* *Bacillus subtillis* *Trametes* spp.
Invertase Invertin Saccharose	3.2.1.26	*Aspergillus niger* *Aspergillus oryzae* *Saccharomyces carlsbergensis* *Saccharomyces cerevisiae*
Lactase		*See* Galactosidase – alpha
Lecithinase		*See* Phospholipase A2
Lipase Fatty acid esterase	3.1.1.3	Calf, kid, lamb Bovine/porcine pancreas

Enzyme type (Common synonyms)	IUB No. (EC numbering 1978)	Natural source
Steapsin		*Aspergillus niger*
		Aspergillus oryzae
		Bacillus spp.
		Candida spp.
		Mucor miehei
		Rhizopus arrhizus
		Rhizopus delemar
Lysozyme Muramidase	3.2.1.17	Hen egg albumen
Macerating enzymes (mixed carbohydrases including pectinases)		*Aspergillus* spp.
		Rhizopus spp.
		Trichoderma viride
		Trichoderma harzianum
Malic acid decarboxylase	1.1.1.38(?)	*Leuconostoc oenos*
Melibiase		*See* Galactosidase – alpha
Muramidase		*See* Lysozyme
Nitrate reductase	1.7.99.4	*Micrococcus* spp.
Penicillin amidase	3.5.1.11	*Bacillus* spp.
		Basidiomycetes spp.
Pentosanase		*See* Hemicellulase
Peroxidase	1.11.1.7	Horseradish root
		Aspergillus spp.
		Penicillium spp.
Phospholipase A2 Lecithinase	3.1.1.4	Porcine pancreas
Plasmin	3.4.21.7	Bovine blood
Pectinases	3.2.1.15	*Aspergillus niger*
		Aspergillus ochraceus
		Aspergillus oryzae
	Anthocyanase	*Aspergillus niger*
	Limonoate dehydrogenase	*Arthrobacter globiformis*
	Naringinase	*Aspergillus niger*

Proteases	Serine		
	Chymotrypsin	3.4.21.1	Bovine/porcine pancreas
	Trypsin	3.4.21.4	Bovine/porcine pancreas
	Subtilisins	3.4.21.14	*Bacillus amyloliquefaciens*
			Bacillus amylosaccharicus
			Bacillus licheniformis
	Aspergillus alkaline		*Aspergillus flavus*
			Aspergillus oryzae

Enzyme type (Common synonyms)	IUB No. (EC numbering 1978)	Natural source
Thiol		
Papain	3.4.22.2	*Papaya latex*
Ficin	3.4.22.3	*Ficus carica*
Bromelain	3.4.22.4	*Bromus* spp.
Carboxyl–acid		
Pepsin	3.4.23.1	Porcine mucosae
Chymosin	3.4.23.4	Abomasum of unweaned calf, kid, lamb
Microbial	3.4.23.6	*Aspergillus niger*
		Aspergillus oryzae
		Aspergillus saitoi
		Endothia parasitica
		Mucor miehei
		Mucor pusillus lindt
		Rhizopus spp.
Microbial metallo (neutral)	3.4.24.4	*Aspergillus oryzae*
		Bacillus subtilis
		Bacillus thermoproteolyticus
		Streptomyces griseus
Pullulanase	3.2.1.41	*Bacillus* spp.
		Klebsiella aerogenes
Rennets	3.4.23.4	Chymosin from calf, kid, lamb
	3.4.23.6	*Endothia parasitica*
		Mucor miehei
		Mucor pusillus lindt
Tannase	3.1.1.20	*Aspergillus niger*
		Aspergillus oryzae
Urease	3.5.1.5	*Canavalia ensiformis* seed
		Bacillus spp.
Uricase	1.7.3.3	*Bacillus fastidiosus*
Xylanase	3.2.1.32	*Aspergillus niger*
Xylose isomerase		*See* Glucose isomerase

DATA INDEX 4B
A GUIDE TO THE USE OF ENZYMES IN THE MORE COMMON INDUSTRIAL AREAS

Alcohol	Amylases, amyloglucosidase, β-glucanase, cellulase, cellobiase, pectinase and protease
Amino acids	Proteases, peptidases and amidases
Analytical	A wide range of purified and also technical grade enzymes, as appropriate to the substance to be determined; continuous assays and electrodes
Biomethanation	Amylases, amyloglucosidase, cellulase, glucanase, lipase, protease, pectinase and macerating enzymes
Biscuits	Amylases and proteases
Bread and flour	Amylases, amyloglucosidase, glucanase, cellulase, cellobiase, lipase and pentosanase
Brewing	Amylases, glucanases, amyloglucosidase, cellulase, protease, pentosanase
Butter	Glucose oxidase and catalase
Cellulose	*For glucose production* – cellulase and cellobiase *For extraction* – pectinase, amylase, lignase
Cheese	*For flavour* – lipase, peptidase, protease *For production* – rennets *For ripening* – protease, peptidase, lipase
Cleaning	*For general use* – amylase and protease *For membranes and filters* – protease, amylase, cellulase, pectinase *For textile printing* – amylase, hemicellulase, protease *For vegetable and fruit* – pectinase, cellulase, and macerating enzymes
Coffee	Pectinase, hemicellulase, cellulase
Colours, natural	*See* 'Plant tissues' and 'Flavours'
Confectionery	*For candies and soft centres* – invertase *For gum products* – amylase *For reworking* – pectinase, amylase, protease, invertase
Dairy products	*For lactose conversion* – lactase *For protein modification* – protease *For heat treated milk* – sulphydryloxidase
Dental hygiene	Dextranase, glucanase, protease, peroxidase, glucoamylase
Detergents	Protease, amylase, lipase

Effluent treatment	Amylase, protease, cellulase, pectinase, lipase
Egg processing	*For liquid* – protease *For whipping and emulsifying* – lipase and phospholipase *For preservation of albumen* – glucose oxidase/catalase
Fats	Lipase, esterase, oxidase
Fermentation feed	*See* 'Alcohol', 'Brewing' and 'Starch'
Fish	*For cleaning, curing and waste utilization* – protease
Flavours	*For extraction* – *See* 'Plant tissues' *For yeasts and proteins* – protease, peptidase, glucanase *For carbohydrates* – amylase, protease *For fats* – lipases, esterases
Fructose	*From starch* – glucose isomerase (*see also* 'Starch') *From inulin* – inulinase
Fruit, candied	Cellulase, amylase, pectinase, protease
Fruit clouds	*For production* – pectinase, amylase *For removal* – pectinase, cellulase, amylase
Fruit colour control	Anthocyanase
Fruit extraction	Pectinase, cellulase, hemicellulase, pentosanase, amylase, amyloglucosidase *For bitterness removal* – naringinase, limonoate dehydrogenase
Fruit juices	Pectinase, amylase, cellulase, amyloglucosidase and macerating enzymes
Fruit pulps	*See* 'Plant tissues'
Gluconic acid	Glucose oxidase/catalase
Glucose	*For production* – *see* 'Starch' *For isomerization* – *see* 'Fructose'
Ice cream	Lactase, lipase
Leather	Protease, lipase, amylase, glucanase
Malt extract	*See* 'Brewing'
Meat tenderization	Protease, peptidase
Nucleic acid	*For removal* – nuclease *For flavour derivatives* – phospho-esterase
Oils	*For plant oils* – *see* 'Plant tissues' *For animal oils* – protease, esterase *For mineral oils* – protease, cellulase, amylase, pectinase and hemicellulase

Oxidation control	Glucose oxidase/catalase
Paper	Amylase, cellulase, lignase, glucanase
Plant tissues	Amylase, amyloglucosidase, cellulase, glucanase, hemicellulase, macerating enzymes, pectinase and protease
Protein	*For extraction* – amylase, cellulase, hemicellulase, glucanase, pectinase *For modification* – protease, peptidase, esterase
Pharmaceuticals	*For production* – *see* 'Fermentation feed', 'Plant tissues', 'Starch', 'Protein modification' *For modification* – amylase, esterase, glucosidase, lipase, oxidase, penicillinamidase
Rubber	Protease
Starch	Amylase, amyloglucosidase, cellulase, hemicellulase, glucanase, pullulanase, lipase, pectinase, protease and isomerase
Sucrose	*For production* – dextranase, amylase, galactosidase *For inversion* – invertase
Tea	*For fermentation* – cellulase, glucanase, pectinase *For solubility* – tannase
Textiles	*For desizing* – amylase, cellulase, glucanase, protease *For printing* – *see* 'Cleaning'
Waste treatments	Amylase, amyloglucosidase, cellulase, pectinase, glucanase, lipase, protease and macerating enzymes
Whey	*See* 'Dairy' and 'Protein'
Wine	*For production* – amylase, cellulase, pectinase *For clarification* – amylase, glucanase, pectinase *For colour control* – glucose oxidase/catalase

Note: This list is intended to be representative but is by no means exhaustive, since many specialized problems may now be tackled in these areas by the application of other enzymes.

DATA INDEX 5

CURRENT INDUSTRIAL ENZYME ASSAYS AND UNIT DEFINITIONS

The following industrial enzyme assay summary has been included by the editors as a guideline only. It is not possible to list every enzyme assay used in detail, as manufacturers and reputable suppliers use different assays and a wide variety of standard conditions. In the case of proteases alone there are more than 20 assays. There are wide disparities in enzyme activity on different substrates, as well as differences in pH and temperature optima making accurate comparison between methods impossible.

Manufacturers and suppliers of industrial enzymes will provide assay protocols on request enabling accurate interpretation of potency that can be directly related to products under test. Assays can also be used to test enzymes of similar type to verify potency and to establish comparative use levels. They can also be used to give an indication of product performance as outlined in Chapter 5.

Enzyme assays	Reaction time	Temperature (°C)	pH	Substrate
Amylase				
Modified Wohlgemuth (MWU)	5–25 min.*	40°	5.4, 5.0, 6.9	Soluble starch – 1%
α-amylase assay (method of SKB)	10–20 min.*	30°	5.0	Soluble starch – 2%
α-amylase 'Falling-Number' – Hagberg	2–10 min.	100°	4.0–6.0	Cereal flours – 28%
Bacterial amylase used in desizing	10–35 min.*	30°	6.6	Soluble starch – 2%
Determination of bacterial α-amylase thermostability	20, 40, 60 min.	70°	6.6	—
Glucoamylase (GAU) (Amyloglucosidase)	60 min.	60°	4.2	Soluble starch – 4%
Catalase				
Keil assay	10 min.	25°	7.0	H_2O_2 – 0.25%
Exhaustion method	—	25°	7.0	H_2O_2 – 1.5%
Cellulase				
Viscometric cellulase	15 min.	40°	4.5	CMC 7HPC – 0.115%
Cellulase	20 min.	50°	4.8	C_x – CMC – 0.4%
				C_1 – Avicel – 1.0%

Cellobiase	15 min.	40°	5.0	D + Cellobiose – 0.2%
β-glucanase	30 min.	30°	7.5	Barley β-glucan – 0.5%
Glucose oxidase.				
Manual assay	15 min.	35°	5.1	Glucose at 3% aerated at 700 ml/min.
Automated assay	5–10 min.	25°	6.5	Glucose – 0.15%
Hemicellulase				
Viscometric hemicellulase	15 min.	40°	4.5	Locust bean gum – 0.17%
Lactase	45 min.	37°	6.5	Lactose monohydrate – 4.75%
Lipase assay	120 min.	37°	7.3	Olive oil – 25%
Glucose isomerase				
Takasaki glucose isomerase	60 min.	70°	7.0	D + Glucose monohydrate
Immobilized glucose isomerase – column	42–48 hrs.	60–65°	7.5	D + Glucose – 40–50%
Pectinase				
Apple juice depectinizing assay	120–180 min.	45°	3.4–3.6	Apple juice – 100%
Pectin methyl esterase	>10 min.	37°	3.5	High methoxyl pectin – 0.5%
Viscometric polygalacturonase	15 min.	30°	4.2	Sodium polypectate – 0.2%
Protease				
Northrop	35 min.	40°	7.4	Casein – 0.2%
Spectrophotometric detergent protease	15 min.	55°	9.2	Casein – 0.6%
Half-hour haemoglobin	30 min.	40°	4.7	Haemoglobin – 1.6%

* Assay is resolved by measuring endpoint in minutes

Enzyme assays	Reaction time	Temperature (°C)	pH	Substrate
Determination of proteolytic activity as haemoglobin units	10 min.	25°	7.5	Haemoglobin – 2%
Spectrophotometric neutral protease	30 min.	37°	7.0	Casein – 0.7%
Acid protease activity	10 min.	30°	2.7	Casein – 2.0%
	30 min.	37°	3.0	Casein – 0.7%
Spectrophotometric protease assay for bromelain	30 min.	30°	5.0	Haemoglobin – 1%
Papain assay (FCC method)	60 min.	40°	6.0	Casein – 1%
Rennet				
Rennet strength	5–6 min.	30°	6.35	Skim milk powder – 12%

DEFINITIONS OF ENZYME ASSAY UNITS

Enzyme assay	*Definition of unit*
Amylase	
Determination of liquefying amylase (Modified Wohlgemuth Method)	One Modified Wohlgemuth Unit (MWU) is that amount of enzyme which will dextrinize one milligramme of soluble starch to a definite size dextrin in 30 minutes under the conditions of the assay
α-Amylase assay (Method of Sandstedt, Kneen and Blish)	One SKB Unit is that amount of enzyme which will dextrinize one gramme of β-limit dextrin to a definite size dextrin in one hour under the conditions of the assay
Assay of bacterial amylase used in desizing	One Bacterial Amylase Unit (BAU) is that amount of enzyme which will dextrinize one milligramme of soluble starch to a definite size dextrin per minute under the conditions of the assay
Determination of bacterial α-amylase thermostability	Thermostability is expressed as per cent relative activity retained with time under the conditions of the assay
Glucoamylase (Amyloglucosidase)	One Glucoamylase Unit (GAU) is that amount of enzyme which will liberate one gramme of reducing sugar as glucose per hour under the conditions of the assay
Catalase	
Keil assay	One Keil Unit (KU) is that amount of enzyme which will decompose one gramme of 100 per cent hydrogen peroxide in ten minutes under the conditions of the assay
Exhaustion method for catalase	One Exhaustion Unit (EU) is that amount of enzyme which will decompose 300 milligrammes of hydrogen peroxide under the conditions of the assay
Cellulase	
Viscometric cellulase assay	One Cellulase Unit (CU) is that amount of enzyme which will produce a change in the relative fluidity of one in a defined sodium carboxymethyl cellulose substrate in five minutes under the conditions of the assay
Cellulase (colorimetric)	One Cellulase Unit is the amount of enzyme which under the given standard conditions forms an amount of reducing carbohydrate equivalent to 1 μmol glucose per minute = 1 International Cellulase Unit, respectively IC_1U and IC_xU
Cellobiase	One Cellobiase Unit (CBU) is the amount of enzyme which under the given standard conditions and with cellobiose as substrate liberates two μmol glucose per minute
β-Glucanase	One β-Glucanase Unit is the amount of enzyme which under standard conditions liberates glucose

Enzyme assay	Definition of unit
	or other reducing carbohydrates with a total reduction power corresponding to one µmol glucose per minute
Glucose oxidase	
Manual assay method for glucose oxidase	One Glucose Oxidase Unit (GOU) is equivalent to that amount of enzyme which will cause the uptake of 10 mm³ oxygen per minute under the conditions of the assay
Hemicellulase	
Viscometric hemicellulase assay	One Hemicellulase Unit (HCU) is that amount of enzyme which will produce a relative fluidity change of one in a defined locust bean gum substrate per five minutes under the conditions of the assay
Lactase	One Lactase Unit (LAU) is defined as the amount of enzymatic activity which releases one µmol of glucose per minute from a solution of 4.75% w/v lactose in pH 6.5 buffer solution at 37 °C
Lipase assay	One Lipase Unit (LU) is that amount of enzyme which will liberate one milli equivalent of fatty acid in two hours under the conditions of the assay
Glucose Isomerase	
Takasaki glucose isomerase	A Takasaki Glucose Isomerase Unit (TGIU) is defined as the amount of an enzyme which, from a 0.1 M D-Glucose solution, under the given assay conditions of pH 7.0 at 70 °C will form one milligramme D(−)Fructose in 60 minutes
Immobilized glucose isomerase – column method	One Immobilized Glucose Isomerase Column Unit (IGICU) is defined as that activity which will produce one µmol of fructose per minute under the conditions of the assay
Pectinase	
Apple juice depectinizing assay	AJDUs are determined by correlating depectinization time of a defined apple juice substrate by the unknown pectinase with depectinization time by a pectinase standard of known activity
Pectinmethylesterase assay	One Pectinmethylesterase Unit (PMEU) is that amount of enzyme which will liberate one µmol of titratable carboxyl groups per minute under the conditions of the assay
Viscometric poly-galacturonase assay	One Polygalacturonase Unit (PGU) is that amount of enzyme which will cause a change in relative fluidity of 0.01 per second in a defined sodium polypectate substrate under the conditions of the assay

Enzyme assay	Definition of unit
Protease	
Colorimetric Northrop assay	One Northrop Unit (NU) is that amount of enzyme which gives 40 per cent hydrolysis of a defined casein substrate under the conditions of the assay
Spectrophotometric detergent protease assay	One Miles Detergent Unit (MDU) is that amount of enzyme which will liberate ten nanomoles of tyrosine per minute under the conditions of the assay
Half-hour colorimetric haemoglobin assay	One Haemoglobin Unit (HU) is that amount of enzyme which will liberate 67.08 milligrammes of non-protein nitrogen under the conditions of the assay
Spectrophotometric neutral protease assay	One Neutral Protease Unit (NPU) is that amount of enzyme which will liberate one µmol of tyrosine per minute under the conditions of the assay
Acid protease assay	One Acid Protease Unit (APU) is that amount of enzyme which will produce a change in absorbance (ΔA) of one at 660 nanometres under the conditions of the assay
Spectrophotometric acid protease assay	One Spectrophotometric Acid Protease Unit (SAPU) is that activity which will liberate one µmol of tyrosine per minute under the conditions of the assay
Spectrophotometric assay for bromelain	One Bromelain Tyrosine Unit (BTU) is that amount of enzyme which will liberate one µmol of tyrosine per minute under the conditions of the assay
Papain assay	One N.F. Papain Unit (N.F. PU) is that amount of enzyme which will liberate the equivalent of one microgram of tyrosine per hour under the conditions of the assay
Rennet	
Rennet strength	The unit of rennet strength is usually as Clotting Units or Rennet Units determined using a rennet standard of known activity as a reference for clotting time under standard conditions

DATA INDEX 6
PRACTICAL BIOCHEMICAL DATA

Editor's notes: Column 1: the enzymes in this index are listed in alphabetical order using the commonly accepted industrial names. *Column 2:* the basis for the International Union of Biochemistry classification and number is given, followed by a description of the reaction catalysed (or special characteristics where appropriate). *Column 3:* representative sources of industrial enzymes are drawn from the preceeding Data Indexes and used to provide examples of data in the subsequent columns. *Columns 4 and 5:* the evident optimum conditions for industrial use have been abstracted from Data Index 3, and extended where necessary by the editor's own sources of data. *Column 6:* brief comments on the known observations mainly regarding inhibitors and activators of industrial relevance. *Column 7:* a listing of the other enzymes that are potentially present and that may be worthy of evaluation when investigating new applications, or seeking explanation of unexpected performance in existing applications. These side activities arise from the intrinsic capability of the source organism and are manifested to varying degrees according to the strain, fermentation medium and methods and degree of purification of the main nominated activity. It is not assumed that any or all of these side activities are present in any particular industrial enzyme preparation. A more detailed discussion of these enzymes will be found in Chapter 5.

Enzyme type	Classification	Example source	Optimum values		Observations	Side activities
			pH	Temp. (°C)		
α-Amylase	1,4-α-D-glucan glucanohydrolase 3.2.1.1 Endo-hydrolysis of polysaccharides with 3 or more 1,4-α-D-glucan units	Cereals	5–6	50	Activated by calcium ions Inhibited by oxidizing agents	β-amylase, β-glucanase, neutral acid protease
		Pancreas	6.5	40	Calcium salts increase heat stability	Esterase, lipase and protease
		Aspergillus niger	5	55	Activated by calcium ions	Cellulase, hemicellulase, acid protease, xylansase

Enzyme	Systematic name / Action	Source	pH	Temperature	Properties	Associated enzymes
		Aspergillus oryzae	5	55	Activated by calcium ions	Glucoamylase, acid protease
		Bacillus subtilis	6–7	70–80	Activated by calcium ions Inhibited by chelating agents	β-glucanase, acid and neutral protease
		Bacillus licheniformis	7–8	90–95	Low calcium dependence, especially in presence of high substrate	β-glucanase, acid and neutral protease; these are thermolabile and rapidly inactivated at amylase use temperatures
α-Galactosidase	α-D-galactoside galactohydrolase 3.2.1.22	Aspergillus niger	4.5	65	Acts on many galactosides	Glucosidase, hemicellulase
	Exo-hydrolysis of terminal nonreducing α-D-galactoside residues of polysaccharides, oligosaccharides, galactomannans, galactolipids	Saccharomyces sp.	5	50	Low galactomannanase activity	Glucosidase, invertase
β-Amylase	1,4-α-D-glucan maltohydrolase 3.2.1.2	Cereals	5.5	55	Activated by reducing agents	α-amylase
	Exo-removal of maltose units from nonreducing end of polysaccharide chains	Soya bean	4–7	55	Acid tolerant	α-amylase, lipoxygenase
		Bacillus sp.	5–7	60	—	α-amylase, β-glucanase, neutral protease

Enzyme type	Classification	Example source	Optimum values		Observations	Side activities
			pH	Temp. (°C)		
β-Galactosidase	β-D-galactoside galactohydrolase 3.2.1.23 Hydrolysis of terminal nonreducing β-D-galactose residues	Aspergillus niger	4.5	55	Product inhibition and the presence of transferase activity is common to all 5 types	α-L-arabinase, glucanase, glucosidase, transferase, invertase, acid protease all common to both Aspergillus sp.
		Aspergillus oryzae	4.5	55		
		Bacillus sp.	7.3	60		Amylase, glucosidase, protease
		Kluyveromyces sp.	6.5	45		Glucosidase, invertase, protease, transferase common to the yeasts
		Saccharomyces sp.	6.5	40		
β-Glucanase	Endo-1,3(4)-β-D-glucanase 3.2.1.6 Endo-hydrolysis of β-D-glucans when the glucose residue whose reducing group is involved, is itself substituted on carbon 3	Aspergillus niger	5	60	Broad range with low specificity	Amylase, glucoamylase, glucosidase
		Bacillus subtilis	7	50–60	Narrow range with higher specificity	Amylase, glucosidase, protease
		Penicillium emersonii	4	70	Tolerant of low pH in some variants	Amylase, dextranase, protease
β-Glucosidase	β-D-glucoside hydrolase 3.2.1.21	Aspergillus niger	5	60	See Cellobiase	Amylase, glucoamylase, protease

Enzyme	Systematic name / EC	Source	pH	Temp.	Characteristics	Impurities
		Aspergillus oryzae	5	65	Broad specificity, unusually high thermotolerance	Amylase, glucoamylase, protease
		Bacillus sp.	7	70	Narrow specificity	Amylase, glucanase, protease
		Clostridium thermocellum	9	60	Very active on β-1,3-bonds	Protease
		Saccharomyces sp.	7	45	Broad specificity	Glucanase, protease
		Sweet almond	7	50	Broad specificity	Usually very pure
		Trichoderma viride	5	65	Broad specificity	Hemicellulase
Catalase	Hydrogen peroxide; hydrogen peroxide oxidoreductase 1.11.1.6	*Aspergillus niger*	5–8	35	Stable at low pH	Usually very pure
		Bovine liver	7	45	Inactivated by alkali	Usually very pure
Cellobiase	β-D-glucoside glucohydrolase 3.2.1.21 Exo-hydrolysis of terminal nonreducing 1,4-α-D-glucose residues	*Aspergillus niger*	5	60	Used to reduce product inhibition of cellobiose when cellulases used	Amylase, glucoamylase, protease, hemicellulase
Cellulase	1,4-(1,3; 1,4)-β-D-glucan 4-glucanohydrolase	*Aspergillus niger*	5	45	Generally low in C_1 type activity	Amylase, cellobiase, glucosidase, glucoamylase

Enzyme type	Classification	Example source	Optimum values		Observations	Side activities
			pH	Temp. (°C)		
	3.2.1.4 Endo-hydrolysis of 1,4-β-glucosidic links of cereal glucans, cellulose, lichenin	Basidiomycetes sp.	4	50	Good C_1 type with broad specificity	Hemicellulase, protease
		Penicillium funiculosum	5	65	Product inhibition is usually low	Amylase, glucoamylase, cellobiase
		Rhizopus sp.	4	45	Broad specificity	Amylase, glucoamylase, protease
		Trichoderma sp.	5	55	High C_1 activity	Hemicellulase
Dextranase	1,6-α-D-glucan 6-glucanohydrolase 3.2.1.11 Endo-hydrolysis of dextrans	Penicillium sp.	5	55	Products are isomaltose and isomaltotriose	Cellulase, hemicellulase
Glucoamylase	1,4-α-D-glucan glucanohydrolase 3.2.1.3 Exo-hydrolysis of terminal 1,4-α-D-glucose residues from nonreducing end of polyglucoside chains	Aspergillus awamori	4–5	60	Acid tolerance and thermotolerance vary widely from source to source	Amylase, glucanase, cellulase, hemicellulase, protease; all common to these fungal sources
		Aspergillus niger	3–5	65		
		Aspergillus oryzae	4.5	60		
		Rhizopus sp.	2.5–5	55		

Enzyme	Description	Source	pH	Temp	Notes	
Glucose isomerase	D-xylose ketol isomerase 5.3.1.5 A true xylose isomerase acting on glucose at high substrate concentration	*Atinoplanes missouriensis*	7.5	60	In the immobilized form, all these are activated by magnesium and cobalt; need for cobalt varies with preparation; magnesium competed for by calcium and must be in excess	All are usually pure
		Bacillus coagulans	8	60		
		Streptomyces sp.	8	63		
		Streptomyces albus	6–7	60–75		
Glucose oxidase	β-D-glucose; oxygen 1-oxidoreductase 1.1.3.4	*Aspergillus niger*	4.5	50	—	Catalase
		Aspergillus sp.	2.5–8	15–70	Acid tolerance, thermotolerance among gp.	Catalase
		Penicillium notatum	3–7	50	—	Catalase
Hemicellulase	Many activities represented; examples include: Endo-1,4-β-D-Mannan hydrolase 3.2.1.78 Exo-α-L-Arabinofuran hydrolase 3.2.1.55 Exo-1,3-β-D-Xylan hydrolase 3.2.1.72	*Aspergillus* sp.	3–6	70	These are very complex enzyme systems that contain many activities that require specific substrates to identify and distinguish them	In common with the *Aspergillus* group, cellulase, glucosidase, pectinase, pentosanase

Enzyme type	Classification	Example source	Optimum values		Observations	Side activities
			pH	Temp. (°C)		
Inulinase	2,1-β-D-Fructan fructanhydrolase 3.2.1.7 Endo-hydrolysis of inulin	Aspergillus sp.	4.5	60	—	Amylase, glucoamylase, invertase, protease, can be anticipated in either type
		Candida sp.	5	40	—	
Invertase	β-D-fructofuranoside fructohydrolase 3.2.1.26	Candida sp.	4.5	50	Maximum activity is shown at low substrate levels	In common with other yeast derived enzymes, the proteases may be present
		Saccharomyces sp.	4.5	55	All contain a bound mannan	
Lactase	See β-Galactosidase					
Lipase	Triacylglycerol acylhydrolase 3.1.1.3	Aspergillus niger	5–7	40	High in esterase	Amylase, cellulase, esterase, hemicellulase, pectinase, protease
		Candida cylindracea	8	50	Active on higher oils and fats	Esterase, protease
		Mucor miehei	7.5	50	High in true lipase activity	Esterase, protease
	Fatty acid esterase Carboxylic-ester hydrolase 3.1.1.1	Pancreatic	7.5–8	40	Preferential action on triglycerides	Amylase, protease

			pH	Temp.	Properties	Associated activities
Aryl-ester hydrolase 3.1.1.2		Pregastric esterase	5.5–7	30–60	High esterase–lipase ratios	Amylase
		Rhizopus sp.	5–8	40	Very varied specificities	Amylase, cellulase, esterase, protease
Lysozyme	Mucopeptide N-acetyl muramoylhydrolase 3.2.1.17 Hydrolysis of 1,4-β-links between N-acetylmuramic acid and 2-acetamido-2-deoxy-D-glucose residues	Hen egg albumen	6–7	35	Inhibited by surfactants	Usually very pure
Pectinase	Poly (1,4-α-D-galacturonide) glycanohydrolase 3.2.1.15	*Aspergillus* sp.	2.5–6	40–60	Wide variety of component activities	Pectin, esterase, pectin lyase, etc (*see* Chapter 5)
		Rhizopus sp.	2.5–5	30–50		
Penicillin amidase	Penicillin amidohydrolase 3.5.1.11	*Bacillus* sp.	7–8	37	—	Usually very pure
		Basidiomycetes sp.	4–6	50	—	Usually very pure
Pentosanase	*See* Hemicellulase					
Peroxidase	Donor; hydrogen peroxide oxidoreductase 1.11.1.7	Horse radish	5–7	45	A specific haem protein enzyme	Catalase

Enzyme type	Classification	Example source	Optimum values pH	Optimum values Temp. (°C)	Observations	Side activities
Phospholipase A_2	Phosphatid-2-acylhydrolase 3.1.1.4	Porcine pancreas	8	45	Stereo-specific for 2 position of lecithins	Usually pure
Plasmin	Fibrinolysin 3.4.21.7 Preferential cleavage of Lys residues	Porcine blood	7.5	37	Converts fibrin to soluble products	Usually very pure
Proteases Serine types	Chymotrypsin 3.4.21.1 Preferential cleavage of Tyr, tryptophan, Phe, Leu residues	Pancreatic	8–9	35	Inhibited by compounds in cereals, beans, potato, egg Bovine more thermotolerant	Amylase, lipase, esterase
	Trypsin 3.4.21.4 Preferential cleavage of Arg and Lys	Pancreatic	8–9	45	Inhibitors as for Chymotrypsin Bovine more thermotolerant	Amylase, lipase, esterase
	Subtilisin 3.4.21.14 Hydrolysis of proteins peptide amides	Bacillus amyloliquefaciens	9–11	60–70	All are subject to inhibition by organo-phosphorus compounds	All are likely to show amylase, glucanase and additional protease
		Bacillus licheniformis	9–11	60–70		

Type	Enzyme	Source	pH	Temp	Properties	Contaminants
		Bacillus subtilis	9–10	55	No peptidase activity	Amylase, glucoamylase cellulase, hemicellulase
		Aspergillus oryzae	8–10	60–70		Lysozyme, glucanase, glucosidase, cellulase
Thiol types	Papain 3.4.22.2 Preferential cleavage of Arg; Lys	*Papaya* sp.	5–7	65	All are activated by reducing compounds. All are inhibited by oxidizing agents	
	Ficin 3.4.22.3 Preferential cleavage of Lys; Ala; Tyr; Gly; Asn; Leu; Val	*Ficus* sp.	5–7	65		Lysozyme, esterase, peroxidase
	Bromelain 3.4.22.4 Preferential cleavage of Lys; Ala; Tyr; Gly	*Ananas* sp.	5–8	55		Cellulase, hemicellulase, pectinase
Carboxyl-acid types	Pepsin 3.4.23.1 Preferential cleavage of Phe; Leu	Porcine mucosa	1.8–2	40–60	Inhibited by aliphatic alcohols	Usually very pure
	Chymosin 3.4.23.4 Specific for 1 bond of kappa casein	Bovine abomasum	4.8–6	30–40	—	Usually pure but may contain pepsin

Enzyme type	Classification	Example source	Optimum values		Observations	Side activities
			pH	Temp. (°C)		
Microbial	3.4.23.6	*Aspergillus niger*	2.5–4	45	Broad specificity	Usually pure but some other proteases are commonly detected
		Aspergillus oryzae	4–7	60	Does not clot milk	
		Endothia parasitica	6–7	40	Narrow specificity and clots milk	These milk coagulant preparations are relatively pure; some lipase, esterase and traces of nonspecific protease may be found
		Mucor miehei	6–7	40	Narrow specificity and clots milk	
		Mucor pusillus lindt	6–7	40	Narrow specificity and clots milk	
Metallo-neutral	Microbial 3.4.24.4 Preferential cleavage; bonds adjacent to hydrophobic residues	*Aspergillus oryzae*	7	50	All are inhibited by reducing agents, chelating agents, halogens	As for most proteases, the side activities are other proteases
		Bacillus thermoproteolyticus	8	65		
		Bacillus sp.	7	50		
Pullulanase	Pullulan-6-glucano hydrolase 3.2.1.41	*Bacillus* sp.	4.5	60	—	Protease, amylase
		Klebsiella aerogenes	5	50	—	Protease, amylase

Hydrolysis of
1,6-α-D-glucosidic
link in pullulan,
amylopectin, glycogen
and limit dextrins

Rennets	*See* Chymosin and Microbial carboxy proteases					
Tannase	Tannin acylhydrolase 3.1.1.20	*Aspergillus niger*	4.5	55	—	Amylase, glucoamylase
		Aspergillus oryzae	3–5	45	—	Amylase, glucoamylase, protease, cellulase
Xylanase	1,3-β-D-Xylan xylanohydrolase 3.2.1.32	*Aspergillus niger*	3–5	45–55	—	Amylase, glucoamylase, glucosidase, cellulase
	Endo-hydrolysis of 1,3-β-D-xylose units of xylans	*Aspergillus oryzae*	4	45	—	Amylase, glucanase, glucoamylase, cellulase
		Bacillus sp.	7–9	55–65	—	Amylase, glucanase protease

INDEX

Abattoirs, 295, 296, 302, 356, 366
Acacia, 120
Acetaldehyde, 198, 208
Acetic acid, 184, 206
Acetic CoA synthetase, 206
Actinomyces missouriensis, 131, 136
Actinoplanes missouriensis, 142, 148, 488, 524, 563
Activated dough method (baking), 212, 213
Activators, 6, 467
Active site, 195
Activity units, 10
Adenosine, 309
Adhesives, 295
Adjuncts (brewing), 226; choice of, 232; cooking, 226, 237; dextrinization, 234; enzymatic processing, 233–4; liquefaction, 234; mash concentrations, 237; processing enzymes, 241, 258; protein content, 233; saccharification, 239
Adsorbant resins, 273
Aflatoxin B$_2$1, 121, 138
Afterwash (textiles), 404
Alcohol (potable), **170–78**, 457, 549; applied cultures and enzymes, 171–3; fermentation, 173, 516, 518; gelatinization, 173; hydrolysis, 173; liquefaction, 173; production size, 177–8; processes involved, 173–7; raw materials, 171; saccharification, 174, 175, 518
Alcohol (power/fuel), **179–93**; amyloglucosidase, 185; cellulose-containing raw materials, 191–3; cooking processes, 185–8; endoamylases, 184; exoamylases, 185; feedstocks containing starch, 182–3; feedstocks containing sugar, 189–91; overall view of production, 181–2
Alcohol dehydrogenase, 198, 208, 209
Alcohol oxidase, 447
Aldehyde oxidase, 447
Alginates, 346, 347
Allergy, 116, 164, 166, 284, 288
All-in-one enzyme systems (brewing), 245
α-aminobenzylpenicillin, 53
α-amylase, 42, 48, 50, 52, 57, 59, 66, 77, 78, 99, 125, 136, 143, 144, 146, 172, 173, 175, 183, 211, 212, 221, 222, 223, 243, 244, 299, 300, 326, 332, 380, 387, 422, 526, 545, 559; assay, 552, 555; bacterial, 174, 175, 215, 227, 231, 232, 241, 246, 307, 343, 379, 384, 422, 448, 514, 516, 517, 520, 521, 523, 526, 528, 537, 540; cereal, 212; fungal, 213, 214, 216 [inactivation], 223, 227 [maltogenic action], 343, 383, 384, 387, 471–2 [inactivation; reaction products], 514, 517, 519, 524, 529, 533, 537, 540, 542; malt, 214 [thermostability], 223, 399; plant, 517, 542; practical

biochemical data, 558; side activities, 558; thermolabile, 184; thermostability determination, assay, 552, 555; thermostable, 174, 184, 246, 422, 470, 471 [calcium dependence; inactivation; inhibitors]
α-galactosidase, 136, 144, 147, 207, 345, 546; practical biochemical data, 558
α-glucosidase, 143, 144, 146, 207
α-ketoglutarate, 196, 198
α-L-arabinase, 560
α-1,4 amylose, 98
AMFEP General Standards for Enzyme Regulations, 139
Amidase, 549
Amino acid acylase, 446
Amino acid oxidase, 447
Amino acids, 208, 370, 372, 448, 549
Aminoacylase, 373, 448
Amino acyl t-RNA synthetase, 66
Aminopenicilloic acid, 53, 98, 446, 516
Amylases, 5, 67, 210, 226, 240, 246, 302, 307, 313, 315, 324, 347, 348, 349, 403, 404, 430, 467–75, 476, 483, 489, 491, 499, 503, 519, 549, 550, 551, 560, 561, 562, 564, 565, 566, 567, 568, 569; action, 467; animal, 397, 541; assays, 552, 555 [unit definitions]; bacterial, 213, 380, 399, 468, 469, [inactivation; inhibitors], 503, 541; content in germinated and ungerminated cereal grain, 211; desizing (textiles), 398–9; fungal, 21, 245, 246 [maltogenic], 346–7, 383, 387, 397, 399, 503; inactivation, 472; liquefying determination of (assay), 555; malt, 213; microbial, 217; pancreatic, 399, 468; plant, 397, 503; supplementation levels, determination of (baking), 216; thermostability, 472; thermostable, 235, 300, 400
Amyloglucosidase, 35, 61, 67, 95, 98, 143, 144, 145, 146, 164, 183, 184, 207, 214, 223, 227, 240, 246, 251, 252, 253, 254, 255, 315, 317, 376, 383, 384, 387, 388, 433, 473, 503, 515, 519, 526, 546, 549, 550, 551; assay, 555; fungal, 239, 241, 473–5 [action, reaction products], 537, 516, 518, 522, 524, 526, 529, 533, 537, 541; thermostability, 253; *see also* 'Glucoamylase'
Amylolysis, 222
Amylopectin, 98, 182, 212, 223, 376, 467, 472, 476; comparision of properties with amylose, 182
Amylose, 182, 223, 375, 379, 467
Amylose–amylopectin complex, 376
Amylose–lipid complex, 375, 378–9, 381
Anaerobic digestion, 302–3; digestive system, 303; methanogenic fermentation, 302–3

570